Chevrolet LUV Automotive Repair Manual

by J H Haynes
Member of the Guild of Motoring Writers

and Ian Coomber

Models covered:
All Series 1 thru 12 with 110.8 cu in (1815 cc) engine
Covers 2-and 4-wheel drive and manual and automatic transmission

ISBN 0 85696 920 6

© Haynes Publishing Group 1977, 1979, 1982, 1983

ABCDE
FGHIJ
KLMNO
PQR

All rights reserved. No part of this book may be reproduced or transmitted in any form or by any means, electronic or mechanical, including photocopying, recording or by any information storage or retrieval system, without permission in writing from the copyright holder.

Printed in England   (319 - 3J3)

Haynes Publishing Group
Sparkford Nr Yeovil
Somerset BA22 7JJ England

Haynes Publications, Inc
861 Lawrence Drive
Newbury Park
California 91320 USA

Acknowledgements

Thanks are due to General Motors (USA) for assistance in providing certain illustrations and technical information; Mike and John Fatz of Two Brothers Automotive in North Hollywood for the loan of the Chevy LUV pick-up used as the project vehicle for this manual, and the Champion Sparking Plug Company who supplied the illustrations showing the various spark plug conditions. The bodywork repair photographs used in this manual were provided by Lloyds Industries Limited who supply 'Turtle Wax', 'Dupli-color Holts', and other Holts range products.

Special thanks are due to all of those people at Sparkford who helped in the production of this manual. Particularly, Brian Horsfall and Les Brazier who carried out the mechanical work and took the photographs respectively. John Rose and David Neilson who edited the text, Lee Saunders and Stanley Randolph who planned the layout of each page and Andy Legg who wrote the Supplement.

About this manual

Its aims

The aim of this book is to help you to get the best value from your vehicle. It can do so in two ways. First it can help you decide what work must be done (even should you choose to get it done by a garage), the routine maintenance, and the diagnosis and of course action when random faults occur. However, it is hoped that you will also use the second and fuller purpose by tackling the work yourself. On the simpler jobs it may even be quicker than booking the vehicle into a garage and going there twice, to leave and collect it. Perhaps most important, considerable amounts of money can be saved by avoiding the costs a garage must charge to cover their labor and overheads.

The book has drawings and descriptions to show the function of the various components so that their layout can be understood. Then the tasks are described and photographed in a step-by-step sequence so that even a novice can cope with complicated work. Such a person is the very one to buy a vehicle needing repair yet be unable to afford garage costs.

The jobs are described assuming only normal tools are available, and not special tools. But a reasonable outfit of tools will be a worthwhile investment. Many special workshop tools produced by the makers merely speed the work, and in these cases guidance is given as how to do the job without them, the oft quoted example being the use of a large hose clip to compress the piston rings for insertion in the cylinder. But on a very few occasions when the special tool is essential to prevent damage to components, then its use is described. Though it might be possible to borrow the tool, such work may have to be entrusted to the official agent.

The manufacturer's official workshop manuals are written for their trained staff, and so assume special knowledge; detail is left out. This book is written for the owner, and so goes into detail.

Using the manual

The book is divided into thirteen Chapters. Each Chapter is divided into numbered Sections which are headed in **bold type** between horizontal lines. Each Section consists of serially numbered paragraphs.

There are two types of illustration: (1) Figures which are numbered according to Chapter and sequence of occurrence in that Chapter and (2) Photographs which have a reference number on their caption. All photographs apply to the Chapter in which they occur so that the reference number pinpoints the pertinent Section and paragraph number.

Procedures, once described in the text, are not normally repeated. If it is necessary to refer to another Chapter the reference will be given.

When the left or right side of the vehicle is mentioned it is as if looking forward from the rear.

Great effort has been made to ensure that this book is complete and up-to-date. However, it should be noted that manufacturers continually modify their vehicles even in retrospect.

While every care is taken to ensure that the information in this manual is correct, no liability can be accepted by the authors or publishers for loss, damage or injury caused by any errors in, or omissions from, the information given.

Contents

	Page
Acknowledgements	2
About this manual	2
Introduction to the Chevrolet LUV	4
Buying spare parts and Vehicle identification numbers	4
Use of English	7
Recommended lubricants	8
Routine maintenance	9
Spare wheel location	12
Jacking points	13
Tools and working facilities	14
Chapter 1 Part A Engine — Series 1 to 4	16
Chapter 1 Part B Engine — Series 5 and 6	39
Chapter 2 Cooling system	52
Chapter 3 Carburetion; fuel, exhaust and emission control systems	56
Chapter 4 Ignition system	85
Chapter 5 Clutch	97
Chapter 6 Part A Manual transmission	103
Chapter 6 Part B Automatic transmission	121
Chapter 7 Driveshaft	125
Chapter 8 Rear axle	128
Chapter 9 Braking system	133
Chapter 10 Electrical system	147
Chapter 11 Suspension and steering	178
Chapter 12 Bodywork and fittings	197
Chapter 13 Supplement — Series 8 thru 12 models, including 4WD	209
Metric conversion tables	278
Index	279

Introduction to the Chevrolet LUV ½-ton Pick-up Truck

This manual covers the Chevrolet LUV pick-up truck Series 1 to 12. One basic body style has been used since the introduction of the Chevy LUV (Light Utility Vehicle) in 1972, with only minor design changes.

Beginning with the Series 9 model in 1979, the LUV was also offered in a 4-wheel drive version (covered in Chapter 13).

All models have a single overhead camshaft in-line engine of 110.8 cu in (1815 cc) capacity. The Series 1 to 4 engine remained basically unchanged up to the introduction of the Series 5 in 1976. Beginning with Series 5, the engine differs in that it has a crossflow cylinder head, with hemispherical combustion chambers and the timing sprockets and chain layout have been considerably modified.

On Series 1 to 4 engines the distributor and oil pump were driven by a layshaft while on the later engines they are driven by a pinion gear in front of the timing sprocket on the crankshaft.

A dry single plate clutch transfers drive to the 4-speed all synchromesh gearbox. Series 5 thru 11 models are available with 3-speed automatic transmission.

Drive to the rear axle is by a propeller shaft having a universal joint at each end; long wheelbase models are equipped with a two section propeller shaft. From the hypoid type differential unit, drive to the roadwheels is by semi-floating axle shafts.

Double acting shock absorbers are fitted front and rear and the front suspension is of the independent type.

Series 1 to 4 models were fitted with drum brakes front and rear, while Series 5 thru 12 models have disc front brakes. A servo unit is incorporated into all models and all brakes are self-adjusting.

Buying spare parts and vehicle identification numbers

Buying spare parts

Spare parts are available from many sources although they generally fall into two categories — those items which are supplied by a Chevrolet dealer and those which are supplied by auto accessory stores. In some cases the two facilities may be combined with an over-the-counter service and a pre-pack display area. In some cases it may be possible to obtain parts on a service-exchange basis but, where this can be done, always make sure that the parts returned are clean and intact. Our advice regarding spare parts purchase is as follows:

Chevrolet dealers: This is the best source of supply for major items such as transmissions, engines, body panels, etc. It is also the only place to obtain parts if your vehicle is still under warranty, since the warranty may be invalidated if non-Chevrolet parts are used.

Auto accessory shops: Auto accessory shops are able to supply practically all of the items needed for repair, maintenance, tune-up and customizing. This is not only true for the vehicle but also for tools and test equipment.

Whichever source of spare parts is used it will be essential to provide information concerning the model and year of manufacture of your vehicle.

Vehicle identification numbers

Modifications are a continuing and unpublicised process in vehicle manufacture quite apart from major model changes. Spare parts manuals and lists are compiled on a numerical basis, the individual vehicle numbers being essential to correct identification of the component required.

Chassis number. The chassis number is located on a plate attached to the left-side pillar and it also gives the date of production.

Engine number. The engine number is located between the distributor and the cylinder head on the cylinder block body.

Paint color code. The identification plate for the paint color code is located on the rear engine compartment panel.

The engine number above the distributor on the cylinder block

The paint color code identification plate in the engine compartment

The vehicle identification number (chassis number) on the left side rear pillar within the cab

A neat looking Chevy LUV pick-up spotted on a Sales Lot in California

A Chevy LUV pick-up

Use of English

As this book has been written in England, it uses the appropriate English component names, phrases, and spelling. Some of these differ from those used in America. Normally, these cause no difficulty, but to make sure, a glossary is printed below. In ordering spare parts remember the parts list may use some of these words:

English	American	English	American
Accelerator	Gas pedal	Locks	Latches
Aerial	Antenna	Methylated spirit	Denatured alcohol
Anti-roll bar	Stabiliser or sway bar	Motorway	Freeway, turnpike etc
Big-end bearing	Rod bearing	Number plate	License plate
Bonnet (engine cover)	Hood	Paraffin	Kerosene
Boot (luggage compartment)	Trunk	Petrol	Gasoline (gas)
Bulkhead	Firewall	Petrol tank	Gas tank
Bush	Bushing	'Pinking'	'Pinging'
Cam follower or tappet	Valve lifter or tappet	Prise (force apart)	Pry
Carburettor	Carburetor	Propeller shaft	Driveshaft
Catch	Latch	Quarterlight	Quarter window
Choke/venturi	Barrel	Retread	Recap
Circlip	Snap-ring	Reverse	Back-up
Clearance	Lash	Rocker cover	Valve cover
Crownwheel	Ring gear (of differential)	Saloon	Sedan
Damper	Shock absorber, shock	Seized	Frozen
Disc (brake)	Rotor/disk	Sidelight	Parking light
Distance piece	Spacer	Silencer	Muffler
Drop arm	Pitman arm	Sill panel (beneath doors)	Rocker panel
Drop head coupe	Convertible	Small end, little end	Piston pin or wrist pin
Dynamo	Generator (DC)	Spanner	Wrench
Earth (electrical)	Ground	Split cotter (for valve spring cap)	Lock (for valve spring retainer)
Engineer's blue	Prussian blue	Split pin	Cotter pin
Estate car	Station wagon	Steering arm	Spindle arm
Exhaust manifold	Header	Sump	Oil pan
Fault finding/diagnosis	Troubleshooting	Swarf	Metal chips or debris
Float chamber	Float bowl	Tab washer	Tang or lock
Free-play	Lash	Tappet	Valve lifter
Freewheel	Coast	Thrust bearing	Throw-out bearing
Gearbox	Transmission	Top gear	High
Gearchange	Shift	Torch	Flashlight
Grub screw	Setscrew, Allen screw	Trackrod (of steering)	Tie-rod (or connecting rod)
Gudgeon pin	Piston pin or wrist pin	Trailing shoe (of brake)	Secondary shoe
Halfshaft	Axleshaft	Transmission	Whole drive line
Handbrake	Parking brake	Tyre	Tire
Hood	Soft top	Van	Panel wagon/van
Hot spot	Heat riser	Vice	Vise
Indicator	Turn signal	Wheel nut	Lug nut
Interior light	Dome lamp	Windscreen	Windshield
Layshaft (of gearbox)	Countershaft	Wing/mudguard	Fender
Leading shoe (of brake)	Primary shoe		

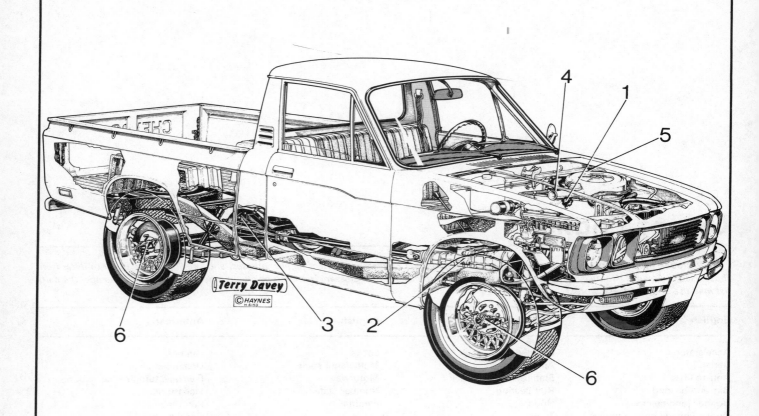

Recommended lubricants

Component	Lubricant
1 Engine	SE engine oil SAE 10W-40
2 Transmission	
Manual	Engine oil
Automatic	Dexron II ATF (automatic transmission fluid)
Manual w/transfer case	Engine oil
3 Differential (front and rear)	GL-5 SAE 90
4 Steering	
Manual steering gear (box)	GL-5 SAE 90
Power steering system	Dexron II ATF (automatic transmission fluid)
5 Brake and clutch hydraulic system	Delco Supreme 11 or DOT-3
6 Wheel bearings	Grease GM part No 1051344
Chassis lubrication	EP chassis grease

Note: *The above are general recommendation only. Different operating conditions require different lubricants. If in doubt as to the correct fluids, consult your driver's handbook or nearest GM dealer.*

Routine maintenance

For information relating to Series 8 thru 12 models, including 4WD, refer to Chapter 13.

Maintenance is essential for ensuring safety and desirable for the purpose of getting the best in terms of performance and economy from the vehicle. Over the years the need for periodic lubrication - oiling, greasing and so on - has been drastically reduced if not totally eliminated. This has unfortunately tended to lead some owners to think that because no such action is required, the items either no longer exist or will last forever. This is a serious delusion. It follows therefore that the largest initial element of maintenance is visual examination. This may lead to repairs or renewals.

In the maintenance summary, certain items are not applicable to all vehicles; as appropriate, these should be ignored.

Routine maintenance - all models

Every 250 miles (400 km) travelled, or weekly - whichever comes first

Check the engine oil level and top-up if necessary (photo).
Check the battery electrolyte level and top-up if necessary.
Check the windshield washer fluid level and top-up if necessary.
Check the tire pressures (when cold).
Examine the tires for wear and damage.
Check the brake and clutch reservoir fluid level and top-up if necessary.
Check the radiator coolant level and top-up if necessary.
Check that the brake operation is satisfactory.
Check the operation of all lights, instruments, warning devices, accessories, controls etc.
Check the cooling fan drivebelt adjustment.

Series 1 to 3 models

Every 3000 miles (5000 kms)

Lubricate the front suspension and steering linkages.
Drain and renew the engine oil. Check for leaks.
Check the tire pressures and their condition (including the spare).
Check that the wheel nuts are tight.
Check the brake lines and hoses for leakages and deterioration.
Check the brake master cylinder fluid level - top-up if required.
Check the clutch reservoir and top-up if required.
Inspect the battery terminal connections and also top-up with distilled water if required.
Check the coolant level and top-up if required. Check for leaks.
Inspect the exhaust system for signs of deterioration, leakage and secure mounting.
Check the engine idling and air fuel mixture - adjust if necessary.
Inspect the fuel lines for leakage, damage or poor connections.
Check the distributor contact breaker points - adjust if necessary.
Check the spark plugs condition and gap. Reset if necessary.

Every 6000 miles (9500 kms)

As per 3000 (5000 kms) miles plus the following:
Check the lubricant level in the rear axle and transmission - top-up as required.
Check propeller shaft flange bolts for tightness and the universal joints for wear.
Lubricate the parking brake pulleys and linkages - ensure that the cable is in good condition.

Topping up the engine oil

The engine oil filter

Raise the rear of the vehicle and check that the parking brake is in good working order.
Check the generator and air pump drivebelts for correct tension and condition - adjust/renew if required.
Check and lubricate the throttle linkages, the brake and clutch pedal springs and bushes.
Check and lubricate the hood and door latch hinges and latches.
Renew the engine oil filter (photo).
Check the throttle control linkage adjustment.
Check and adjust if necessary the ignition timing.

Outside Temperature	Viscosity Lubricant To Be Used
BELOW 50°F	SAE 80
0° TO 90°F	SAE 90
ABOVE 50°F CONSISTENTLY	SAE 140

The recommended oil viscosity chart for the rear axle

Every 12000 miles (19000 kms)

As per 6000 miles (9500 kms) plus the following:
Lubricate the front wheel bearings with wheel bearing grease.
Inspect the steering gear and check the lubricant level.
Inspect the brake linings and renew if required.
Check all electrical wiring, light bulbs and headlight alignment for malfunction and signs of deterioration - adjust and renew as required.
Drain the cooling system, flush and refill the system - check for leaks.
Lubricate the seat adjusters and seat track.
Remove the door trim panels and lubricate the window winder and door latch linkages.

Emission Control System
Carburetor/cylinder head
Check the carburetor and manifold retaining nuts for the correct torque specified.
Check the carburetor choke mechanism for correct adjustment and operation. Use a CO meter if available.
Check the cylinder head retaining bolts for correct torque specified.
Compression test each cylinder. If any cylinder registers less than 70% than the highest cylinder reading, the engine condition is suspect and should be repaired as required.
Check the valve clearances, and adjust if required.
Check the carburetor idle speed and adjust if required.

Distributor/timing
Renew the distributor points and condenser. Lubricate the cam.
Inspect the distributor cap and rotor and renew if worn or cracked.
Check the timing and advance mechanism, and adjust if required.

Coasting richer system
Check the carburetor solenoid operation (see Chapter 3).

Fuel tank and supply lines
Inspect the fuel tank and supply lines for signs of leakage or damage.
Check that the fuel filler cap is correctly sealed. Renew as required.
Renew the fuel line filters.

Air cleaner
If vehicle operates in dusty conditions renew the air cleaner element.
Check the installation of the thermostat control and ensure that the hoses and ducting are securely connected. Test the check valve for correct operation.

Spark plugs
Renew the spark plugs and check the gaps.

Evaporation Control System (ECS)
Inspect all fuel and vapor lines/hoses to ensure good connections.
Remove the relief valve and check valve and test for correct operation. Renew as required.

Positive Crankcase Ventilation System (PCV)
Renew the PCV valve and ensure good connections.
Blow the hose out with compressed air.

Vacuum system and fittings
Inspect all fittings and connections for damage or deterioration. Renew as required.

Exhaust Gas Recirculation System (EGR)
Check the operation of the EGR valve. Run the engine at idle and open the throttle slightly. The valve should open and close on returning to idle.

Every 15000 miles (24000 kms)

Under extreme conditions of usage, such as heavy city traffic or very hot climatic conditions, drain the oil from the automatic transmission and replenish using the correct grade of lubricant.

Every 24000 miles (38500 kms)

Check the AIR system hoses and fittings and the mixture control valve operation. Renew as required.
Drain the transmission oil and renew with recommended grade.
Drain the rear axle oil and renew with recommended grade.
Renew the distributor cap and rotor.
Check the ignition HT cables and renew if needed.
Renew the air cleaner element.
Clean and repack the wheel bearings with the recommended grease.

Every 30000 miles (48000 kms)

Renew the automatic transmission fluid.

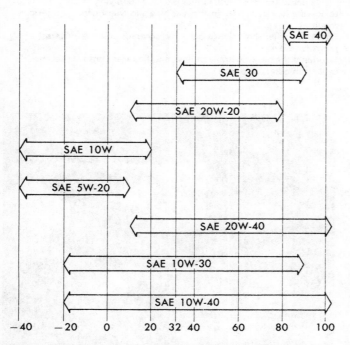

The recommended oil viscosity chart for the engine lubricant
Note: SAE 5W - 20 oils are not recommended for continued high speed driving.

Every 36000 miles (58000 kms)

Drain and renew the steering gear lubricant with the recommended grade.

Series 4 and 5 models

Every 7500 miles (12000 kms) or 6 months

Chassis lubrication
Check and lubricate the front suspension and steering linkages.
Lubricate the parking brake pulley, cables and linkages.
Lubricate the hood latch and hinges, the tailgate latch and hinges.
Check condition of latches and hinges and adjust if required.

Lubricant and fluid levels
Check the brake master cylinder, clutch master cylinder, battery, rear axle, engine and transmission lubricant/fluid levels. Replenish as required.
A higher lubricant or fluid loss than normal in any of the above items is indicative of a malfunction, and should therefore be checked and rectified accordingly.

Propeller shaft
Ensure that the propeller shaft flange bolts are tightened to the correct torque - 18 - 20 lb f ft (2.4 - 2.7 kg fm).

Engine
Drain the engine oil and replenish with the correct grade.

Wheels and tires
Change the respective wheels and tires to alternative positions
Check the tires for wear, cuts or deterioration.
Check the wheel nuts for tightness. Torque to 60 - 80 lb f ft (8.2 - 11 kg fm).

Exhaust system
Inspect the exhaust system for signs of leakage, deterioration and security. Repair or renew as required.

Suspension and Steering
Inspect the steering and suspension joints for signs of wear or lack of lubrication. Renew immediately any suspect components.

Brake lines and hoses
Clean and inspect the respective brake lines and hoses for any signs of cracks, chafing, leakage or general deterioration. Renew any suspect parts immediately.
On Series 5 models, inspect the disc pads on the front brakes - renew if required. Lubricate the caliper sliding parts.

General checks
Inspect the lap and shoulder safety belts, buckles and retractors for deterioration and efficient operation. Check that the anchorage points are secure.
Inspect the windshield wipers and fit new wiper blades if required.
Check the washer unit fluid level and ensure that the washer unit is fully operational.
Ensure that the parking brake is fully operational.
Check the side marker lights, headlamps, parking lights, indicators, back-up lights and hazard warning flashers.
Check the generator and air pump drivebelts for condition and tension.
Remove the spark plugs and clean and adjust the gaps.
Check the oil level in the rear axle and gearbox.

Every 15000 miles (24000 kms)

General
Service as per 7500 miles (12000 kms) plus the following:
Inspect the fan, generator and air pump drivebelts for signs of cracking, wear or deterioration. Check the tension and adjust if required.
Clean, inspect and lubricate the throttle and parking brake linkages and the brake and clutch pedal springs, bushes and pins.
Check the headlight alignment (see Chapter 10) and adjust if necessary.
Remove the rear brake drums and inspect the brake linings and wheel cylinders on each wheel.
Check the parking brake adjustment.
Steam clean or hose the underbody and chassis. Inspect for damage or signs of rust and repair as required.
Change the engine oil filter.
Drain and flush the cooling system. Replenish and check for leaks.
Check the steering gear lubricant - drain and renew lubricant every 37500 miles (60000 kms).

Emission control system

Timing
Adjust the carburetor idle speed and mixture - check with a CO meter if possible - see Chapter 3.
Check and adjust if required the engine timing and dwell.
Renew the distributor points and condenser, and inspect the distributor cap and rotor for signs of damage, wear or cracking. Renew as required.
Lubricate the cam.
Check and lubricate the distributor advance mechanism.

Carburetor/Cylinder head
Check the carburetor and manifold retaining nuts for tightness and tighten to the specified torque if required.
Check the choke adjustment and operation, in particular inspect the choke shaft which may be coated in petroleum gum or damaged.
Clean and repair/renew as required.

Fuel tank and fuel system
Check the fuel tank, fuel cap and supply lines for signs of leakage or damage. Ensure that the fuel cap is able to seal correctly.
Renew the fuel line.

Air cleaner
Inspect the thermostatically controlled air cleaner to ensure that all hoses, ducts and connections are in good condition. Check that the valve operates correctly.

Evaporation Control System
Inspect the fuel and vapor lines and hose connections.
Remove the relief and check valve and check that they are fully operational - renew damaged or worn parts as required.

Positive Crankcase Ventilation (PCV) system
Renew the PCV valve.
Inspect the hose and connections to ensure they are in good condition. Blow the hose clean with compressed air, and if the hose is damaged or has deteriorated renew it.

Vacuum system and Air Injection Reactor (AIR) System
Inspect the fittings and connections for damage. If the hoses are cracked, worn or show signs of deterioration, renew them as required.

Valve clearance
Remove the cam cover to check the valve clearances.
Check that the camshaft carrier/rocker shaft bracket nuts are to the correct torque before checking the clearances. Adjust as required.

Spark plugs
Renew the spark plugs - check the clearances before assembly.
Check the HT cables and coil wires for condition - renew as required.

Air conditioning system
Inspect the air conditioning system hoses for secure connections and condition.
Inspect the refrigerant charge at sight glass (if fitted).
Renew the hoses or refrigerant if required. If any hoses show signs of deterioration they must be renewed by a GM Dealer. See Chapter 10, Section 41.

Every 30000 miles (48000 kms)

Air cleaner
Remove and replace the engine air filter element, but do not use the vehicle with the filter removed - backfiring can occur causing a fire in the engine compartment.
Change the filter more frequently if vehicle operates in dusty areas.

Air Injector Reactor System
Check the hoses and fittings.
Check the mixture control valve operation and if defective renew.

Rear axle and Transmission
Drain and replenish the rear axle and transmission oils with the correct grade of lubricant.

Wheel bearings
Clean and repack the wheel bearings with the specified lubricant.

Cooling System
Drain, flush and refill the cooling system.
Renew any cracked or deteriorated hoses.
Have the filler cap pressure tested.

Spare wheel location

Jacking points

Jacking points: side/rear

Jacking points: front

Jacking points: rear

Chassis stand lift point

Tools and working facilities

Introduction

A selection of good tools is a fundamental requirement for anyone contemplating the maintenance and repair of a motor vehicle. For the owner who does not possess any, their purchase will prove a considerable expense, offsetting some of the savings made by doing-it-yourself. However, provided that the tools purchased are of good quality, they will last for many years and prove an extremely worthwhile investment.

To help the average owner to decide which tools are needed to carry out the various tasks detailed in this manual, we have compiled three lists of tools under the following headings:
Maintenance and minor repair, Repair and overhaul, and *Special.*
The newcomer to practical mechanics should start off with the *Maintenance and minor repair* tool kit and confine himself to the simpler jobs around the vehicle. Then, as his confidence and experience grow, he can undertake more difficult tasks, buying extra tools as, and when, they are needed. In this way, a *Maintenance and minor repair* tool kit can be built-up into a *Repair and overhaul* tool kit over a considerable period of time without any major cash outlays. The experienced do-it-yourselfer will have a tool kit good enough for most repair and overhaul procedures and will add tools from the *Special* category when he feels the expense is justified by the amount of use these tools will be put to.

It is obviously not possible to cover the subject of tools fully here. For those who wish to learn more about tools and their use there is a book entitled *How to Choose and Use Car Tools* available from the publishers of this manual.

Maintenance and minor repair tool kit

The tools given in this list should be considered as a minimum requirement if routine maintenance, servicing and minor repair operations are to be undertaken. We recommend the purchase of combination wrenches (ring one end, open-ended the other); although more expensive than open-ended ones, they do give the advantages of both types of wrench.

Combination wrenches - 3/8 to 11/16 in AF or 10 to 17 mm
Adjustable wrench - 9 inch
Engine sump/gearbox/rear axle drain plug key (Where applicable)
Spark plug wrench (with rubber insert)
Spark plug gap adjustment tool
Set of feeler gauges
Brake adjuster wrench (where applicable)
Brake bleed nipple wrench
Screwdriver - 4 in long x ¼ in dia (flat blade)
Screwdriver - 4 in long x ¼ in dia (cross blade)
Combination pliers - 6 inch
Hacksaw, junior
Tire pump
Tire pressure gauge
Grease gun (where applicable)
Oil can
Fine emery cloth (1 sheet)
Wire brush (small)
Funnel (medium size)

Repair and overhaul tool kit

These tools are virtually essential for anyone undertaking any major repairs to a motor vehicle, and are additional to those given in the *Maintenance and minor repair* list. Included in this list is a comprehensive set of sockets. Although these are expensive they will be found invaluable as they are so versatile - particularly if various drives are included in the set. We recommend the ½ inch square-drive type, as this can be used with most proprietary torque wrenches. If you cannot afford a socket set, even bought piecemeal, then inexpensive tubular box wrenches are a useful alternative.

The tools in this list will occasionally need to be supplemented by tools from the *Special* list.

Sockets (or box wrenches) to cover range 6 to 27 mm
Reversible ratchet drive (for use with sockets)
Extension piece, 10 inch (for use with sockets)
Universal joint (for use with sockets)
Torque wrench (for use with sockets)
Self-grip wrench - 8 inch
Ball pein hammer
Soft-faced hammer, plastic or rubber
Screwdriver - 6 in long x 5/16 in dia (flat blade)
Screwdriver - 2 in long x 5/16 in square (flat blade)
Screwdriver - 1½ in long x ¼ in dia (cross blade)
Screwdriver - 3 in long x 1/8 in dia (electricians)
Pliers - electricians side cutters
Pliers - needle nosed
Pliers - circlip (internal and external)
Cold chisel - ½ inch
Scriber (this can be made by grinding the end of a broken hacksaw blade)
Scraper (this can be made by flattening and sharpening one end of a piece of copper pipe)
Center punch
Pin punch
Hacksaw
Valve grinding tool
Steel rule/straight edge
Allen keys
Selection of files
Wire brush (large)
Axle stands
Jack (strong scissor or hydraulic type)

Special tools

The tools in this list are those which are not used regularly, are expensive to buy, or which need to be used in accordance with their manufacturers' instructions. Unless relatively difficult mechanical jobs are undertaken frequently, it will not be economic to buy many of these tools. Where this is the case, you could consider clubbing together with friends (or an automobile club) to make a joint purchase, or borrowing the tools against a deposit from a local repair station or tool hire specialist.

The following list contains only those tools and instruments freely available to the public, and not those special tools produced by the vehicle manufacturer specifically for its dealer network. You will find occasional references to these manufacturers' special tools in the text of this manual. Generally, an alternative method of doing the job without the vehicle manufacturer's special tool is given. However, sometimes, there is no alternative to using them. Where this is the case and the rele-

Tools and working facilities

vant tool cannot be bought or borrowed you will have to entrust the work to a franchised dealer.

- Valve spring compressor
- Piston ring compressor
- Balljoint separator
- Universal hub/bearing puller
- Impact screwdriver
- Micrometer and/or vernier gauge
- Carburetor flow balancing device (where applicable)
- Dial gauge
- Stroboscopic timing light
- Dwell angle meter/tachometer
- Universal electrical multi-meter
- Cylinder compression gauge
- Lifting tackle
- Trolley jack
- Light with extension lead

Buying tools

For practically all tools, a tool factor is the best source since he will have a very comprehensive range compared with the average repair station or accessory store. Having said that, accessory stores often offer excellent quality tools at discount prices, so it pays to shop around.

Remember, you don't have to buy the most expensive items on the shelf, but it is always advisable to steer clear of the very cheap tools. T There are plenty of good tools around at reasonable prices, so ask the proprietor or manager of the shop for advice before making a purchase.

Working facilities

Not to be forgotten when discussing tools, is the workshop itself. If anything more than routine maintenance is to be carried out, some form of suitable working area becomes essential.

It is appreciated that many an owner mechanic is forced by circumstances to remove an engine or similar item, without the benefit of a garage or workshop. Having done this, any repairs should always be done under the cover of a roof.

Wherever possible, any dismantling should be done on a clean flat workbench or table at a suitable working height.

Any workbench needs a vise: one with a jaw opening of 4 in (100 mm) is suitable for most jobs. As mentioned previously, some clean dry storage space is also required for tools, as well as the lubricants, cleaning fluids, touch-up paints and so on which soon become necessary.

Another item which may be required, and which has a much more general usage, is an electric drill with a chuck capacity of at least 5/16 in (8 mm). This, together with a good range of twist drills, is virtually essential for fitting accessories such as wing mirrors and back-up lights.

Last, but not least, always keep a supply of old newspapers and clean lint-free rags available, and try to keep any working area as clean as possible.

Care and maintenance of tools

Having purchased a reasonable tool kit, it is necessary to keep the tools in a clean serviceable condition. After use, always wipe off any dirt, grease and metal particles using a clean, dry cloth, before putting the tools away. Never leave them lying around after they have been used. A simple tool rack on the garage or workshop wall, for items such as screwdrivers and pliers is a good idea. Store all normal spanners and sockets in a metal box. Any measuring instruments, gauges, meters, etc., must be carefully stored where they cannot be damaged or become rusty.

Take a little care when the tools are used. Hammer heads inevitably become marked and screwdrivers lose the keen edge on their blades from time-to-time. A little timely attention with emery cloth or a file will soon restore items like this to a good serviceable finish.

Wrench jaw gap comparison table

Jaw gap (in)	Wrench size
0.250	¼ in AF
0.275	7 mm AF
0.312	5/16 in AF
0.315	8 mm AF
0.340	11/32 in AF; 1/8 in Whitworth
0.354	9 mm AF
0.375	3/8 in AF
0.393	10 mm AF
0.433	11 mm AF
0.437	7/16 in AF
0.445	3/16 in Whitworth; ¼ in BSF
0.472	12 mm AF
0.500	½ in AF
0.512	13 mm AF
0.525	¼ in Whitworth; 5/16 in BSF
0.551	14 mm AF
0.562	9/16 in AF
0.590	15 mm AF
0.600	5/16 in Whitworth; 3/8 in BSF
0.625	5/8 in AF
0.629	16 mm AF
0.669	17 mm AF
0.687	11/16 in AF
0.708	18 mm AF
0.710	3/8 in Whitworth; 7/16 in BSF
0.748	19 mm AF
0.750	¾ in AF
0.812	13/16 in AF
0.820	7/16 in Whitworth; ½ in BSF
0.866	22 mm AF
0.875	7/8 in AF
0.920	½ in Whitworth; 9/16 in BSF
0.937	15/16 in AF
0.944	24 mm AF
1.000	1 in AF
1.010	9/16 in Whitworth; 5/8 in BSF
1.023	26 mm AF
1.062	1.1/16 in AF; 27 mm AF
1.100	5/8 in Whitworth; 11/16 in BSF
1.125	1.1/8 in AF
1.181	30 mm AF
1.200	11/16 in Whitworth; ¾ in BSF
1.250	1¼ in AF
1.259	32 mm AF
1.300	¾ in Whitworth; 7/8 in BSF
1.312	1.5/16 in AF
1.390	13/16 in Whitworth; 15/16 in BSF
1.417	36 mm AF
1.437	1.7/16 in AF
1.480	7/8 in Whitworth; 1 in BSF
1.500	1½ in AF
1.574	40 mm AF; 15/16 in Whitworth
1.614	41 mm AF
1.625	1.5/8 in AF
1.670	1 in Whitworth; 1.1/8 in BSF
1.687	1.11/16 in AF
1.811	46 mm AF
1.812	1.13/16 in AF
1.860	1.1/8 in Whitworth; 1¼ in BSF
1.875	1.7/8 in AF
1.968	50 mm AF
2.000	2 in AF
2.050	1¼ in Whitworth; 1.3/8 in BSF
2.165	55 mm AF
2.362	60 mm AF

Chapter 1 Part A: Engine - Series 1 to 4

Contents

Ancillary components - removal ... 8	General ancillary items - reassembly ... 44
Camshaft carrier - installation ... 42	General description ... 1
Camshaft - examination and renovation ... 28	Layshaft - examination and renovation ... 29
Camshaft - removal ... 17	Lubrication system - description ... 22
Clutch and flywheel - removal ... 15	Major operations possible with the engine installed ... 2
Connecting rods and connecting rod/big-end bearings - examination and renovation ... 25	Major operations requiring engine removal ... 3
	Methods and equipment for engine removal ... 4
Crankshaft and main bearings - examination and renovation ... 24	Oil pan (sump) - installation ... 39
	Oil pan (sump) - removal ... 11
Crankshaft and main bearings - reassembly ... 35	Oil pump - examination ... 33
Crankshaft and main bearings - removal ... 21	Oil pump - installation ... 38
Cylinder bores - examination and renovation ... 26	Oil pump - removal ... 12
Cylinder head and valves - servicing and decarbonising ... 31	Piston pin (gudgeon pin) - removal ... 20
Cylinder head - installation ... 41	Piston rings - removal ... 19
Cylinder head - removal ... 10	Pistons and connecting rods - removal ... 18
End plates, layshaft and primary timing sprocket - assembly ... 37	Pistons and rings - examination and renovation ... 40
Engine/Transmission - installation ... 46	Pistons, rings and connecting rods - reassembly and installation ... 36
Engine and Transmission - removal ... 6	
Engine dismantling - general ... 7	Primary timing sprocket and layshaft - removal ... 14
Engine - initial start-up after overhaul ... 47	Secondary timing sprocket - installation ... 43
Engine reassembly - general ... 34	Timing case - removal ... 13
Engine - removal ... 5	Timing chain, sprockets and camshaft - removal ... 9
Examination and renovation - general ... 23	Timing chain, sprockets and chain tensioners - examination and renovation ... 30
Flywheel and ring gear - inspection and renovation ... 32	
Flywheel, crankshaft pulley and oil filter - installation ... 39	Valve clearances - adjustment ... 45
Front and rear plates - removal ... 16	

Specifications

Engine	4-cylinder, 4-stroke in-line, ohc, water cooled
Firing order	1, 3, 4, 2
Combustion chambers type	Wedge
Bore	3.31 in
Stroke	3.23 in
Piston displacement	110.8 cu in
Compression ratio	8.2 : 1
Compression pressure	163.56 psi @ 300 rpm
Compression pressure maximum variance	8.53 psi
Maximum output	75 HP @ 5000 rpm (SAE net)
Maximum torque	88 ft lb @ 3000 rpm

Chapter 1 Part A: Engine - Series 1 to 4

Engine weight-dry
Dry ... 342 lb
With oil/coolant ... 355 lb

Exhaust gas control ... Air Injection Reactor system (AIR) (Federal)
Exhaust Gas Recirculation (EGR) (California)

Cylinder head
Inlet valves open at ... 31° BTDC
   close at ... 67° ATDC
Exhaust valves open at ... 59° BTDC
   close at ... 23° ATDC
Valve clearance (between rocker and cam) at low temperature
   inlet ... 0.004 in
   exhaust ... 0.006 in
Valve head diameter
   inlet ... 1.69 in
   exhaust ... 1.30 in
Valve stem to guide clearance
   inlet (standard) ... 0.0016 in
      maximum wear allowance ... 0.0079 in
   exhaust (standard) ... 0.0020 in
      maximum wear allowance ... 0.0098 in
Valve spring tension
   outer spring at length of 1.58 in ... 47.4 lb nominal
      wear limit ... 41.8 lb
      maximum deviation from vertical ... 0.08 in
   inner spring at length of 1.50 in ... 17.8 lb nominal
      wear limit ... 15.4 lb
      maximum deviation from vertical ... 0.08 in

Camshaft
Camshaft
   nominal height of cam ... 1.534 in
   minimum height ... 1.514 in
Camshaft journals
   nominal diameter ... 1.89 in
   minimum diameter allowance ... 1.87 in
   minimum wear allowance ... 0.002 in
Camshaft endplay
   standard ... 0.032 in
   maximum end play allowance ... 0.008 in
Camshaft journal to bearing clearance
   standard ... 0.0024 in
   maximum wear allowance ... 0.0059 in
Layshaft end play
   standard ... 0.0032 in
   maximum wear allowance ... 0.008 in
Layshaft clearance between journal and bearing
   journal outside diameter ... 1.772 in nominal
   maximum wear allowance ... 1.756 in
   maximum unever wear allowance ... 0.002 in
   journal to bearing clearance (standard) ... 0.0024 in

Pistons
Type ... Trunk (no T slot)
Material ... LO-EX
Number of rings
   compression ... 2
   oil control ... 1
Piston to cylinder wall clearance ... 0.0018 to 0.0026 in
Piston size
Standard:
   Grade 'A' ... 3.3049 to 3.3053 in
   Grade 'B' ... 3.3053 to 3.3057 in
   Grade 'C' ... 3.3507 to 3.3061 in
   Grade 'D' ... 3.3061 to 3.3065 in
Oversize pistons
   Standard to cylinder diameter ... 3.3071 to 3.3087 in
   0.005 in oversize to cylinder diameter ... 3.3120 to 3.3136 in
   0.010 in oversize to cylinder diameter ... 3.3169 to 3.3185 in
   0.020 in oversize to cylinder diameter ... 3.3268 to 3.3283 in
   0.030 in oversize to cylinder diameter ... 3.3366 to 3.3382 in
   0.040 in oversize to cylinder diameter ... 3.3465 to 3.3480 in
   0.050 in oversize to cylinder diameter ... 3.3563 to 3.3579 in
   0.060 in oversize to cylinder diameter ... 3.3661 to 3.3677 in

Piston ring gap
- 1st compression ring ... 0.008 to 0.016 in
- 2nd compression ring ... 0.008 to 0.016 in
- Oil ring ... 0.039 to 0.012 in

Piston ring to groove clearance
- standard compression rings ... 0.0012 to 0.0028 in
- maximum limit ... 0.012 in
- Oil control ring ... 0.0008 to 0.0024 in
- maximum limit ... 0.006 in

Gudgeon pin
- nominal outside diameter ... 0.87 in
- maximum wear allowance ... 0.8650 in

Connecting rods end play
Connecting rod to crankpin
- standard ... 0.011 in
- maximum wear allowance ... 0.014 in

Connecting rod to piston boss ... 0.069 in

Gudgeon pin to connecting rod small end bush clearance
- standard ... 0.00024 in
- maximum wear allowance ... 0.002 in

Connecting rod small end bush inside diameter (after reaming) ... 0.8662 to 0.8665 in

Crankshaft

	Journal outside diameter	Crankpin outside diameter
Standard	2.2016 to 2.2022 in	1.9262 to 1.9268 in
−0.010	2.1917 to 2.1923 in	1.9163 to 1.9169 in
−0.020	2.1819 to 2.1825 in	1.9065 to 1.9071 in
−0.030	2.1720 to 2.1726 in	1.8967 to 1.8972 in
−0.040	2.1622 to 2.1628 in	1.8868 to 1.8874 in
−0.050	2.1524 to 2.1529 in	1.8770 to 1.8776 in

Crankshaft end float
- standard setting ... 0.0059 in
- maximum allowance ... 0.012 in

Torque settings

	lb f ft	kg f m
Cylinder head bolts		
Series 1 - initial	43	5.9
Final	58	8.0
Series 2 to 4 - initial	43	5.9
Final	70	9.6
Crankshaft main bearing cap bolt	72	9.9
Crankcase bolts (Series 1)	15	2.0
Camshaft sprocket retaining nut	33	4.5
Connecting rod big-end bolts	43	5.9
Camshaft carrier mounting bolts	15	2.0
Layshaft sprocket retaining nut	33	4.5
Layshaft thrust plate bolts		
Series 1	6	0.8
Series 2 to 4	9	1.2
Flywheel retaining bolt -		
Series 1	36	4.9
Series 2 to 4	69	9.5

1 General description

The engine is of the four cylinder in-line type, the valves being operated by an overhead camshaft.

Both the crankcase (cylinder block) and cylinder head are manufactured in cast aluminium alloy. To increase rigidity, the crankshaft is supported by five main bearings which also enable the engine to operate smoothly under its varied operating conditions and prolong the life of the crankshaft bearings and allied components.

Both the crankshaft main bearings and the connecting rod 'big-end' bearings are of the shell type. The connecting rod small ends are fitted with bushes reamed to suit the piston pin (gudgeon pin) size.

In the cylinder head the valve system incorporates double valve springs unequally pitched to minimize valve bounce at high engine revolutions.

Wedge type combustion chambers are used.

The engine components are lubricated by means of a trochoid type oil pump driven from the engine layshaft (auxiliary shaft). A full flow oil filter is fitted so that oil, delivered from the pump, is filtered prior to being forced into circulation to the various components.

The engine cooling system incorporates a radiator, thermostat, cooling fan and connecting pipes and hoses.

Special notes

Special precautions must be taken when handling the refrigerant lines of the air conditioning system (when fitted), and these are as follows:

a) *Flexible hose lines must not be bent to a radius of less than 4 inches (101.6 mm).*

b) *Flexible hose lines must be located at least 2½ inches (63.5 mm) from the exhaust manifold when refitted.*

Chapter 1 Part A: Engine - Series 1 to 4

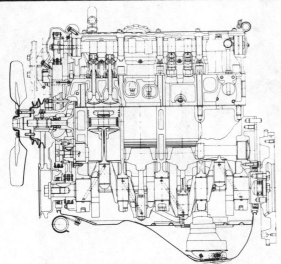

Fig. 1.1, Sectional drawing of the engine

5.7a Disconnect the carburetor control cable

Fig. 1.2. Remove the air filter unit

c) All metal tubing lines must be free of kinks and must therefore be handled with care.
d) **Do not disconnect any of the air conditioning supply lines.**
e) Do not weld or apply heat in the vicinity of the air conditioning lines.

2 Major operations possible with the engine installed

The following major tasks can be achieved with the engine installed in the vehicle. The degree of difficulty varies according to the tools, lifting tackle and personal mechanical experience.

a) *The removal and installation of the cylinder head.*
b) *The removal and installation of the camshaft and bearings.*
c) *The removal and installation of the engine mountings.*
d) *The removal and installation of the flywheel.*
e) *The removal and installation of the clutch pilot bearing.*
f) *The removal and installation of the gear case and seal.*
g) *The removal and installation of the timing chains and sprockets.*
h) *The removal and installation of the oil and water pumps.*
i) *The removal and installation of the oil pan (sump).*
j) *The removal and installation of the pistons and connecting rods.*
k) *The removal and installation of the connecting rod bearings.*

3 Major operations requiring engine removal

The only items which cannot be removed and installed without the engine being removed are the crankshaft and main bearings.

4 Methods and equipment for engine removal

1 The engine can either be removed complete with the transmission, or the two units can be removed separately. If the transmission is to be removed, it is preferable to remove it while still coupled to the engine.
2 Essential equipment includes a suitable jack or support(s) so that the vehicle can be raised whilst working underneath, and a hoist or lifting tackle capable of taking the engine (and transmission, if applicable) weight. If the lifting tackle demands a reduction of weight, such items as the cylinder head, manifolds and engine ancillaries can be removed before lifting out the engine, provided that a suitable sling can be used for lifting purposes. If an inspection pit is available some problems associated with jacking will be alleviated, but at some time during the engine removal procedure a jack will be required beneath the transmission (a trolley jack is very useful for this application).

5 Engine - removal

1 Prior to commencing work it is necessary to drive the vehicle over an inspection pit, onto ramps or jack it up for access to the exhaust and transmission.
2 Remove the hood (bonnet) as described in Chapter 12.
3 Detach the ground cable from its battery terminal.
4 Drain the cooling system and remove the radiator as described in Chapter 2.
5 Drain the engine oil into a suitable container (and transmission if the gearbox is to be removed also).
6 Disconnect the hoses from the air cleaner and unscrew the air cleaner retaining clamp. Remove the air cleaner unit from the carburetor.
7 Disconnect the carburetor control cable (photo) and the fuel supply hoses. Disconnect the carburetor wiring (photo).
8 Unscrew the exhaust pipe to manifold flange retaining nuts and disconnect the pipe from the manifold.
9 Detach the connecting wires from the generator and starter unit. Remove the starter motor retaining bolts and withdraw the starter motor (if removing engine only).

5.7b Disconnect the carburetor wiring from the multi plug connector

5.8 Unscrew the exhaust pipe to manifold nuts

5.12 The ground cable and chain tensioner oil feed pipe

6.3 Remove the gearshift lever

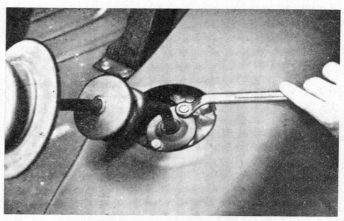

Fig. 1.3. Remove the gearshift lever retainer

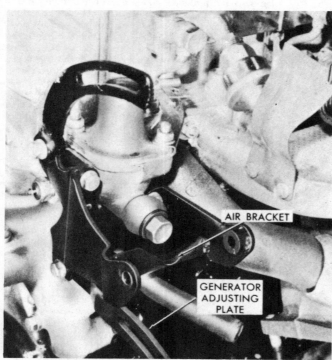

Fig. 1.4. Remove the air pump adjustment plates

6.7 Removing the engine and transmission units. Note the angle of tilt and take care not to damage surrounding bodywork or fittings when lifting

8.8 Remove the generator retaining bolts and note the hose clip attached to the lower bolt at the rear

10 Detach the heater hose from the fender side and the vacuum hose.
11 On the right-hand engine side, detach the distributor high tension cables and the ground cable from the cylinder head cover to dashboard on the cylinder head cover side. Detach the distributor ground cable from the three retaining clips on the engine unit.
12 Disconnect the thermometer unit and oil pressure unit cords, and detach the ground cable from the timing gear case on the engine side (photo).
13 Detach the check and relief valve hoses.
14 Locate the lifting sling into the engine hangers and slightly lift the engine. Undo and remove the rear and front engine mountings.
15 Place a jack under the gearbox to support it and then undo and withdraw the respective engine to clutch housing retaining bolts.
16 Check that all of the surrounding engine to body connections and fittings are disconnected.
17 Carefully draw the engine forwards, away from the transmission, taking care that the engine weight is not transferred to the transmission input shaft (manual transmission), then lift the engine upwards and away from the vehicle.
18 It may be necessary to detach and remove the fan and pulley and also the crankshaft drive pulley to allow the engine extra forward movements in order to clear the transmission input shaft, and prevent damage to the matrix just forward of the radiator position. Take care not to let the engine sway or damage accessories, wiring or vehicle paintwork during removal.
19 With the engine removed, do not attempt to move the vehicle with the transmission installed unless the transmission is supported by a suitable means from the vehicle frame.

6 Engine and transmission - removal

Follow the instructions given in Section 5, paragraphs 1 to 13 and then proceed as follows:
1 Undo the clutch slave cylinder retaining bolts and detach the cylinder unit. Disconnect the exhaust pipe bracket from the clutch housing.
2 Withdraw the speedo drive cable from the transmission unit and detach the transmission to body ground cable.
3 From inside the cab, lift the gearshift lever boot clear of the transmission tunnel and unscrew the two gearshift lever retaining bolts (Fig. 1.3). Remove the lever (photo).
4 Detach the propeller shaft from the differential unit end (see Chapter 7) and withdraw the propeller shaft from the transmission unit. Insert a plug into the end of the transmission extension housing to prevent oil spillage.
5 On some models, it will be necessary to remove the gearbox rear mounting plate and crossmember to facilitate the gearbox downwards movement during removal. Support the gearbox with a jack and then remove the two bolts on top of the crossmember and two at the front.
6 Locate the lifting sling into the engine hangers and slightly lift the engine. Undo and remove the rear and front engine mountings.
7 Check that the surrounding engine and gearbox fittings and accessories are detached, lift the engine and transmission unit forwards and upwards (at an approximate angle of 60°). Take care not to let the engine and transmission unit sway or damage the surrounding accessories and paintwork (photo).
8 Transfer the engine and transmission to a suitable clean work area so that the engine and transmission ancillaries can be detached as necessary.

7 Engine dismantling - general

1 It is best to mount the engine on a dismantling stand but if one is not available, then stand the engine on a strong bench so as to be at a comfortable working height. Failing this, the engine can be stripped down on the floor.
2 During the dismantling process the greatest care should be taken to keep the exposed parts free from dirt. As an aid to achieving this, it is sound advice to thoroughly clean down the outside of the engine, removing all traces of oil and congealed dirt.
3 Use kerosene or a good grease solvent. The latter compound will make the job much easier, as, after the solvent has been applied and allowed to stand for a time, a vigorous jet of water will wash off the solvent and all the grease and filth. If the dirt is thick and deeply embedded, work the solvent into it with a wire brush.
4 Finally wipe down the exterior of the engine with a rag and only then, when it is quite clean should the dismantling process begin. As the engine is stripped, clean each part in a bath of kerosene.
5 Never immerse parts with oilways in kerosene, (eg. the crankshaft) but to clean, wipe down carefully with a kerosene dampened rag. Oilways can be cleaned out with wire. If an air-line is present all parts can be blown dry and the oilways blown through as an added precaution.
6 Re-use of old engine gaskets is false economy and can give rise to oil and water leaks, if nothing worse. To avoid the possibility of trouble after the engine has been reassembled always use new gaskets throughout.
7 Do not throw the old gaskets away as it sometimes happens that an immediate replacement cannot be found and the old gasket is then very useful as a template for making up a replacement. Hang up the old gaskets as they are removed on a suitable hook or nail.
8 To strip the engine it is best to work from the top down. The oil pan provides a firm base on which the engine can be supported in an upright position. When the stage where the oil pan must be removed is reached, the engine can be turned on its side and all other work carried out with it in this position.
9 Wherever possible install nuts, bolts and washers fingertight from wherever they were removed. This helps avoid later loss and muddle. If they cannot be installed then lay them out in such a fashion that it is clear from where they came.

8 Ancillary components - removal

1 If you are stripping the engine completely or preparing to install a reconditioned unit, all the ancillaries must be removed first. If you are going to obtain a reconditioned 'short' motor (block, crankshaft, pistons and connecting rods) then obviously the cam box, cylinder head and associated parts will need retention for installing to the new engine. It is advisable to check just what you will get with a reconditioned unit as changes are made from time to time.
2 If the transmission is attached, unscrew the retaining bolts from the clutch housing and separate the engine and gearbox units.
3 Unscrew the oil feed pipe joint bolt and remove the pipe.
4 Detach the respective hoses and cables from the engine, taking note of their respective positions for reassembly. Attach identity labels to ensure correct refitting.
5 Unscrew the distributor retaining bolt and mark the distributor bowl and engine in line with the rotor segment to ensure correct reassembly (see Chapter 4). Disconnect the vacuum line and withdraw the distributor from the engine.
6 Unscrew the oil filter retaining bolts and remove the filter. Withdraw the engine oil dipstick.
7 Slacken the air pump adjustment bolt and mounting bolt. Pivot the air pump and remove the drive belt, then remove the bolts to withdraw the pump unit.
8 Slacken the generator adjustment bolt and mounting bolt, pivot the generator and remove the drive belt, then remove the bolts to withdraw the generator (photo).
9 If it has not already been removed during the engine removal, undo the four securing bolts and remove the fan, spacer and fan pulley.
10 The air pump brackets are now removed after unscrewing the retaining bolts. Also remove the air pump and generator adjustment plates as in Fig. 1.4.
11 Disconnect the air cleaner bracket and air duct.
12 From the carburetor, unclip the throttle return spring and then unscrew the four nuts retaining the carburetor to the manifold. Remove the carburetor unit, and also the heat insulator and heat protector.
13 Unscrew the air injection nozzle joint bolt in the air manifold and remove the nozzle from the manifold. Slacken the air injection sleeve nut, unscrew the two retaining bolts from the air manifold bracket and remove the air manifold.

8.9 Remove the fan, spacer and pulley

Fig. 1.5. Unscrew the retaining bolts to remove the front cover

Fig. 1.6. Unscrew the timing sprocket retaining bolt

Fig. 1.7. The locating dowel

Fig. 1.8. Slacken the air injection nozzle sleeve nuts

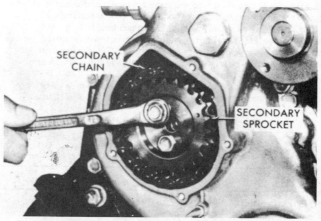
Fig. 1.9. Method of removing the secondary timing sprockets

Chapter 1 Part A: Engine - Series 1 to 4

14 From the crankshaft pulley unscrew the retaining bolt. To stop the crankshaft from turning, wedge the flywheel ring gear with a cold chisel. Remove the bolt and pulley.
15 Unscrew the four bolts retaining the water pump and remove the pump unit.

9 Timing chain, sprockets and camshaft - removal

1 Undo and remove the screws retaining the camshaft timing sprocket front cover and remove the cover (Fig. 1.5).
2 From the secondary chain tensioner plug remove the oil line. Withdraw the tensioner plug and tensioner spring.
3 Unscrew the timing sprocket retaining bolt and plate washer (Fig. 1.6) and then unscrew and remove the two damper bolts from the front of the cylinder head which retain the upper secondary timing chain.
4 Slacken the lower timing chain damper bolts and then remove the timing sprocket from the camshaft with the chain. The timing sprocket can now be carefully removed from the chain, but ensure that the timing sprocket pin does not fall out. It must be noted that when the camshaft is removed from the timing sprocket, the pin should be positioned at the top. Mark the pin position on the timing sprocket before dismantling.
5 Locate the chain with a suitable piece of wire or cord, and then unscrew the securing bolts and remove the camshaft cover.
6 The camshaft carrier bolts can now be unscrewed in an alternate and progressive manner. This is essential as the camshaft carrier is tensioned by the valve springs and therefore the tension must be evenly distributed and reduced to all of the carrier bolts. On removal of the bolts the camshaft carrier and cam can be withdrawn, but take care not to misplace the camshaft carrier locating dowel (Fig. 1.7).
7 Slacken each of the air injection nozzle sleeve nuts and rotate the nozzles 180° to remove them (Fig. 1.8).
8 Unscrew the timing gear cover retaining bolts and detach the cover. Unscrew the secondary timing sprocket securing bolt and remove the sprocket from the layshaft. This is easily achieved by locating and tightening alternatively two suitable bolts into the timing sprocket threaded holes as in Fig. 1.9.
9 On removal of the sprocket, extract the chain from the top and then remove the secondary chain tensioners from the cylinder head. Remove the bolts from the sprocket.

10 Cylinder head - removal

1 Undo and remove the timing gear case to cylinder head retaining bolts and then using an extension bar wrench progressively unscrew the cylinder head retaining bolts in the sequence shown in Fig. 1.10.
2 Having removed the cylinder head bolts the cylinder head can now be detached from the block.
3 The thermostat housing and air pump bracket can now be removed and also the inlet and exhaust manifolds from the head.

11 Oil pan (sump) and crankcase - removal

1 Invert the cylinder block and place on a clean flat surface.
2 Unscrew and remove the oil pan securing bolts in the sequence shown in Fig. 1.11.
3 The oil pan can now be removed. If it is reluctant to be separated from the base of the block, lightly tap it with a wooden or hide mallet.
4 On Series 1 models the crankcase is removed by unscrewing the 18 retaining bolts and prised from the cylinder block using a screwdriver.

12 Oil pump - removal

1 Unscrew the oil pipe connecting nut from the cylinder block.
2 Unscrew and remove the two oil pump body retaining bolts and withdraw the oil pump unit (photo).

13 Timing case - removal

1 Unscrew the timing case retaining bolts.
2 Remove the access plug and then unscrew the inner timing case securing bolt as in photo.
3 With a screwdriver, apply pressure to the cut-away sections of the timing case outer rim and prise the case from the block.

14 Primary timing sprocket and layshaft - removal

1 Unscrew the primary chain tensioner securing nuts (Fig. 1.12) and carefully remove the tensioner, ensuring that the shoe is not dislodged by the spring tension.
2 Remove the timing sprockets and chain. To do this, insert screw rods alternately and evenly into the layshaft sprocket holes.
3 The layshaft can now be withdrawn from the cylinder block after removing the two thrust plate to layshaft bolts. Take care not to dislodge the layshaft bearings during removal.

15 Clutch/flywheel - removal

1 Mark the relative positions of the clutch unit to the flywheel to ensure that they are reassembled in the exact position.

12.2 Removing the oil pump and feed pipe

Fig. 1.10. Cylinder head bolt removal and tightening sequence

Chapter 1 Part A: Engine - Series 1 to 4

13.2 The access plug removed and wrench to remove the inner securing bolt

2 Slacken the clutch cover to flywheel securing bolts evenly one turn at a time to release the spring tension.
3 Supporting the clutch pressure plate, unscrew fully the securing bolts and remove the clutch unit.
4 Bend the flywheel retaining bolt locking tabs over. Earlier engines had four retaining bolts with two locking tabs (photo), whilst the later engines have six with three locking tabs.
5 Undo the retaining bolts and remove the flywheel.

16 Front and rear plates - removal

1 Unscrew the retaining bolts and remove the front plate by tapping lightly with a mallet.
2 Unscrew the retaining bolts and remove the rear plate by tapping with a mallet.

17 Camshaft - removal

1 Unscrew the thrust plate retaining bolts from their position on the front of the camshaft carrier, as in Fig. 1.13.
2 With the bolts removed, detach the thrust plate and withdraw the camshaft from the front of the engine.

18 Pistons/connecting rods - removal

1 Stand the block on end with the flywheel side down.
2 Examine the respective big-end bearing caps and connecting rods. They should be match-marked from 1 to 4 from the front of the engine. If they are not, numerically dot punch the caps and rods at adjacent points, noting which side of the engine the numbers or punch marks face so that they can be installed in their original relative positions.
3 Unbolt number 1 big-end (connecting rod bearing) cap and remove it.
4 Before removing the pistons and connecting rods, carefully remove the ridge of carbon at the top of the cylinder bores with a scraper or emery cloth.
5 Using a wooden hammer handle, carefully tap the piston and connecting rod assembly upwards and out of the bore, taking care not to score the cylinder wall with the connecting rod as it passes through.
6 It is unlikely that the original shell bearings will be used again but should this be the case, retain them in exact order, identifying them in sequence with the connecting rod and cap sections.
7 Extract the remaining pistons and rods in the same manner.

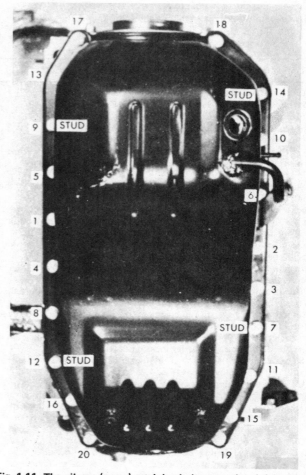

Fig. 1.11. The oil pan (sump) retaining bolt removal and tightening sequence

Fig. 1.12. Unscrew the primary chain tensioner securing nuts

Chapter 1 Part A: Engine - Series 1 to 4

15.4 The early flywheel showing four retaining bolts and locking tabs

Fig. 1.13. Unscrew the thrust plate nuts

Fig. 1.14. Method of removing the piston pin

19 Piston rings - removal

1 Each ring should be sprung open just sufficiently to permit it to ride over the lands of the piston body, but take care not to scratch the land outer surfaces with the ends of the rings. Broken rings congealed in their grooves with carbon can be carefully extracted using a hooked tool. An old hacksaw blade ground to shape is ideal, but take care not to damage the ring lands and grooves.
2 If a piston ring expander tool is not available and the 2nd compression and oil control rings are out of their grooves, it is helpful to cut three ¼ in (6 mm) wide strips of tin and slip them under the ring at equidistant points.
3 Using a twisting motion, this method of removal prevents the ring dropping into the empty groove(s) as it is being removed from the piston. Always remove the rings over the top of the piston.

20 Piston pin (gudgeon pin) - removal

1 The piston pins are an interference fit in the pistons and finger pressure fit in the connecting rod small end bush.
2 With a pair of snap-ring pliers remove the snap-rings from the piston.
3 Immerse the piston into hot water and heat to approximately 158 to 212°F (70° to 100°C)..
4 Locate the connecting rod in a vise as in Fig. 1.14 and using a soft drift, drive the pin from the piston and connecting rod.
5 Ensure that each piston and piston pin are kept with its respective connecting rod, and note which way round the piston is fitted on the connecting rod.

21 Crankshaft and main bearings - removal

1 Position the cylinder block with the crankshaft side upwards.
2 Unscrew the crankshaft rear oil seal retaining bolts and remove the seal and housing and note direction of seal.
3 Examine the main bearing caps for numbers and directional fitting arrows. If they are not marked, dot punch them in order from the front.
4 Unbolt and remove each of the bearing caps, and note that the number 3 journal incorporates the thrust washers.
5 The crankshaft can now be lifted from the cylinder block. If the original shell bearings are to be used again, which is unlikely, retain them in exact order identifying them in relation to the respective crankcase and cap sections.

22 Lubrication system - description

The engine lubricating oil is drawn from the oil pan (sump) by a trochoid type oil pump which is fitted with a filter. From the pump the oil is forced into a disposable element type full flow filter. The oil is then transferred by force to the various engine bearings and components via oil galleries and feed pipes.
At normal operating temperature the trochoid oil pump will deliver 4.57 gallons of SAE30 oil per minute at a pump speed of 1400 rpm. The system is developed to supply oil at a pressure of 57 lbs per square inch.
The oil pressure relief valve is incorporated into the oil filter mounting and pressure unit body (Fig. 1.16). The opening pressure of the relief valve is 61 to 67 lbs per square inch. The overflow valve opening pressure is 11 to 17 lbs per square inch.

23 Examination and renovation - general

1 With the engine stripped, and all components thoroughly cleaned, it is now time to examine everything for wear and damage. Parts and assemblies should be checked and, where possible, renovated or otherwise renewed as described in the following Sections.
2 Oilseals and gaskets must always be renewed. Oilseals can be prised out using a screwdriver or in the case of the timing gear case oilseal, it can be drifted out via the two holes as shown in the photo. Refit the seals by tapping in with a wooden block (Fig. 1.18).

Fig. 1.15. Remove the rear oil seal

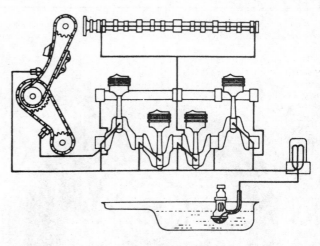
Fig. 1.16. The lubrication system

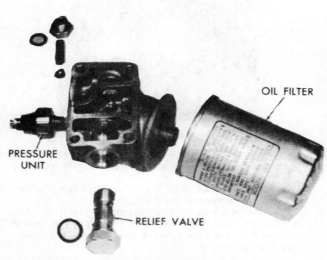
Fig. 1.17. The oil filter and mounting showing the pressure relief and pressure unit removed

23.2 Remove the gearcase oil seal

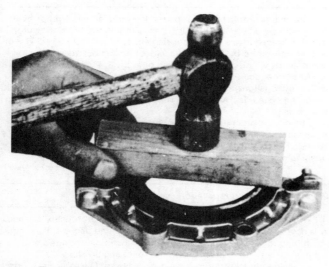
Fig. 1.18. Replacing an oil seal using a wooden block

Fig. 1.19. Compare the plastigage strip with the index to calculate the clearance

Chapter 1 Part A: Engine - Series 1 to 4

3 Clean and inspect all casings for signs of damage and cracks. Renew if required. Clean all old gaskets and sealing compound from the mating surfaces.
4 If for any reason paper gaskets are not available, they can be made by smearing the surface to be sealed with engineers marking blue and a suitable sheet of gasket paper is then pressed carefully against the surface. The gasket can then be cut out round the blue marking left on the paper. A set of hole punches are useful if an extensive amount of gaskets are to be cut.
5 This system is especially useful when only one gasket is required as it saves buying a complete set.

24 Crankshaft and main bearings - examination and renovation

1 Examine the crankpin and main journal surfaces for signs of scoring or scratches. Check the ovality of the crankpins at different positions with a micrometer. If out of round by more than the specified amount, the crankpin will have to be reground. It will also have to be reground if there are any scores or scratches present. Also check the journals in the same fashion.
2 If it is necessary to regrind the crankshaft and fit new bearings your local garage or engineering works will be able to decide how much metal to grind off and the size of new bearing shells required.
3 Full details of crankshaft regrinding tolerances and bearing undersizes are given in Specifications.
4 The main bearing clearances may be established by using a strip of Plastigage between the crankshaft journals and the main bearing/shell caps. Tighten the bearing cap bolts to a torque of 72 lb f ft (9.9 kg fm). Remove the cap and compare the flattened Plastigage strip with the index provided. The clearance should be compared with the tolerances in the Specifications. **Note:** Do not rotate the crankshaft with plastigage installed.
5 Temporarily install the crankshaft to the crankcase having positioned the other halves of the shell main bearings in their locations. Install the No. 3 main bearing cap only, complete with shell bearing and tighten the securing bolts to between 72 lb f ft (9.9 kg fm) torque. Using a feeler gauge, check the end-play by pushing and pulling the crankshaft. Where the end-play is outside the specified tolerance, the center bearing shells will have to be renewed (photo).
6 Finally examine the clutch pilot bearing (bush) which is located in the center of the flywheel mounting flange at the rear end of the crankshaft. If it is worn, renew it by tapping a thread in it and screwing in a bolt. Carefully press in the new bush so that it is fully located.

25 Connecting rods and connecting rod (big-end) bearings - examination and renovation

1 Big-end bearing failure is indicated by a knocking from within the crankcase and a slight drop in oil pressure.
2 Examine the bearing surfaces for pitting and scoring. Renew the shells in accordance with the sizes specified in Specifications. Where the crankshaft has been reground, the correct undersize shell bearings will be supplied by the repairer.
3 Should there be any suspicion that a connecting rod is bent or twisted or the small end bush no longer provides an interference fit for the gudgeon pin then the complete connecting rod assembly should be exchanged for a reconditioned one but ensure that the comparative weight of the two rods is equal.
4 Measurement of the big-end bearing clearances may be carried out in a similar manner to that described for the main bearings in the previous Section. The running clearances are given in Specifications.
5 Finally check the big-end thrust clearance which should be between 0.008 and 0.013 in (0.2 and 0.3 mm).

26 Cylinder bores - examination and renovation

1 The cylinder bores must be examined for taper, ovality, scoring and scratches. Start by carefully examining the top of the cylinder bores. If they are at all worn, a very slight ridge will be found on the thrust side. This marks the top of the piston ring travel. The owner will have a good indication of the bore wear prior to dismantling the engine, or removing the cylinder head. Excessive oil consumption accompanied by blue smoke from the exhaust is a sure sign of worn cylinder bores and piston rings.
2 Measure the bore diameter just under the ridge with a micrometer and compare it with the diameter at the bottom of the bore, which is not subject to wear. If the difference between the two measurements is more than 0.008 in (0.2 mm) then it will be necessary to install special pistons and rings, or to have the cylinders rebored and install oversize pistons.
3 The standard clearance between a piston and the cylinder walls is between 0.0018 and 0.0026 in (0.045 and 0.066 mm). The easiest way to check this is to insert the piston into its bore with a feeler blade 0.001 in (0.04 mm) in thickness inserted between it and the cylinder wall. Attach the feeler blade to a spring balance and note the force required to extract the blade while pulling vertically upwards. This should be between 1.1 and 2.2 lbs.
4 Where less than specified force is required to withdraw the feeler blade, then remedial action must be taken. Oversize pistons are available as listed in Specifications.
5 These are accurately machined to just below the indicated measurements so as to provide correct running clearances in bores bored out to the exact oversize dimensions.
6 If the bores are slightly worn but not so badly worn as to justify reboring them, then special oil control rings and pistons can be installed, which will restore compression and stop the engine burning oil. Several different types are available and the manufacturer's instructions concerning their installation must be followed closely.
7 If new pistons are being installed and the bores have not been reground, it is essential to slightly roughen the hard glaze on the sides of the bores with fine glass paper so that the new piston rings will have a chance to bed in properly.

27 Pistons and piston rings - examination and renovation

1 Where new pistons have been supplied to match the rebore diameter, new sets of piston rings will also be provided but it is worthwhile checking the ring clearances, as described in the following paragraphs.
2 If the original pistons are being used, carefully remove the piston rings as described in Section 19.
3 Clean the grooves and rings free from carbon, taking care not to scratch the aluminium surfaces of the pistons.
4 If new rings are being fitted to old pistons (cylinders not rebored) then order the top compression ring to be stepped to prevent it impinging on the 'wear ring' which will almost certainly have been formed at the top of the cylinder bore.
5 Before installing the rings to the pistons, push each ring in turn down its cylinder bore (use an inverted piston to do this to keep the ring square) and then measure the ring end gap. The gaps must be as given in the Specifications and should be measured with a feeler blade.
6 The piston rings should now be tested in their respective grooves for side clearance. The clearances must be as listed in the Specifications.
7 Piston ring end gaps can be increased by rubbing them carefully with carborundum stone.
8 Where necessary, a piston ring which is slightly tight in its groove may be rubbed down holding it perfectly squarely on a carborundum or a sheet of fine emery cloth laid on a piece of plate glass. Excessive tightness can only be rectified by having the grooves machined out.

28 Camshaft - examination and renovation

1 Clean and inspect the camshaft lobes and journals for signs of wear or pitting. If these are readily apparent then the camshaft must be renewed.
2 Slight score marks on the cams can be removed by carefully and lightly rubbing down with a very fine emery cloth or oil stone.
3 Measure with a micrometer the outside diameter of the camshaft journals and cam profiles. Check them against the dimensions quoted in the Specifications and if necessary renew the camshaft.
4 To measure the camshaft end play, install the thrust plate into position. If outside the specified limits, renew the thrust plate.

31.5a The rocker arm and springs showing the location of the springs

31.5b Unclip the spring ...

31.5c ... remove the rocker arm and ...

31.5d ... the rocker guide

31.7a Compress the valve spring and remove the collets ...

31.7b ... and withdraw the springs and seats

31.11 Remove the valve

33.3 Remove the filter screen to gain access to the four bolts retaining the case to the pump

Fig. 1.20. Depress the rocker spring (1) and remove from rocker spring (2). Detach the rocker arm (3) and extract the rocker guide (4)

29 Layshaft - examination and renovation

1 Clean and carefully inspect the layshaft journals and oil pump drive gear for signs of wear, and renew the shaft if necessary.
2 Check the layshaft end play by fitting the thrust plate and timing sprockets to the layshaft and tighten the retaining bolts to 33 lb f ft (4.5 kg f m).
3 Move the thrust plate fully towards the timing sprocket and measure the distance between the end of the journal and the thrust plate. Renew the thrust plate if the clearance is beyond that specified.
4 Inspect the clearance between the bearing and journal and if beyond that specified, renew either the layshaft or the bearing.
5 If the bearing is being renewed it is a job best left to your local Chevrolet dealer who has the necessary tool for removing the bearing.

30 Timing chain, sprockets and chain tensioners - examination and renovation

1 If the timing sprockets show signs of excess wear or damage they must be renewed.
2 Lay the timing chain out straight and pull on it with a force of 40 ft. lbs. Now measure the length 40 links (at random) from the pin centers. The standard length is 15 inches and the maximum allowance for wear is 15.16 in. If the chain has stretched beyond this length it must be renewed.
3 Inspect the primary and secondary chain tensioners and shoes. Renew the tensioners if they show signs of wear or damage and if the tension spring is broken or distorted it must be renewed.
4 Inspect the primary and secondary chain side dampers for signs of wear or damage and renew if required.

31 Cylinder head and valves - servicing and decarbonising

1 With the cylinder head removed, use a blunt scraper to remove all traces of carbon deposits from the combustion spaces and ports. Remember that the cylinder head is aluminium alloy and can be damaged easily during the decarbonising operations. Scrape the cylinder head free from scale or old pieces of gasket or jointing compound. Clean the cylinder head by washing it in kerosene and take particular care to pull a piece of rag through the ports and cylinder head bolt holes. Any dirt remaining in these recesses may well drop onto the gasket and cylinder block mating surfaces as the cylinder head is lowered into position and could lead to a gasket leak after reassembly is complete.
2 With the cylinder head clean, test for distortion if a history of coolant leakage has been apparent. Carry out this test using a straight edge and feeler gauges or a piece of plate glass. If the surface shows any warping in excess of 0.0079 in (0.2 mm) then the cylinder head will have to be resurfaced which is a job for a specialist engineering company.
3 Clean the pistons and top of the cylinder bores. If the pistons are still in the block then it is essential that great care is taken to ensure that no carbon gets into the cylinder bores as this could scratch the cylinder walls or cause damage to the piston and rings. To ensure this does not happen, first turn the crankshaft so that two of the pistons are at the top of their bores. Stuff rag into the other two bores or seal them off with paper and masking tape. The waterways should also be covered with small pieces of masking tape to prevent particles of carbon entering the cooling system and damaging the water pump.
4 Before scraping the carbon from the piston crowns, press grease into the gap between the cylinder walls and the two pistons which are to be worked on. With a blunt scraper carefully scrape away the carbon from the piston crown, taking great care not to scratch the aluminum. Also scrape away the carbon from the surrounding lip of the cylinder wall. When all carbon has been removed, scrape away the grease which will be contaminated with carbon particles, taking care not to press any into the bores. To assist prevention of carbon build-up the piston crown can be polished with a metal polish. Remove the rags or masking tape from the other two cylinders and turn the crankshaft so that the two pistons which were at the bottom are now at the top. Place rag or masking tape in the cylinders which have been decarbonised and proceed as just described.

5 The rocker arms can be removed from the cylinder head by the following method, refer to Fig. 1.20. Depress the rocker spring (1) and remove the arm from the adjuster ball (2). Detach the rocker arm (photo) (3), and extract the rocker guide (4).
6 On removal, the rockers must be kept in their correct sequence unless they are badly worn and are to be renewed.
7 With a valve spring compressor, compress each valve in turn until the two halves of the collets can be removed (photo). Release the compressor and remove the inner and outer springs and spring seats (retaining caps) (photo).
8 If, when the valve spring compressor is screwed down, the valve spring retaining cap refuses to free to expose the split collet, do not continue to screw down the compressor as there is a likelihood of damage occuring.
9 Gently tap the top of the tool directly over the cap with a light hammer. This will free the cap. To avoid the compressor jumping off the valve spring retaining cap when it is tapped, hold the compressor firmly in position with one hand.
10 When the valve springs are being compressed ensure that the valve spring compressor screw contacts the center of the valve head.
11 Remove the valve (photo). Examine the heads of the valves for pitting and burning, especially the heads of the exhaust valves. The valve seatings should be examined at the same time. If the pitting on valve and seat is very slight the marks can be removed by grinding the seats and valve together with coarse, and then fine, valve grinding paste.
12 Where bad pitting has occured to the valve seats it will be necessary to recut them and fit new valves. If the valves seats are so worn that they cannot be recut, then it will be necessary to fit new valve seat inserts. These latter two jobs should be entrusted to the local Chevrolet agent or engineering works. In practice it is very seldom that the seats are so badly worn that they require renewal. Normally., it is the valve that is too badly worn for replacement, and the owner can easily purchase a new set of valves and match them to the seats by valve grinding.
13 Valve grinding is carried out as follows:
   Smear a trace of coarse carborundum paste on the seat face and apply a suction grinder tool to the valve head. With a semi-rotary motion, grind the valve head to its seat, lifting the valve occasionally to redistribute the grinding paste. When a dull matt even surface finish is produced on both the valve seat and the valve, wipe off the paste and repeat the process with fine carborundum paste, lifting and turning the valve to redistribute the paste as before. A light spring placed under the valve head will greatly ease this operation. When a smooth unbroken ring of light grey matt finish is produced, on both valve and valve seat faces, the grinding operation is complete.
14 Scrape away all carbon from the valve head and the valve stem. Carefully clean away every trace of grinding compound, taking great care to leave none in the ports or in the valve guides. Clean the valves and the valve seats with a kerosene soaked rag then with a clean rag, and finally, if an air-line is available, blow the valves, valve guides and valve ports clean.
15 Test each valve in its guide for wear. After a considerable mileage, the valve guide bore may wear oval. This can be tested by inserting a new valve in the guide and moving it from side to side. If the tip of the valve stem deflects by about 0.0080 in (0.2 mm) inlet or 0.010 in (0.25 mm) exhaust then it must be assumed that the tolerance between the stem and guide is greater than the permitted maximum.
16 If replacement valve guides are to be fitted this is a job best left to your local GM dealer or auto engineer, who will have the correct tool for removal and refitting of the guide.
17 The pivot should not be removed from the head unless it is being renewed. When fitting a new pivot, tighten it to a torque of 90 lb f ft (12.4 kg f m).

32 Flywheel and ring gear - inspection and renovation

1 If the teeth on the flywheel starter ring are badly worn or some are missing, it will be necessary to renew the ring gear.
2 This can be achieved by splitting the ring with a cold chisel, but the greatest care must be taken not to damage the flywheel during this process.
3 To fit the new ring heat it evenly with an oxy-acetylene flame until the ring has expanded sufficiently to allow it to be located on the

flywheel. Allow the ring gear to cool naturally, and it will contract and should be a tight and permanent fit on the flywheel.
4 Care must be taken to ensure that the ring is not overheated or the temper of the ring will be lost.
5 Inspect the flywheel for signs of cracks or damage and renew if necessary. Check also the thickness of the friction face which is 1.46 in standard hub with a maximum wear limit of 1.42 in.

33 Oil pump - examination and renovation

1 The oil pump drive gear is located on its shaft by a taper pin. At the other end of the shaft the drive rotor in the oil pump is located with a pin.
2 Remove the oil pump shaft drive gear to obtain access to the pin on the oil pump drive rotor.
3 The oil pump rotors are accessible on removing the filter screen and four bolts retaining the case to the pump (photo).
4 Clean the components of the pump and inspect for signs of excessive wear (photo).
5 Inspect the rotor and lobes. Any deep score marks or pittings will necessitate renewal of the pump.
6 With a feeler gauge measure the clearance between the tip of the rotor and vane. The normal clearance is 0.0012 to 0.0059 in and any wear beyond these limits necessitates renewal of the pump, rotor or vane (Fig. 1.21.).
7 Measure the clearance between the vane and pump body inner wall. Renew the pump body or vane if the clearance exceeds 0.008 to 0.011 in.
8 Check the clearance between the rotor, vane and pump body cover as in Fig. 1.23 using a straight edge rule and feeler gauge. A clearance beyond 0.006 in means the pump must be renewed.
9 With a micrometer measure the pump rotor shaft diameter and the inside diameter of the rotor shaft bore in the body. If the clearance exceeds 0.008 in renew the pump.

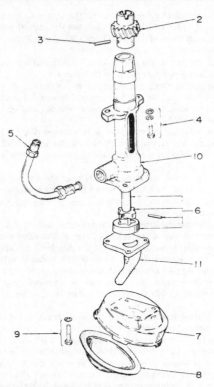

Fig. 1.21. The oil pump component parts

2	Pinion	7	Case
3	Pin	8	Screen
4	Bolt, lockwasher, plain washer	9	Bolt, lockwasher
5	Pipe assembly	10	Body
6	Rotor set	11	Cover

33.4 The rotor and vanes cleaned for inspection

35.2 Insert the main bearing shells into position

Fig. 1.22. Insert feeler gauge between rotor and vane tip

Fig. 1.23. Method to check the rotor, vane and pump body clearance using straight edge and feeler gauge

Chapter 1 Part A: Engine - Series 1 to 4

34 Engine reassembly - general

1 To ensure maximum life with minimum trouble from a rebuilt engine, not only must every part be correctly assembled, but everything must be spotlessly clean, all the oilways must be clear, locking washers and spring washers must always be fitted where indicated and all bearings and other working surfaces must be thoroughly lubricated during assembly. Before assembly begins, renew any bolts or studs whose threads are in any way damaged; whenever possible use new spring washers and cotter pins.
2 Apart from your normal tools, a supply of lint-free cloths, an oil can filled with engine oil (an empty washing-up fluid plastic bottle thoroughly clean and washed out will invariably do just as well), a supply of new spring washers, a set of new gaskets and a torque wrench should be collected together.
3 Carefully check the specified clearances of components during reassembly, and where given tighten all nuts and bolts to the recommended torque values.

35 Crankshaft and main bearings - reassembly

1 Invert the cylinder block and ensure that it is perfectly clean and that the oil ways are clear.
2 Insert the main bearing shells into their seats (photo) and lubricate them with clean engine oil.
3 On Series 1 engines locate the seal into its groove in the rear end of the block and cut off any protruding seal flush with the end cap mating face (photo). Insert the seal into the rear main bearing cap groove and cut off flush any protruding seal, see Fig. 1.24.
4 Place the crankshaft carefully into position in the block (photo), and then install the thrust bearings into their location on both sides of the No. 3 journal as in Fig. 1.27. The thrust bearing smooth face is positioned towards the crankshaft on later models. On Series 1 engines the smooth face is towards the crankcase.
5 Place the bearing shells into their caps and lubricate them with oil. Place them in position. The marking on the back face of the caps point to the front of the engine. No. 1 bearing cap differs in contour from the other caps. Nos. 2 and 3 caps are marked with an 'A' on the back face as in Fig. 1.25.
6 Alternatively tighten the bearing caps starting at number 3, to a torque of 72 lb f ft (9.9 kg f m). (photo).
7 With all the bearing caps tightened, rotate the crankshaft to ensure that it turns smoothly and does not drag. Check the crankshaft and play which should be 0.0059 in (photo). If on Series 1 engines the crankshaft is tight in rotation, recheck the location of the rear oil seal. The crankshaft on this engine series should be able to turn freely under a torque loading of 15 to 22 lb f ft (2 to 3 kg f m).
8 On Series 2 to 4 models the rear oil seal is now positioned in its retainer. The seal is an interference fit in the retainer and can be inserted by locating squarely on the retainer and tapped into position using a wooden block as in Fig. 1.18.
9 Grease the lips of the oil seal and position the gasket to cylinder block, smearing a jointing compound over the two faces. Align the dowel rods and holes and fit the oil seal and retainer, securing with the bolts and washers.

36 Pistons, rings and connecting rods - reassembly and installation

1 Position the cylinder block with the flywheel end downwards.
2 To assemble the respective pistons to their connecting rods, heat the piston to 158° to 212°F (70° to 100°C), the piston can be suspended in a bowl of hot water to achieve this. Fit a snapring into a piston pin bore groove and then align the connecting rod small end bore with the piston pin bore.
3 Press the piston pin, using finger pressure, into the piston and through the small end and locate the pin against the previously inserted snapring (photo).
The piston and connecting rod must be assembled with the cylinder number mark on the righthand side of the groove in the front of the piston crown.
4 Assemble the rings to the pistons. The rings must be fitted with their markings NPR or TOP facing upwards. Use a ring expander for fitting if available or alternatively reverse the dismantling procedure

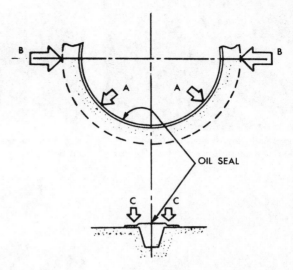

Fig. 1.24. Series 1 models rear oil seal location

A must be free from projection and depression
B cut off excess seal
C cut off excess seal

With the seal correctly located the crankshaft can be rotated using a torque of 15 to 22 ft. lbs. (2.0 to 3.0 kg fm).

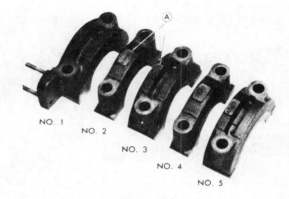

Fig. 1.25. The respective main bearing caps

Fig. 1.26. Using a ring expander to fit the piston rings

35.3 Insert and trim off excess rear seal into recess in rear main of block and cap (Series I models)

35.4 Place the crankshaft carefully into position

35.5 Position the main bearing caps

35.6 Tighten the main bearing caps to the recommended torque

35.7 Check the crankshaft end play with a feeler gauge

36.3 Insert the piston pin, pressing it flush against the snap ring

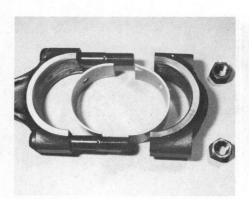
36.8 Locate the respective connecting rods, caps and bearing shells prior to assembly

36.9 Use a ring compressor to fit the pistons and compress the rings

36.12a Fit the big end (connecting rod) bearing and cap ...

36.12b ... but ensure that the identification numbers are adjacent to each other

36.13 Tighten the connecting rod end cap bolts to the specified torque

37.2 The front end plate bolted into position

Chapter 1 Part A: Engine - Series 1 to 4

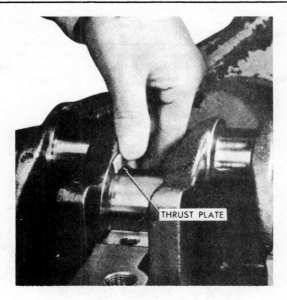

Fig. 1.27. Insert the thrust bearings

Fig. 1.28. Align the sprockets with the timing marks adjacent

37.4 The layshaft thrust plate bolted in position

37.6a Refit the timing chain tensioner. Note the timing marks

5 When all the rings are assembled onto the pistons, rotate each top compression ring so that its gap faces the front of the piston and then rotate and position the 2nd compression and oil ring gaps at 120° intervals.
6 Rotate the crankshaft to the TDC position.
7 Lubricate with clean engine oil the bore of the first piston to be fitted and also lubricate the piston and rings.
8 Position the connecting rod and big-end cap bearing shells into position (photo) and lubricate them.
9 Compress the piston rings with a suitable compressor and locate in the cylinder bore (photo).
10 Carefully tap the piston crown with a wooden hammer handle and insert the piston into its cylinder.
11 Engage the connecting rod with the crankshaft when the crankpin is at its lowest point of rotation and travel. Lubricate the exposed part of the crankpin.
12 Install the big-end cap (photo) complete with shell bearing ensuring that the numbers are adjacent in their sequence (photo).
13 Tighten the big-end bolts to the specified torque figure of 43 lb f ft (5.9 kg f m) (photo) and then check the connecting rod end play which should be 0.0079 to 0.0130 in (0.2 to 0.3 mm).
14 Repeat the operations to install the remaining three pistons/connecting rod assemblies.

37 End plates layshaft and primary timing sprocket - assembly

1 Relocate the rear plate to the cylinder block and secure with bolts.
2 Smear the front plate gasket with jointing compound and refit the front plate. Secure with bolts, (photo).
3 Rotate the crankshaft to position number 1 and 4 pistons at TDC (top dead centre). To check that TDC position is correct, position the timing cover over the dowels and place the crankshaft pulley in position. The TDC mark should align with the timing pointer.
4 Lubricate the layshaft and insert it into position. Locate the thrust plate on the layshaft and tighten the securing bolts to a torque of 9 lb f ft (1.2 kg f m), on Series 2 to 4. Series 1 models tighten to 6 lb f ft (0.8 kg f m) (photo).
5 Insert the crankshaft and layshaft primary timing sprockets into position in their drive chain and ensure that the timing marks (Fig. 1.28) are adjacent to each other as shown. When fitting the primary sprocket to the layshaft the timing mark on the layshaft timing mark must align with that of the crankshaft sprocket.
Note: The layshaft keyway is not the timing mark. The layshaft keyway and sprocket are aligned and then the layshaft is turned and the sprocket positions for timing set by lightly tapping each sprocket alternately. Support the layshaft from within the crankcase to prevent it turning when locating the sprockets.
6 Refit the primary chain tensioner (photo), and install the water pump and gasket (photo).

38 Oil pump - installation

1 With the marks on the timing sprockets aligned, fit the oil pump into position (photo) with the groove in the end of the pump drive pinion at a right-angle to the layshaft, and the oil pump rotor slot towards the rear of the cylinder block.

Fig. 1.29. Fit the oil pan/cylinder block rubber gasket

41 Cylinder head - installation

1 Ensure that the cylinder block and gear case upper faces are parallel using a straight edge as in Fig. 1.30. The gearcase must be relocated or renewed if the distortion is in excess of 0.007 in (0.17 mm).

37.6b Refit the water pump and gasket

2 The oil pump drive pinion must mesh with the helical drive gear on the layshaft.
3 Insert the pump retaining bolts and tighten.
4 Reconnect the oil pipe from the pump to the cylinder block, and tighten the pipe nut.

39 Crankcase oil pan (sump) - installation

Series 1
1 Apply a liberal quantity of jointing compound to the sealing gaskets. Insert the gaskets into position in the No. 1 and 5 main bearing caps. The gaskets must fit fully with no end clearances.
2 Fit the crankcase and insert the retaining bolts. Tighten to 15 lb f ft torque (2.0 kg f m).

Series 2 to 4
3 Smear a liberal quantity of jointing compound to the front and rear portions of the jointing face and install the rubber gasket as in Fig. 1.29.
4 Reposition the oil pan and gasket, and locate the retaining bolts and nuts.
5 Tighten the retaining bolts and nuts evenly in the sequence shown in Fig. 1.11 to a torque of 50 lbf in (0.57 kgf m).

38.1 Relocate the oil pump - observe the instructions in paragraphs 1 and 2

40 Flywheel/Crankshaft pulley/Oil filter - installation

1 Locate the flywheel onto the crankshaft and fit the lock plates and retaining bolts.
2 Tighten the bolts to a torque of 36 lb f ft (4.9 kg f m) on Series 1 models, and 69 lb f ft (9.5 kg f m) on Series 2 to 4.
3 Lock the bolts by bending over the lock plate tabs. Reassemble the clutch unit as described in Chapter 5.
4 Refit the under cover to the rear plate.
5 Lubricate with oil the face of the crankshaft pulley to oil seal face and refit the pulley onto the shaft.
6 Secure the pulley with bolt and lock washer.
7 Refit the oil filter unit (photo). If the pressure unit or oil pressure relief valve have been removed they must be installed to the filter unit but ensure that the respective components are perfectly clean.

38.4 Reconnect the oil pipe

39.3 The oil pan gasket in position

40.7 Refit the oil filter unit using a new sealing gasket

41.3 The cylinder head gasket in position

41.6 Tighten the cylinder head bolts to the torque specified

41.7 Locate and tighten the timing case bolts

41.8 Relocate the air injection nozzles but do not tighten

43.2 Align the camshaft and thrust plate

43.5 Locate the timing chain onto the timing sprocket

43.6 The layshaft timing sprocket located and with the timing mark at the 2 o'clock position

43.7 The camshaft timing sprocket located with the timing mark at approximately 12 o'clock position (exact position if correct 6° 20')

44.2 The thermostat housing, air pump bracket and generator adjuster plate

44.3 Fit the heater hose nipple

Fig. 1.30. Check cylinder block and gear case upper faces for alignment using a straight edge

Fig. 1.31. The cylinder head bolt tightening sequence

Fig. 1.32. The cam carrier in position showing the PVC hose clips and air manifold bracket locations

44.5 Locate the generator

44.6 Attach the air nozzles to the air manifold and tighten securely

44.7 The inlet and exhaust manifolds refitted

44.9 The secondary chain tensioner oil pipe location

44.10 Refit the air pump

45.1 The crankshaft pulley and timing case pointer indicating top dead center (TDC)

Chapter 1 Part A: Engine - Series 1 to 4

2 Fit the secondary chain tensioner with its thick shoe side facing up.
3 Ensure that the cylinder head and cylinder block mating surfaces are perfectly clean and position the cylinder head gasket in position on the block with the side marked 'TOP' uppermost (photo).
4 Locate the 'O' rings into the oil ports and place the gear case to cylinder head gasket in position on the gear case.
5 Carefully place the cylinder head into position and insert the retaining bolts. Refer to Fig. 1.31 and tighten all of the bolts in the sequence shown to a torque figure of 43 lb f ft (5.9 kg f m). Then loosen them.
6 On Series 1 models retighten the bolts to 58 lb f ft (8.01 kg f m) and on Series 2 to 4 further tighten the bolts marked 1, 2, 3 and 6 to 70 lb f ft (9.6 kg f m) and the remaining 6 bolts to a torque figure of 60 lb f ft (8.2 kg f m) (photo).
7 The timing gear case retaining bolts are now tightened (photo).
8 Relocate the air injection nozzles but do not fully tighten (photo).

42 Camshaft carrier - installation

1 Align the camshaft thrust plate mark with the corresponding alignment mark on the camshaft.
2 Fit the 'O' rings to the camshaft carrier and install in position, but do not fully tighten the dowel retaining bolts. The longest bolt is installed in the dowel position.
3 Insert the camshaft carrier mounting bolts and tighten in progression alternately to 15 lb f ft torque (2.0 kg f m).
4 It should be noted that the camshaft carrier mounting bolts also retain two parts of the air manifold bracket and the PVC hose clips (Fig. 1.32).
5 Reconnect the air injection nozzle to the air manifold and tighten.

43 Secondary timing sprocket - installation

1 Rotate the crankshaft and position number 4 piston to its top dead centre (TDC) on compression stroke.
2 Ensure that the alignment marks on the camshaft and thrust plate are in alignment (photo). If they are not aligned, position the camshaft sprocket on the camshaft and fit the pin into the camshaft sprocket. Now turn the crankshaft and align the marks.
3 Remove the camshaft sprocket and reposition the number four piston to TDC position on compression stroke. If to make adjustment the crankshaft was turned in the reverse direction, the final adjustment must be made in the normal direction to align the relative marks, and the chain must be tensioned (in direction of rotation).
4 It is essential that the camshaft timing sprocket and the mating surface on the camshaft are perfectly clean. This is necessary as the drive torque from the camshaft to sprocket is by means of frictional contact.
5 The timing chain is now inserted through the aperture in the top and the layshaft timing sprocket located into it (photo). The punched timing mark on the sprocket must align with the key on the layshaft.
6 When installed the sprocket punch mark must be approximately at the 2 o'clock position (photo).
7 Insert the camshaft sprocket into the chain with its punch mark at the 12 o'clock position, and fit the sprocket to the camshaft (photo).
8 The camshaft timing sprocket position must be adjusted in relation to the camshaft. The punch mark on the sprocket must be up when the timing chain drive side is tensioned. To tension for adjustment, apply pressure to the tensioner shoe thru the plug hole in the secondary chain tensioner.
9 Correctly installed, the camshaft sprocket punch mark will be at a position of 6° 20' from the top center (in direction of rotation).
10 Insert the locating pin into the hole in the camshaft flange which aligns with a sprocket hole.
11 Tighten the layshaft retaining bolt to a torque of 33 lb f ft (4.5 kg f m)
12 Fit the camshaft sprocket plate washer and tighten the retaining bolt to 33 lb f ft (4.5 kg f m).
13 Prior to fitting the gearcase front cover and camshaft carrier cover ensure that the alignment marks on the sprockets are correct then fit the covers.

44 General ancillary items - reassembly

1 Fit the secondary chain tensioner.
2 Fit the thermostat housing, the air pump bracket, adjusting plate and also the generator adjustment plate (photo).
3 Fit the heater hose joint nipple and locate the air cleaner bracket and air duct into position (photo).
4 Insert the distributor into its location and ensure that with the No. 4 piston on TDC (compression stroke) that the scribed marks on the distributor bowl aligns with the rotor segment.
5 Refit the generator but do not fully tighten the adjustment bolts (photo).
6 Reconnect the air injection nozzles to the air manifold and ensure they are firmly tightened.
7 Refit the inlet and exhaust manifolds using new gaskets (photo).
8 Refit the carburetor heat protector, heat insulator, and then the carburetor unit.
9 Reconnect the secondary drain tensioner oil pipe (photo).
10 Refit the air pump (photo) and then install the spacer fan pulley and fan.
11 Refit the fan belt and air pump drive belts and tension both to a deflection of 0.4 in.
12 Attach the distributor primary cable and meter unit cable.

45 Valve clearances - adjustment

1 Rotate the crankshaft so that No. 1 or No. 4 piston is on top dead center (TDC) on compression stroke. In this position the timing marks on the gear case and crankshaft pulley are in alignment (photo).
2 The valve clearances are now adjusted with a feeler gauge in the following manner.
3 Slacken the locknut and insert the feeler gauge between the rocker arm and cam (photo). Turn the rocker arm adjustment screw to take up the clearance and lock in position by retightening the locknut.
4 The correct clearances are:
   Inlet — 0.004 in (0.101 mm) at low temperature
   Exhaust — 0.006 in (0.152 mm) at low temperature
5 Refit the camshaft side cover.
6 Refit the respective hoses to their connections ensuring that they are in good condition and also their securing clips. Position the hoses so that they are not in contact with adjoining components.

46 Engine/transmission - installation

1 Whether the engine is to be installed on its own, or together with the transmission, the procedure is basically the reverse of the removal procedure.

45.3 Method of checking valve clearance

2 Raise the engine (and transmission, where appropriate), using a hoist, and lower it into position in the vehicle engine bay. A jack will be required to support the transmission until the rear support is installed if the transmission is connected to the engine.
3 Install the engine and transmission mounts.
4 Where applicable, install the starter motor and the flywheel housing bolts.
5 Install the exhaust pipe.
6 Install the oil filter (where applicable).
7 Connect the electrical, ground, fuel and air lines to the engine.
8 Install the carburetor controls and linkages.
9 Connect the heater and emission control hoses.
10 Install the fan, radiator and hoses.
11 If applicable, connect the gearshift lever, speedometer drive cable and propeller shaft.
12 Top-up the engine oil and coolant (and transmission oil, where applicable) (photo).
13 Install the air cleaner.

47 Engine - initial start-up after overhaul

1 Make sure that the battery is fully charged and that all lubricants, coolant and fuel are replenished.
2 If the fuel system has been dismantled it will require several revolutions of the engine on the starter motor to pump the petrol up to the carburetor on models fitted with a mechanical fuel pump.
3 As soon as the engine fires and runs, keep it going at a fast idle only, (no faster) and bring it up the normal working temperature.
4 As the engine warms up there will be odd smells and some smoke from parts getting hot and burning off oil deposits. The signs to look for are leaks of water or oil which will be obvious if serious. Check also the exhaust pipe and manifold connections, as these do not always 'find' their exact gas tight position until the warmth and vibration have acted on them, and it is almost certain that they will need tightening further. This should be done of course, with the engine stopped.
5 When normal running temperature has been reached adjust the engine idling speed as described in Chapter 3.
6 Stop the engine and wait a few minutes to see if any lubricant or coolant is dripping out when the engine is stationary.
7 Road test the car to check that the timing is correct and that the engine is giving the necessary smoothness and power. Do not race the engine - if new bearings and/or pistons have been fitted it should be treated as a new engine and run in at a reduced speed.

Chapter 1 Part B: Engine - Series 5 and 6

For information relating to Series 8 thru 12 models, including 4WD, refer to Chapter 13.

Contents

Ancillary components - removal ... 51	Examination and renovation - general ... 64
Cam cover - removal ... 52	Flywheel and ring gear - servicing ... 72
Camshaft - examination and renovation ... 69	Front cover - removal ... 58
Connecting rods and connecting rod bearings - examination and renovation ... 66	General description ... 47
	Initial start-up after major overhaul ... 80
Crankshaft and main bearings ... 62	Inlet and exhaust manifolds - removal ... 55
Crankshaft and main bearings - examination and renovation ... 65	Lubrication system - description ... 63
	Oil pan (sump) - removal ... 56
Cylinder bores - examination and renovation ... 67	Oil pump - examination and renovation ... 73
Cylinder head and valves - servicing and decarbonising ... 71	Oil pump - removal ... 57
Cylinder head - assembly and installation ... 76	Pistons and connecting rods - removal ... 60
Cylinder head - removal ... 53	Pistons and rings - examination and renovation ... 68
Engine and transmission - removal ... 49	Rear main oil seal ... 61
Engine and transmission - installation ... 79	Rocker arms/valves and springs - removal ... 54
Engine dismantling - general ... 50	Timing chain, pinion gear, sprockets and chain adjuster/tensioner and guides - examination and renovation ... 70
Engine - final assembly after major overhaul ... 78	
Engine - reassembly ... 75	Timing chain, pinion gear, sprockets and chain adjuster/tensioner and guides - removal ... 59
Engine reassembly - general ... 74	
Engine - removal ... 48	Valve clearance - adjustment ... 77

Specifications

Engine type	4-cylinder, 4-stroke in-line, ohc, water cooled
Firing order	1, 3, 4, 2
Combustion chamber type	Hemispherical
Bore	3.31 in
Stroke	3.23 in
Piston displacement	110.8 cu in
Compression ratio	8.5:1
Compression pressure	170.64 psi @ 3000 rpm
Maximum compression pressure variance	8.53 psi
Maximum output	80 HP @ 4,800 rpm
Maximum torque	95 ft lb @ 3,000 rpm
Engine weight	
Dry	316 lb
With oil/coolant	334 lb
Lubrication method	Forced circulation

Oil pump
Type	Trochoid
Oil pressure @ 2,800 rpm	56 to 71 psi @ 122°F
Drive rotor to driven rotor clearance	0.0005 to 0.0059 in
Maximum permissible clearance	0.0079 in
Driven rotor to inner pump body wall clearance	0.0063 to 0.0087 in
Maximum permissible clearance	0.0098 in
Drive rotor/driven rotor to oil pump cover clearance	0.0012 to 0.0035 in
Maximum permissible clearance	0.0079 in
Drive shaft O/D to pump cover bore I/D clearance	0.0028 to 0.0043 in
Maximum permissible clearance	0.0098 in
Maximum oil capacity	5.7
Oil pan capacity	Maximum 5.3 liq/qt
Oil filter type	Full-flow paper (cartridge type)
Oil pressure switch actuation pressure	4.27 - 7.11 psi
Relief valve opening pressure	56.88 - 71.10 psi
Overflow valve opening pressure	11.38 - 17.06 psi

Cylinder head
Inlet valves	
open at	21° BTDC
close at	65° ATDC
Exhaust valves	
open at	55° BBDC
close at	20° ATDC
Valve clearance (between rocker and cam) - at low temperature	
inlet	0.006 in
exhaust	0.010 in
Valve head diameter	
inlet	1.67 in
exhaust	1.34 in
Valve spring free height	
outer spring nominal height	1.8465 in
limit for use	1.7874 in
Inner spring nominal height	1.7835 in
limit for use	1.7244 in
Valve spring tension	
outer spring compressed to 1.614 in (standard)	32.2 to 37.0 lbs
limit for use	30.9 lbs
inner spring compressed to 1.516 in (standard)	18.7 to 21.5 lbs
limit for use	16.5 lbs
Maximum valve spring inclination allowance	0.0787 in
Camshaft	
nominal height	1.4508 in
minimum height	1.4311 in
Camshaft journals	
nominal diameter	1.3362 to 1.3368 in
minimum wear allowance	0.0020 in
minimum diameter allowance	1.3307 in
Camshaft run out	
nominal	0.0020 in
maximum run out allowance	0.0038 in
Camshaft end play	
standard	0.0020 to 0.0059 in
maximum end play allowance	0.0078 in
Camshaft journal diameter	1.3362 to 1.3370 in
Camshaft bearing inside diameter	1.3386 to 1.3398 in
Standard camshaft journal/bearing clearance	0.0016 to 0.0035 in
Maximum journal to bearing clearance allowance	0.0059 in
Rockers and shaft	
nominal shaft diameter	0.8071 in
maximum wear allowance	0.8012 in
maximum shaft run out	0.0079 in
rocker arm shaft diameter to rocker arm inside diameter maximum clearance allowable	0.0078 in

Pistons
Type	Track (No T slot)
Number of rings	
compression	2
oil	1

Chapter 1 Part B: Engine - Series 5 and 6

Standard size
- 'A' grade ... 3.3049 to 3.3053 in
- 'B' grade ... 3.3053 to 3.3057 in
- 'C' grade ... 3.3057 to 3.3061 in
- 'D' grade ... 3.3061 to 3.3065 in

0.020 in oversize A, B ... 3.3246 to 3.3254 in
0.040 in oversize A, B ... 3.3443 to 3.3451 in

Piston ring gaps in bore
- 1st compression ... 0.008 to 0.016 in
- 2nd compression ... 0.008 to 0.016 in
- oil control ... 0.008 to 0.035

Maximum limit for use 1st, 2nd and oil control ... 0.059 in

Crankshaft

Journal diameter
- nominal ... 2.205 in
- wear limit ... 0.002 in or more
- limit for use ... 2.1555 in

Crankpin diameter
- nominal ... 1.929 in
- wear limit ... 0.002 in or more
- limit for use ... 1.8799 in

Crankshaft journal diameter
- standard bearing size ... 2.2016 to 2.2022 in
- 0.010 in undersize ... 2.1917 to 2.1923 in
- 0.020 in undersize ... 2.1819 to 2.1825 in

Crankpin diameter
- standard ... 1.9262 to 1.9268 in
- 0.010 in undersize ... 1.9163 to 1.9169 in
- 0.020 in undersize ... 1.9065 to 1.9071 in

Cylinder bore grades (Stamped on cylinder block upper face)
- 'A' grade ... 3.3071 to 3.3075 in
- 'B' grade ... 3.3075 to 3.3079 in
- 'C' grade ... 3.3079 to 3.3083 in
- 'D' grade ... 3.3083 to 3.3087 in

Crankshaft end play
- crankshaft to thrust bearing clearance ... 0.0024 to 0.0094 in
- maximum allowance ... 0.0117 in

Torque wrench settings

	lb f ft	kg f m
Cylinder head bolts		
initial	61	8.4
final	72	9.9
Main bearing cap bolts	72	9.9
Connecting rod bearing cap bolts	33	4.5
Flywheel bolts	69	9.5
Rocker arm shaft bracket nuts	16	2.2
Camshaft sprocket bolt	58	8.0

47 General description

The engine is a four-cylinder in-line type, the valve being operated by overhead camshaft.

The aluminum alloy cylinder head has hemispherical combustion chambers with inclined valves, transversely opposed giving a crossflow head design for maximum efficiency.

The camshaft is located in five main bearings. The bearing caps are incorporated in the rocker arm shaft brackets which in turn are bolted to the cylinder head.

The crankshaft is supported in five main bearings.

A double rocker arm shaft is used to operate the valves, the inlet and exhaust valves being operated independently. The camshaft, operated by the timing chain, actuates the rocker arms to open and close the valves. The valves have double return springs to improve the valve performance at high engine speeds.

The timing chain, driven from the crankshaft pulley, is adjusted by means of an automatic tensioner and mechanical adjuster.

The engine lubricant is circulated by means of a trochoid type oil pump and a full flow oil filter with disposable cartridge element is also fitted.

The cooling system is of the pressurised circulation type incorporating an impeller type water pump and wax-pellet type thermostat.

A comprehensive emission control system is fitted comprising the following:

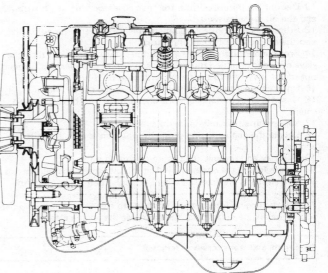

Fig. 1.33. Sectional view of the Series 5 and 6 engines

a) Positive Crankcase Ventilation System (PCV).
b) Evaporative Control System (ECS).
c) Air Injection Reactor System (AIR).
d) Coasting Richer System (CRS).
e) Exhaust Gas Recirculation System (EGR).
f) Controlled Combustion System (CCS).
g) Oxidizing Catalytic Converter (OCS) (California only).
h) Over Temperature Control System (OTC) (California only).

The carburetor fitted is a Stromberg down-draft two barrel type incorporating an electric automatic choke system.

48 Engine - removal

1 Before commencing work, drive the vehicle over an inspection pit, onto ramps or jack it up, for access to the exhaust and transmission.
2 Remove the hood (bonnet) as described in Chapter 12.
3 Detach the ground cable from the battery terminal.
4 Drain the cooling system as described in Chapter 2 and remove the radiator. Remove the fan blade unit.
5 Drain the engine oil into a suitable container (and transmission if gearbox is to be removed).
6 Remove the air cleaner unit by detaching:

a) *The Positive Crankcase Ventilation System (PCV) hose from the cam cover.*
b) *The Evaporative Control System (ECS) hose from the air cleaner unit body.*
c) *The Air Injection Reactor System (AIR) hose from the thermosensor to inlet manifold.*
d) *Undo the air cleaner unit clamping bolt, lift the cleaner unit sufficiently to detach the Controlled Combustion System (CCS) hose from the thermosensor to inlet manifold.*

Now remove the air cleaner unit.
7 Detach the Controlled Combustion System hose and also the manifold cover.
8 Disconnect the generator wires, the cylinder block to frame ground cable, the coil high tension cable and the back-up lamp switch and top/third switch wires to the connector at the rear of the engine.
9 Disconnect the exhaust downpipe to manifold connection.
10 Disconnect the following hoses:

a) *The heater hoses at heater core tubes.*
b) *The carburetor fuel hoses.*
c) *The inlet manifold vacuum hose.*
d) *The Evaporative Control System hose from the oil pan.*
e) *The radiator hoses.*

11 Detach the accelerator cable from the carburetor.
12 Disconnect the wires from the thermosensor unit, the distributor and the oil pressure switch.
13 From the carburetor, disconnect the leads to the solenoid valve and the auto choke at the connector.
14 From underneath remove the clutch return spring and then the slave cylinder retaining bolts. Detach the slave cylinder and suspend out of the way - tie to the frame with cord or wire.
15 Unscrew the retaining bolts and remove the starter motor.
16 Unscrew the bellhousing bolts and remove the flywheel splash shield.
17 Support the transmission with a jack.
18 Locate the engine lifting sling into position and slightly lift the engine.
19 Unscrew and remove the engine mounting nuts.
20 Before lifting the engine from the vehicle, check that all fittings and attachments are free and where possible, tied back out of the way.
21 Pull the engine forward and upwards to clear the transmission and tilting the engine upwards at the front lift the engine clear.

49 Engine and transmission - removal

Manual transmission
1 To remove the engine and manual transmission as a unit, follow the instructions given in Part 'A' of this Chapter, Section 6.

Automatic transmission
2 Slacken the throttle valve cable adjusting nuts, withdraw the cotter pin and joint pin from the carburetor level. Detach the control cable (Fig. 6.38).
3 Disconnect the transmission oil level gauge.
4 From underneath, remove the lower converter housing dust cover.
5 Detach the control rod from the shift lever by removing the cotter pin (Fig. 6.38).
6 Disconnect the exhaust pipe bracket.
7 Slacken the oil cooler pipe retaining nuts (on the right-hand side), detach the pipe location clips and place the pipes towards the body sides.
8 Remove the speedo cable from the transmission.
9 Unscrew and remove the three nuts and bolts retaining the frame bracket to the transmission rear mounting.
10 Lift the engine and transmission accordingly to remove the four bolts securing the frame bracket and crossmember. Remove the rear mounting.
11 The engine and transmission units can now be removed from the vehicle. Take care not to damage the transmission oil cooler pipes and surrounding fittings.

50 Engine dismantling - general

Refer to Part 'A', Section 7.

51 Ancillary components - removal

1 Refer to the procedures given in paragraphs 1, 2 and 4 in Section 8, and in addition remove the oil filter.
2 Remove the clutch unit as in Chapter 5.
3 To remove the flywheel bend over the lock tabs and remove the retaining bolts.
4 Remove the rear plate.

52 Cam cover - removal

1 Remove the air cleaner unit (see Chapter 3).
2 Detach the HT cables from the cam cover brackets.
3 Unscrew and remove the four cam cover securing nuts and washers.
4 Remove the cover.

53 Cylinder head - removal

1 Remove the cam cover as described in the previous Section.
2 Detach the Exhaust Gas Return (EGR) pipe clamp bolt from the cylinder head at the rear.
3 If engine is still in the vehicle, disconnect the exhaust manifold to downpipe connection and drain the cooling system (see Chapter 2). Detach the heater hoses from the inlet manifold and rear of the cylinder head. Remove the sparking plugs, all electrical wiring connections to the cylinder head (noting their respective positions), the vacuum lines and sparking plug cables.
4 Turn the engine to position the No. 1 piston at TDC (firing position). Lift off the distributor cap and mark the rotor to housing relative position. Remove the distributor (see Chapter 4).
5 Depress and turn the automatic timing chain adjuster slide pin 90° clockwise (Fig. 1.34) to lock the chain adjuster.
6 Unscrew and remove the camshaft timing sprocket retaining bolt. Remove the sprocket from the camshaft, but do not remove the sprocket from the chain - retain it on the damper and tensioner.
7 Detach the air injector reactor (AIR) hose and the check valve from the exhaust manifold.
8 Unscrew and remove the cylinder head to timing cover bolts.
9 Using an extension bar wrench unscrew and remove the cylinder head bolts. Progressively unscrew the bolts in reverse sequence, starting with the outer bolts (Fig. 1.36).
10 The cylinder head and manifolds can now be lifted clear. The help of an assistant will be useful at this point.
11 Remove the cylinder head to the workbench for cleaning and inspection/dismantling.

54 Rocker arm/valves and springs - removal

1 With the cam cover removed, slacken each rocker arm shaft bracket nut, in sequence, a little at a time, Start with the outer brackets (Fig. 1.35).
2 Remove the nuts from the brackets.
3 Remove the spring from the rocker arm shaft and dismantle the rocker arm brackets and arms, keeping in sequence of removal.
4 The valves can be removed from the cylinder head by the following method. Compress each spring in turn with a valve spring compressor until the spring retainers can be removed. If available, use special spring compressor J-26513 (Fig. 1.36). Release the compressor and remove the spring and spring retainer.
5 If, when the valve spring compressor is screwed down, the valve spring retaining cap refuses to free, do not continue to screw down on the compressor as there is a likelihood of damaging it.
6 Gently tap the top of the tool directly over the cap with a light hammer. This will free the cap. To avoid the compressor jumping off the valve spring retaining cap when it is tapped, hold the compressor firmly in position with one hand.
7 Slide the rubber oil control seal off the top of each valve stem and then drop out each valve through the combustion chamber.
8 It is essential that the valves are kept in their correct sequence unless they are so badly worn that they are to be renewed.

55 Inlet/exhaust manifolds - removal

Inlet manifold

1 If the engine is still in the vehicle, disconnect the following:

   a) The EGR pipe clamp bolt from the rear of the cylinder head.
   b) The EGR pipe from the inlet and exhaust manifolds.
   c) The EGR valve from the lower side of the inlet manifold complete with bracket.
   d) Drain the cooling system (Chapter 2) and disconnect the radiator top hose at the water outlet and the heater hose from the manifold upper side.
   e) Detach the accelerator linkage, vacuum lines and the electrical connections from the carburetor. Disconnect the fuel feed line to the carburetor.

2 Unscrew the manifold securing nuts and remove the manifold/carburetor from the studs and position on cam cover temporarily. Detach the lower heater hoses and then remove the manifold and carburetor unit.
3 From the manifold disconnect the carburetor, the temperature sender switch and vacuum fittings.

Exhaust manifold

4 If the engine is still in vehicle, follow instructions 1a and b for inlet manifold removal and detach the exhaust pipe to manifold connection.
5 Unscrew the manifold shield bolts and detach the shield.
6 Detach the heat stove hose from the air cleaner unit and remove the heat stove.
7 Unscrew and remove the exhaust manifold to cylinder head retaining nuts. Remove the manifold and clean off any gasket material from the manifold or head flanges.

56 Oil pan - removal

1 If engine is still in vehicle, drain the oil into a suitable container and detach the front splash shield.
2 Remove the front crossmember. Detach the relay rod from the idler arm and lower relay rod.
3 Remove the bellhousing brace from the left-hand side and disconnect the vacuum line from the oil pan.
4 Remove the bolts and detach the oil pan and gasket.

57 Oil pump - removal

1 The oil pump can be removed with the engine in place in the vehicle or out of it as required.
2 To remove it in the vehicle, remove the cam cover, the distributor unit and oil pan as described in Sections 52, 53 and 56, respectively.
3 Unscrew and remove the oil pick up tube to block bolt and detach the tube from the pump (Fig. 1.37).
4 Unscrew the pump unit mounting bolts and withdraw the pump unit.
5 Detach the rubber hose and relief valve assembly from the pump unit.

Fig. 1.35. The rocker arm assemblies

Fig. 1.34. The locking timing chain adjuster positions

Fig. 1.36. Using the special valve spring compressor J-26513

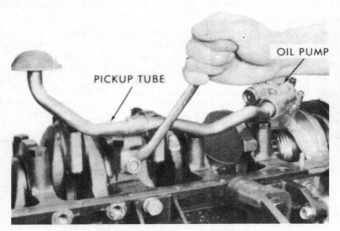

Fig. 1.37. Remove the oil pump pick up tube

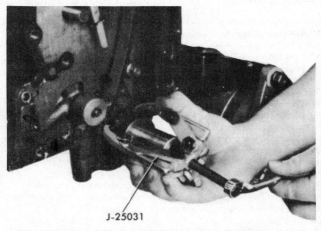

Fig. 1.38. Remove the crankshaft timing sprocket using a puller

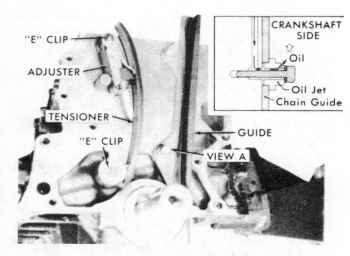

Fig. 1.39. The timing chain tensioner, guide and adjuster and inset oil jet position

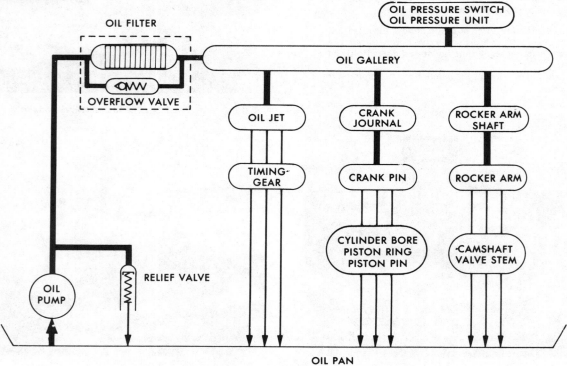

Fig. 1.40. The engine lubrication system

58 Front cover - removal

1 Remove the cylinder head as given in Section 53 and the oil pan, see Section 56.
2 Detach the oil pick up tube from the oil pump.
3 Remove the generator and air conditioning drive belts and fan unit.
4 Unscrew and remove the crankshaft pulley center bolt and remove the pulley and harmonic balancer unit.
5 Undo the air pump mounting bolts, hinge the pump towards the engine and remove the drive belt.
6 On vehicles fitted with air conditioning, undo the mounting bolts and remove the compressor and brackets.
7 Remove the distributor as described in Chapter 4.
8 Unscrew and remove the front cover retaining bolts.
9 Remove the front cover, carefully tapping with a soft faced mallet if it sticks.

59 Timing chain, pinion gear, sprockets and chain adjusters, tensioner and guides - removal

1 With the front cover removed, see previous Section, remove the timing chain from the lower timing sprocket and the camshaft sprocket.
2 Withdraw the pinion gear and timing sprocket using a puller as in Fig. 1.38. If available use special puller J-25031.
3 Refer to Fig. 1.39 and 'E' clip on the automatic chain adjuster.
4 Remove the 'E' clip and the chain tensioner.
5 Remove the guide bolts and withdraw the guide and oil jet.

60 Pistons and connecting rods - removal

1 If the engine is in the vehicle, remove the cylinder head and oil pan as described in Sections 53 and 56.
2 Remove the pistons and connecting rods in the same manner as described in Part A, Section 18.

61 Rear main oil seal - removal

1 This can be removed from the vehicle, with the engine installed providing that the oil pan and transmission are removed. Refer to Section 56 and Chapter 6, respectively for the relevant removal instructions.
2 On manual transmission models detach the clutch cover and pressure plate assembly.
3 With the battery ground cable disconnected, remove the starter motor unit.
4 Bend over the bolt locking tabs and remove the flywheel bolts. Remove the flywheel and cover.
5 Unscrew the main seal retainer bolts and remove the seal and retainer. Prise the seal from the container assuming it is to be renewed.

62 Crankshaft and main bearings - removal

1 Rotate the crankshaft and position it vertically to the top of the cylinder block.
2 Unscrew the crankshaft main bearing cap bolts in sequence. Start from the end caps.
3 Before removal note the respective positions of the caps. The arrow mark on the rear face of each cap points to the front of the engine and each journal and cap number are in alignment.
4 On removing the caps, lay them out in sequence with their respective bearing shells ready for cleaning and inspection.
5 Remove the crankshaft and thrust bearings from the cylinder block.

63 Lubrication system - description

The pressurised circulation type lubrication system consists of a Trochoid type oil pump, a full flow oil filter with cartridge element, and the various front cover/cylinder block/cylinder head oilways for supplying oil to the engine components.

Oil is drawn from the oil pan by the oil pump and is fed to the full flow oil filter. From the filter the oil is forced into the oil gallery then onto the crankshaft journals via the oil ports. The number 3 crankshaft oil port has a diversion port which feeds oil to the cylinder head. This then lubricates the rocker assemblies via oilways in the inlet and exhaust valve rocker shafts.

The camshaft is lubricated from an oil well in the upper cylinder head face.

The timing sprockets and chain are lubricated by oil which is delivered from the number 1 crankshaft journal oil port which delivers the oil to the oil jet on the timing chain guide.

A relief valve is incorporated in the oil pump and under excess pressure the valve opens and allows excess oil to be diverted back into the oil pan.

The oil filter body contains an overflow valve in order that a difference of oil pressure on each side of the filter maintains the specified level. When the overflow valve opens the oil is by-passed direct to vital engine parts.

64 Examination and renovation - general

Refer to Section 23.

65 Crankshaft and main bearings - examination and renovation

Refer to the procedure given in Section 24 but where reference to Specifications are concerned refer to those given at the start of Part B for the Series 5 and 6 models.

66 Connecting rods and connecting rod (big end) bearings - examination and renovation

Refer to Section 25, and the Specifications in Part B.

67 Cylinder bores - examination and renovation

Refer to the procedure given in Section 26 and the Specifications in Part B.

68 Pistons and rings - examination and renovation

Refer to the procedure given in Section 27 and the Specifications in Part B.

69 Camshaft - examination and renovation

Refer to the procedures given in Section 28 and the Specifications in Part B.

70 Timing chain, pinion gear, sprockets and chain adjuster and tensioner - examination and renovation

1 Check the timing chain sprockets for signs of excessive wear or damage. Renew if required.
2 Inspect the timing chain for excessive wear or damage. Refer to Fig. 1.41 and calculate the distance (L) with the chain stretched to a tension of 22 lbs (9.9 kg). If distance (L) exceeds 15.16 inches (38.4 cm) the chain must be renewed.
3 Check the automatic chain adjuster pin, arm, wedge and rack teeth of the locking plate. If worn or damaged, renew.
4 Check the chain adjuster and renew if required.
5 Inspect the chain guide and oil jet. If worn or damaged, renew.

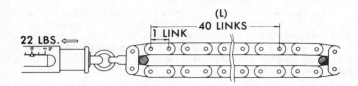

Fig. 1.41. Check the timing chain for wear - see Sec. 70, paragraph 2

71 Cylinder head and valves - servicing and decarbonising

Refer to the procedures given in Section 31 and the Specifications in Part B. Note the following differences:

Maximum valve stem to guide allowable clearance
0.1298 in (3.27 mm)   inlet
0.0097 in (0.23 mm)   exhaust

72 Flywheel and ring gear - servicing

Refer to the procedures given in Section 32 but note that the minimum friction face thickness of the flywheel is 1.22 in (3.1 cm).

73 Oil pump - examination and renovation

1 Thoroughly clean and dry the pump unit.
2 Insert a feeler gauge between the drive rotor and driven rotor. The maximum allowable clearance is 0.0079 in (0.19 mm). If the clearance exceeds this tolerance or the respective parts are damaged or scored, renew the pump unit.
3 Insert the feeler gauge between the driven rotor and pump body inner wall. The maximum allowable clearance is 0.0098 in (0.25 mm). Renew the pump unit if the clearance exceeds this tolerance or if damaged or badly pitted or scored.
4 Insert a feeler gauge between the drive rotor and driven rotor end faces and a straight edge placed across the pump body end faces (Fig. 1.42). If the clearance exceeds 0.0079 in (0.19 mm) or the faces are damaged in any way, renew the pump unit.
5 Measure the driveshaft outside diameter and the driveshaft bore diameter. If the clearance exceeds 0.0098 in (0.25 mm) renew the pump.
6 Inspect the oil pump rubber hose and renew if damaged or deterioration is apparent.
7 Inspect the relief valve and spring. If worn, weak or damaged, renew the oil pump unit.
8 Inspect the oil pump body, cover and drive gear for signs of wear, cracks or damage. If apparent, renew the oil pump unit.

74 Engine reassembly - general

Refer to Section 34.

75 Engine - reassembly

1 In general, the engine reassembly is a reversal of the dismantling process. The crankshaft, piston and connecting rod assembly procedures are basically the same as those given in Part A for the Series 1 to 4 models, although the clearance and torque tightening specifications may differ. Careful reference should therefore be made to the Specifications quoted in Part B during assembly.
2 The following instructions are the principle assembly details which differ to those in Part A and where special care must be taken.
3 Lubricate all crankshaft journals, main bearings and connecting rod bearings with clean engine oil as the respective components are assembled. Also lubricate the pistons, rings and cylinder bores in the same manner.
4 Insert the crankshaft thrust bearing with the oil groove face turned outwards (Fig. 1.43).
5 Fit the crankshaft bearing caps with the marking arrow on the rear face of each cap pointing to the front of the engine, and the corresponding cap and journal numbers aligned.
6 Half tighten the main bearing cap bolts in the sequence 3 - 4 - 2 - 5 - 1. Then fully tighten the bolts to 72 lb f ft (9.9 kg fm) torque and ensure that the crankshaft rotates freely.
7 When fitting the piston rings, ensure that they are positioned with their gaps located as shown in Fig. 1.44.
8 When fitting the piston and connecting rods, ensure that the face of the connecting rod with the cylinder number mark is located on the starter motor side with the piston facing the front (Fig. 1.45). Tighten the connecting rod cap bolts to 33 lb f ft (4.5 kg fm). Rotate the crankshaft to ensure that it turns without binding.
9 Torque the flywheel retaining bolts to 69 lb f ft (9.4 kg fm) and lock by bending over the lock tabs.
10 Refit the timing sprocket and pinion gear with the grooved side to the front cover.
11 When refitting the timing chain align the mark plate on the chain with the crankshaft timing sprocket mark. The mark plate is on the front side of the chain with the greater number of links between mark plates on the chain guide side. The crankshaft sprocket is fitted with the marked side forward and the triangular mark in alignment with the chain mark plate (Fig. 1.46).
12 Ensure that the camshaft sprocket and timing chain are in mesh until the sprocket is fitted to the camshaft.
13 When fitting the front cover, the oil pump drive gear punch mark must be aligned with the oil filter side of the cover. The dowel pin center and the alignment mark on the oil pump casing must be in line as in Fig. 1.47.
14 Before fitting the cover, ensure that the no. 1 and 4 pistons are at TDC, then install the cover engaging the pinion gear with the oil pump drive gear, and check that the oil pump drive gear punch mark faces the rear side. This can be seen through the front cover to cylinder block clearance as in Fig. 1.48.
15 The oil pump shaft end split must be parallel to the front cylinder block face but with the offset forward (Fig. 1.49).
16 When installing the oil pump, align the notch on the crankshaft pulley with the 'O' marking on the front cover. If the cylinder head is in position, also align the camshaft and number 1 rocker arm marks. This is to ensure that the number 4 piston is on TDC on compression stroke.
17 Insert the driven rotor with the alignment mark in line with the drive rotor mark as in Fig. 1.50.
18 The oil pump unit is installed so that the drive gear and pinion gear are engaged. The drive gear alignment mark faces rearwards and at an angle of about 20° clockwise from the crankshaft line as in Fig. 1.51. The oil relief valve, pump cover, oil pipe and hose can then be refitted. When refitting the oil pan, a thin coating of Permatex No. 2 or equivalent should be applied to the positions indicated by the arrows in Fig. 1.52.

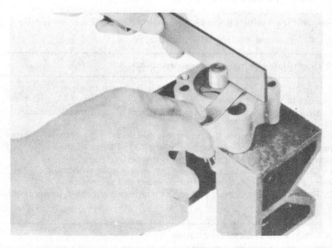

Fig. 1.42. Measuring the rotor to pump cover clearance with a feeler gauge and straight edge

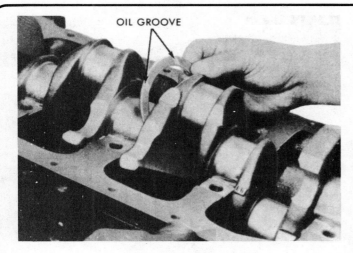

Fig. 1.43. Installing the thrust bearing

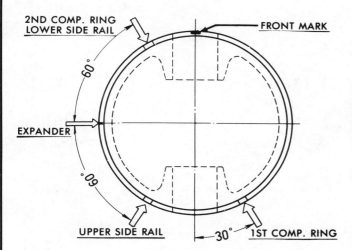

Fig. 1.44. The correct piston ring positions prior to assembly

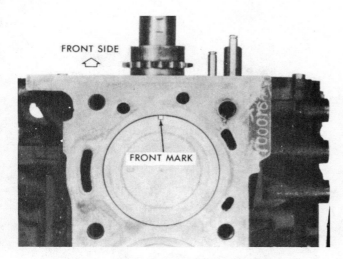

Fig. 1.45. The piston in position showing the front mark

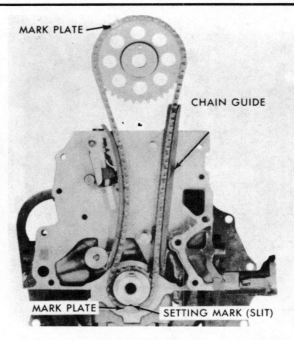

Fig. 1.46. The timing chain position showing the alignment marks

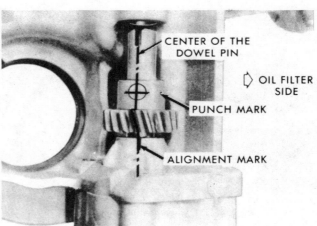

Fig. 1.47. The oil pump alignment position

Fig. 1.48. Check the oil pump drive gear punch mark

Fig. 1.49. The correct oil pump shaft alignment position

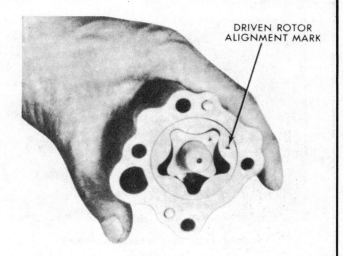

Fig. 1.50. Align the drive and driven rotors punch marks

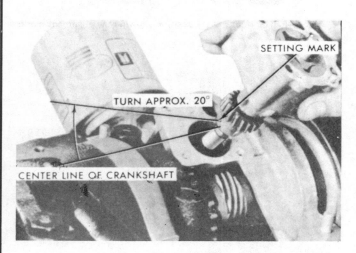

Fig. 1.51. Insert the oil pump

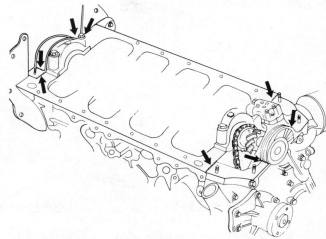

Fig. 1.52. The sealer positions on the oil pan gasket surface

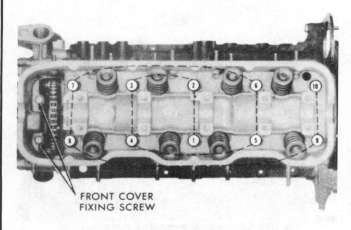

Fig. 1.53. The cylinder head bolt tightening sequence

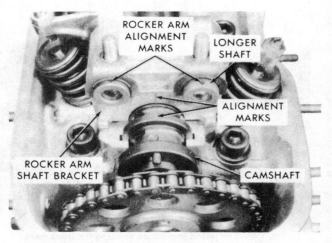

Fig. 1.54. The rocker arm and camshaft alignment marks

Chapter 1 Part B: Engine - Series 5 and 6

76 Cylinder head - assembly and installation

1 Position the head gasket over the dowel pins with the 'TOP' side upwards.
2 Insert each valve into its respective guide, applying a little engine oil to the stems.
3 Reassemble the respective valves, springs, retainers and seals to the cylinder head using the valve spring compressor. Ensure that each retainer is fully seated in the valve stem groove.
4 Place the cylinder head into position on the cylinder block and fit the retaining bolts. Retain the camshaft sprocket and chain with a piece of wire or rod. Tighten the head bolts progressively in sequence (Fig. 1.53), to 61 lb f ft (8.4 kg fm), then further tighten to 72 lb f ft (9.9 kg fm).
5 Lubricate the camshaft and place it into position.
6 Install the rocker arm shafts and note that the longer shaft is fitted to the exhaust valve side.
7 Fit the rocker arm shaft brackets and rocker arms to the shafts with the cylinder number marked on the bracket upper faces pointing forwards.
8 The mark on the number 1 rocker arm shaft bracket must align with that of the inlet and exhaust valve side rocker arm shafts.
9 When the rocker arm shaft stud holes are in line with the rocker arm shaft bracket stud holes, ensure that the number 1 rocker arm shaft bracket is longer on the exhaust side as can be seen by the amount of projection.
10 Insert the rocker arm springs into position between the shaft bracket and rocker arm. Ensure that the rocker arm shaft is turned upwards as indicated by the punch marks (see Fig. 1.54) and then install the rocker arm shaft bracket assembly onto the head studs.
11 Now align the camshaft and number 1 rocker arm shaft bracket marks, and then tighten the rocker arm shaft bracket nuts to 16 lb f ft (2.2 kg fm).
12 To prevent damaging the rocker arm springs, retain them with an adjustable wrench when tightening the bracket stud nuts.
13 Align the camshaft sprocket with the pin on the camshaft and assemble the sprocket to the shaft, but do not remove the sprocket from the chain. Retain the sprocket with bolt and washer and refit the half moon seal in the front of the head. Torque the retaining bolt to 58 lb f ft (8.0 kg fm).
14 Reset the automatic adjuster with a screwdriver by rotating the slide pin 90° counter clockwise, and check the timing chain tension.
15 To check the valve timing the rotor and mark on the distributor housing must be aligned with the number 1 piston at TDC on firing position. The crankshaft pulley timing mark must align with the 'O' mark on the front cover.

77 Valve clearance - adjustment

1 Remove the rocker arm cover.
2 Prior to adjusting the valves make sure that the rocker arm shaft bracket nuts are tight and at their proper torque.
3 Rotate the crankshaft so that either the number 1 or 4 piston is at top dead center (TDC) on the compression stroke. In this position the timing marks on the gear case and crankshaft pulley are in alignment.
4 When the number 1 piston is at TDC, the number 1 cylinder intake and exhaust valves, the number 2 intake valve and the number 3 exhaust valve can all be adjusted. When the number 4 piston is at TDC the number 2 exhaust valve, the number 3 intake valve and the number 4 intake and exhaust valves can be adjusted.
5 Use a feeler gauge to measure the clearance between the rocker arm and the valve stem. There should be a slight drag on the gauge.
6 If the valve needs adjusting, loosen the lock nut and turn the adjusting screw until the proper clearance is obtained. Then retighten the lock nut, and recheck the clearance to be sure it hasn't changed.

78 Engine - final assembly after major overhaul

1 Fit the inlet manifold gasket over the studs and reverse the inlet manifold removal sequence as given in Section 55. Tighten the manifold retaining nuts progressively from the inner studs outwards.
2 Reconnect the hoses to the inlet manifold with the PCV hose to the upper left side, the hot idle hose compensation hose on the upper right and the air vacuum hose on the lower side.
3 Fit the exhaust manifold gasket into position over the studs and reverse the manifold removal sequence given in Section 55. Tighten the retaining nuts working from the inside-out in a progressive manner.
4 Refit the crankshaft pulley and harmonic balancer. Tighten the retaining bolt and washer to a torque of 87 lb f ft (12.027 kg fm).
5 Refit the flywheel splash shield.
6 Re-install the starter motor if engine and transmission are being refitted.

79 Engine/transmission - installation

1 Locate lifting slings to the engine and lift and lower into position in the vehicle, taking care not to damage the surrounding fittings.
2 With the engine transmission in position, refit the engine mounting nuts and then refit the respective ancillary components in the reverse order of removal given in Section 51.
3 When all hoses, wires and attachments are refitted, refill the engine cooling system and check for leaks.
4 Refill the engine/transmission oil.
5 On manual transmission models check the clutch pedal free play.
6 Adjust the fan belt and air pump drivebelts for tension.

80 Initial start-up after major overhaul

Follow the instructions given in Section 47.

81 Fault diagnosis - engine

Symptom	Reasons	Remedy
Engine fails to turn over when starter control operated		
No current at starter motor	Flat or defective battery	Charge or renew battery. Push start car.
	Loose battery leads	Tighten both terminals and ground ends of earth leads.
	Defective starter solenoid or broken wiring	Run a wire direct from the battery to the starter motor or by-pass the solenoid.
	Engine ground strap disconnected	Check and retighten strap.
Current at starter motor	Jammed starter motor drive pinion	Place car in gear and rock to and fro.
	Defective starter motor	Remove and recondition.
Engine turns over but will not start		
No spark at spark plug	Ignition damp or wet	Wipe dry the distributor cap and ignition leads.
	Ignition leads to spark plugs loose	Check and tighten at both spark plug and distributor cap ends.
	Grounded or disconnected low tension leads	Check the wiring on the coil and to the distributor.

Symptom	Reasons	Remedy
	Dirty, incorrectly set, or pitted contact breaker points	Clean, file smooth, and adjust.
	Faulty condenser	Check contact breaker points for arcing, remove and install new.
	Defective ignition switch	By-pass switch with wire.
	Ignition leads connected wrong way round	Remove and refit leads to spark plugs in correct order.
	Faulty coil	Remove and install new coil.
	Contact breaker point spring earthed or broken	Check spring is not touching metal part of distributor. Check insulator washers are correctly placed. Renew points if the spring is broken.
No fuel at carburetor float chamber or at jets	No gas in gas tank	Refill tank!
	Vapor lock in fuel line (in hot conditions or at high altitude)	Blow into gas tank, allow engine to cool, or apply a cold wet rag to the fuel line.
	Blocked float chamber needle valve	Remove, clean and refit.
	Fuel pump filter blocked	Remove, clean and refit
	Choked or blocked carburetor jets	Dismantle and clean.
	Faulty fuel pump	Remove, overhaul and refit
Engine stalls and will not start		
Excess of gas in cylinder or carburetor flooding	Too much choke allowing too rich a mixture to wet plugs	Remove and dry spark plugs or with wide open throttle, push start the car.
	Float damaged or leaking or needle not sealing	Remove, examine, clean and refit float and needle valve as necessary.
	Float damaged or	
	Float lever incorrectly adjusted	Remove and adjust correctly.
No spark at spark plug	Ignition failure - sudden	Check over low and high tension circuits for breaks in wiring.
	Ignition failure - misfiring precludes total stoppage	Check contact breaker points, clean and adjust. Renew condenser if faulty.
	Ignition failure - in severe rain or after traversing water splash	Dry out ignition leads and distributor cap.
No fuel at jets	No gas in gas tank	Refill tank!
	Sudden obstruction in carburetor	Check jets, filter, and needle valve in float chamber for blockage.
	Water in fuel system	Drain tank and blow out fuel lines.
Engine misfires or idles unevenly		
Intermittent spark at spark plug	Ignition leads loose	Check and tighten as necessary at spark plug and distributor cap ends.
	Battery leads loose on terminals	Check and tighten terminal leads.
	Battery earth strap loose on body attachment point	Check and tighten ground lead to body attachment point.
Intermittent sparking at spark plug	Engine ground lead loose	Tighten lead.
	Low tension leads to coil loose	Check and tighten leads if found loose.
	Dirty, or incorrectly gapped plugs	Remove, clean and regap.
	Dirty, incorrectly set, or pitted contact breaker points	Clean, file smooth and adjust.
	Tracking across inside or distributor cover	Remove and install new cover.
	Ignition too retarded	Check and adjust ignition timing.
	Faulty coil	Remove and install new coil.
Fuel shortage at engine	Mixture too weak	Check jets, float chamber needle valve, and filter for obstruction. Clean as necessary.
	Carburetor incorrectly adjusted	Remove and overhaul carburetor.
	Air leak in carburetor	Remove and overhaul carburetor.
	Air leak at inlet manifold to cylinder head, or inlet manifold to carburetor	Test by pouring oil along joints. Bubbles indicate leak. Renew manifold gasket as appropriate.
Lack of power and poor compression		
Mechanical wear	Incorrect valve clearances	Adjust rocker arms to take up wear.
	Burnt out exhaust valves	Remove cylinder head and renew defective valves.
	Sticking or leaking valves	Remove cylinder head, clean, check and renew valves as necessary.
	Weak or broken valve springs	Check and renew as necessary.
	Worn valve guides or stems	Renew valve guides and valves.
	Worn pistons and piston rings	Dismantle engine, renew pistons and rings.

Chapter 1 Part B: Engine - Series 5 and 6

Symptom	Reason	Remedy
Fuel/air mixture leaking from cylinder	Burnt out exhaust valves	Remove cylinder head, renew defective valves.
	Sticking or leaking valves	Remove cylinder head, clean, check, and renew valves as necessary.
	Worn valve guides and stems	Remove cylinder head and renew valves and valve guides.
	Weak or broken valve springs	Remove cylinder head, renew defective springs.
	Blown cylinder head gasket (accompanied by increase in noise)	Remove cylinder head and install new gasket.
	Worn pistons and piston rings	Dismantle engine, renew pistons and rings.
	Worn or scored cylinder bores	Dismantle engine, rebore, renew pistons and rings.
Incorrect adjustments	Ignition timing wrongly set. Too advanced or retarded	Check and reset.
	Contact breaker points incorrectly gapped	Check and reset contact breaker points.
	Incorrect valve clearances	Check and reset rocker arm to valve stem gap.
	Incorrectly set spark plugs	Remove, clean and regap.
	Carburation too rich or too weak	Tune carburetor for optimum performance.
Carburation and ignition faults	Dirty contact breaker points	Remove, clean and refit.
	Fuel filters blocked causing poor top end performance through fuel starvation	Dismantle, inspect, clean and refit all fuel filters.
	Distributor automatic balance weights or vacuum advance and retard mechanisms not functioning correctly	Overhaul distributor.
	Faulty fuel pump giving top end fuel starvation	Remove, overhaul, or install exchange reconditioned fuel pump.
Excessive oil consumption	Excessively worn valve stems and valve guides	Remove cylinder head and fit new valves and valve guides.
	Worn piston rings	Install oil control rings to existing pistons or purchase new pistons.
	Worn pistons and cylinder bores	Install new pistons and rings, rebore cylinders.
	Excessive piston ring gap allowing blow-up	Install new piston rings and set gap correctly.
	Piston oil return holes choked	Decarbonise engine and pistons.
Oil being lost due to leaks	Leaking oil filter gasket	Inspect and install new gasket if necessary.
	Leaking rocker cover gasket	Inspect and install new gasket as necessary.
	Leaking timing gear cover gasket	Inspect and install new gasket as necessary.
	Leaking oil pan gasket	Inspect and install new gasket as necessary.
	Loose oil pan plug	Tighten and install new gasket as necessary.
Unusual noises from engine Excessive clearances due to mechanical wear	Worn valve gear (noisy tapping from rocker box)	Inspect and renew rocker shaft, rocker arms, and ball pins as necessary.
	Worn big-end bearing (regular heavy knocking)	Drop oil pan, if bearings broken up clean out oil pump and oilways, install new bearings. If bearings not broken but worn install bearing shells.
	Worn timing chain and sprockets (rattling from front of engine)	Remove timing cover, install new timing sprockets and timing chain.
	Worn main bearings (rumbling and vibration)	Drop oil pan, remove crankshaft, if bearing worn but not broken up, renew. If broken up strip oil pump and clean out oilways.
	Worn crankshaft (knocking, rumbling and vibration)	Regrind crankshaft, install new main and big-end bearings.

Chapter 2 Cooling system

For information relating to Series 8 thru 12 models, including 4WD, refer to Chapter 13.

Contents

Antifreeze and rust inhibitors ... 6	Fault diagnosis - cooling system ... 11
Cooling system - draining ... 3	General description ... 1
Cooling system - filling ... 5	Radiator - removal and installation ... 8
Cooling system - flushing ... 4	Thermostat - removal, testing and installation ... 9
Cooling system - routine maintenance ... 2	Water pump - removal and installation ... 10
Fan/alternator drivebelt - tension adjustment ... 7	

Specifications

System type ...	Pressurized circulation with pump assistance
Radiator type ...	Flat tubes with corrugated fins
Cap pressure valve opening pressure ...	15 psi
Vacuum valve opening pressure ...	0.6 - 0.7 psi
Radiator leakage test specified pressure ...	28.4 psi
Coolant capacity	
Series 1 to 4 ...	1.5 US gallons
Series 5 and 6 ...	1.6 US gallons
Water pump	
Type ...	Centrifugal impeller
Delivery volume at 3000 rpm ...	13.2 gal/min or more (Series 1 to 4)
Delivery volume at 6000 rpm (68°F water temperature) ...	42.3 gal/min (Series 5)
Seal type ...	Mechanical
Bearing type ...	Ball bearing
Pump pulley ratio ...	1.10 (1.18 with A/C)
Number of impellers ...	6
Impeller diameter ...	2.756 in
Cooling fan	
Number of blades ...	4
Series 5 and 6 with A/C ...	7
Cooling fan diameter -	
Series 1 to 4 ...	12.6 inch
Series 5 and 6 ...	13 in without A/C
Series 5 and 6 ...	15.4 in with A/C
Fan belt deflection ...	0.4 in
Thermostat	
Type ...	Wax pellet
Valve opening temperature ...	177° - 182°F (80° - 83°C)
Valve lift 203°F (96°C) ...	0.32 inches or more

Chapter 2/Cooling system

1 General description

The cooling system comprises the radiator, top and bottom water hoses, the water pump, cylinder block and head water jackets and a thermostat. The radiator cap has a pressure relief valve.

The principle of the system is that cold water in the bottom of the radiator circulates upwards through the lower radiator hose to the water pump, where the pump impeller pushes the water round the cylinder block and head through the various cast-in passages to cool the cylinder bores, combustion surfaces and valve seats. When sufficient heat has been absorbed by the cooling water, the engine has reached an efficient working temperature, the water moves from the cylinder head past the now open thermostat into the top radiator hose and into the radiator header tank.

The water then travels down the radiator tubes where is is rapidly cooled by the in-rush of air, when the vehicle is in forwards motion. A multi-bladed fan, mounted on the water pump pulley, assists this cooling action. The water, now cooled, reaches the bottom of the radiator and the cycle is repeated.

When the engine is cold the thermostat remains closed until the coolant reaches a pre-determined temperature (see Specifications); this assists rapid warming-up.

2 Cooling system - routine maintenance

1 The cooling system requires very little routine maintenance but in view of its important nature, this maintenance must not be neglected.
2 The maintenance intervals are given in the Routine Maintenance Section at the beginning of this manual.
3 Apart from regular checking of the coolant level, and inspection for leaks and deterioration of hose connections, the only other items of major importance are the use of antifreeze solutions or rust inhibitors suitable for aluminium engines, and renewal of the coolant. These items are covered separately in this Chapter.
4 It must be remembered that the cooling system is pressurized. This means that when the engine is hot, the coolant will be at a temperature in excess of 212°F (100°C). Great care must therefore be taken if the radiator pressure cap has to be removed when the engine is hot since steam and boiling water will be ejected. If possible, let the engine cool down before removing the pressure cap. If this is not possible, place a cloth on the cap and turn it slowly counter-clockwise to the first notch. Keep it in this position until all the steam has escaped then turn it further until it can be removed.

3 Cooling system - draining

1 Park the vehicle on level ground. If the coolant is to be re-used, place a suitable clean container under the drain taps or plug for its collection.

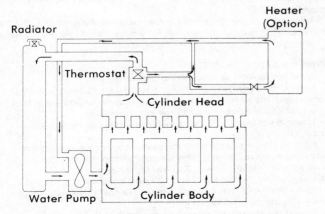

Fig. 2.1. The cooling system circulation

2 Remove the radiator cap. If the engine is hot see paragraph 4 of the previous Section.
3 Open the radiator and engine drain taps/plugs and drain off all of the coolant.
4 When coolant ceases to flow from the drain taps, probe them with a stiff wire to ensure that no sediment is blocking the drain orifices.
5 On completion refer to Section 4 or 5 as applicable.

4 Cooling system - flushing

1 With the passage of time deposits can build up in an engine which will lead to engine overheating and possibly serious damage.
2 It is a good policy, whenever the cooling system is drained, to flush the system with cold water from a hosepipe. This can conveniently be done by leaving a hosepipe in the radiator filler orifice for about 15 minutes while water is allowed to run thru. This will usually be sufficient to clear any sediment which may be present.
3 If there appears to be a restriction, first try reverse flushing; this is the application of the hose to the drain orifices and forcing water back thru the radiator tubes and out of the filler.
4 If the radiator flow is restricted by something other than loose sediment, then no amount of flushing will shift it and it is then that a proprietary chemical cleaner, suitable for aluminium engines, is needed. Use this according to the directions and make sure that the residue is fully flushed out afterwards. If leaks develop after using a chemical cleaner, a proprietary radiator sealer may cure them, but the signs are that the radiator has suffered considerable chemical corrosion and that the metal is obviously getting very thin in places.

5 Cooling system - filling

1 When draining (and flushing if applicable) has been accomplished, close the drain taps and top-up the cooling system with water which contains the correct proportion of antifreeze or inhibitor (see Section 6). Ensure that the heater water valve is open to prevent airlocks from occurring in the heater matrix.
2 When the radiator is full, run the engine for 5 to 10 minutes at a fast idle. As the water circulates and the thermostat opens, the coolant level will be seen to fall. Top-up the radiator to about halfway up the fill elbow, install the cap, then run the engine for a few more minutes and check carefully for water leaks.
3 Allow the system to cool, then recheck the coolant level. Top-up if necessary so that, with the engine idling, the coolant is within 1½ in of the top of the filler neck and then refit the radiator cap.

6 Antifreeze and rust inhibitors

1 Tap water alone should not be used in an aluminium engine except in an emergency. If it has to be used, it should be drained off at the earliest opportunity and the correct coolant mixture used instead.
2 Generally speaking, the basis of the coolant mixture can be tap water, except where this has a high alkali content or is exceptionally hard. If these conditions exist, clean rainwater or distilled water should be used.
3 Antifreeze must be of a type suitable for use with aluminium engines (ethylene glycol based antifreeze is suitable) and many proprietary products will be available for use. The fact that all products tend to be expensive should not deter you from using them, since they are a good insurance against freezing and corrosion.
4 The following table gives suitable concentrations of antifreeze. Do not use concentrations in excess of 55% except where protection to below −35°F (−37°C) is required as there is a possibility of overheating in very hot weather.

Coolant freezing water point	Mixture percentage (volume)		Specific gravity of mixture at 68°F (20°C)
	Anti-freeze	Water	
−4°F (−20°C)	35	65	1.051
−49°F (−45°C)	55	45	1.078

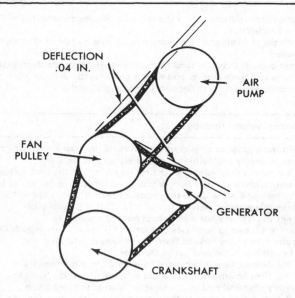

Fig. 2.2. The drive belt layout and their adjustment

8.3 Lift the radiator clear taking care not to damage the core

Fig. 2.3. Loosen but do not remove the bolt arrowed

5 Antifreeze mixtures are normally suitable for use for a period of two years (even so-called permanent antifreeze) after which they should be discarded and a fresh mixture used.
6 Antifreeze normally contains suitable corrosion inhibitors for protection of the engine. However, if antifreeze is not used for some reason, the vehicle manufacturers market suitable inhibitors which will give satisfactory protection to the engine. Any inhibitors which are used must be mixed in accordance with the instructions on the container.

7 Fan/alternator drivebelt - tension adjustment

1 In order that the engine cooling and alternator charge rate are efficient, the drivebelt tension must be correct.
2 To check the belt tension, apply thumb pressure to the belt at its midway point of its longest run between pulleys.
3 If the distance between the pulleys is 13 to 16 inches, the belt deflection should be ½ inch. If the distance between the pulleys is 7 to 10 inches the belt deflection must be ¼ inch.
4 If adjustment is required, loosen the alternator mounting bolts and pivot bolt and move the alternator as necessary. Avoid prying against the side of the alternator or irreparable damage may result.
5 Tighten the bolts after adjustment has been made, then recheck the tension. After running the vehicle for about 150 miles (250 km), recheck the belt tension and adjust if necessary.

8 Radiator - removal and installation

1 Drain the cooling system - refer to Section 3 if necessary.
2 Loosen the hose clamps and then disconnect the top and bottom hoses.
3 Undo and remove the four radiator retaining bolts, then lift the radiator clear of the vehicle taking care not to damage the core (photo).
4 Installation is a direct reversal of removal but be sure to shut the drain tap before refilling with coolant. After filling check the hose connections for any signs of leakage.

9 Thermostat - removal, testing and installation

Series 1 to 4
1 Drain the radiator by means of the drain tap. Any coolant to be re-used may be drained into a suitable clean container.
2 Undo the top hose clamp from the engine connection.
3 Disconnect the outlet pipe to remove the thermostat.
4 To test whether the unit is serviceable, suspend the thermostat on a piece of string in a pan of water being heated. Using a thermometer, with reference to the opening and closing temperature in Specifications, its operation may be checked. The thermostat should be renewed if it is stuck open or closed or if it fails to operate at the special temperature. The operation of the thermostat is not instantaneous and sufficient time must be allowed for the movement during test. Never install a faulty unit - leave it out if no replacement is available immediately.
5 Installation of the thermostat is the reverse of the removal procedure, but be sure to check for leaks when fully reassembled and the engine is running.

Series 5 and 6
6 Drain the radiator as described in Section 3.
7 Detach the crankshaft ventilation hose, the emission control system hose, the air hose and the controlled combustion system hose from the hot idle compensator to the inlet manifold.
8 Unscrew the two air cleaner attachment bolts and slacken the clamp bolt.
9 Remove the air cleaner from the carburetor and disconnect the controlled combustion hose from the thermosenser to the inlet manifold from the air cleaner. Remove the air cleaner unit.
10 Unscrew and remove the two outlet pipe attachment bolts and withdraw the outlet pipe and water hose. The thermostat can now be removed from the inlet manifold.
11 To test the thermostat refer to paragraph 4 of this Section.
12 Refitting is a direct reversal of removal but be certain to renew any defective hoses or clips. On refilling the system check for leaks.

10 Water pump - removal and installation

Series 1 to 4
1. Detach the battery ground cable.
2. Drain the cooling system as described in Section 3.
3. Remove the radiator and shroud assembly.
4. Disconnect the generator and the air pump drivebelts by slackening their respective adjustment bolts.
5. Disconnect the fan with pulley and spacer.
6. The water pump unit can now be removed from the engine complete with gasket. When disconnecting the water pump, the bolt to the rear of the timing gear cover (Fig. 2.3) must be loosened but not removed.
7. The heater hose and radiator bottom hose can now be detached from the water pump.
8. Installation of the water pump is a direct reversal of removal but always fit a new gasket to the pump. Readjust the generator and air pump belts on reassembly and finally check for leaks when the system has been refilled.

Series 5 and 6
9. Disconnect the ground cable from the battery and drain the cooling system as described in Section 3.
10. On models without air conditioning equipment, undo and remove the four nuts securing the engine fan and detach the fan. On models with air conditioning equipment fitted, unscrew the generator and air pump mounting bolts and remove the drive belts. Detach the engine fan and pulley by unscrewing the four retaining bolts. Unscrew the four fan set plate and pulley attachment bolts to remove the set plate and pulley.
11. The water pump can now be removed by unscrewing the six retaining bolts.
12. Installation is a direct reversal of removal but be certain to readjust the generator and air pump drivebelts and when the system has been refilled, check for any signs of leakage.

11 Fault diagnosis - cooling system

Symptom	Reason/s	Remedy
Loss of coolant	Leak in system	Examine all hoses, hose connections, drain taps and the radiator and the heater for signs of leakage when the engine is cold, then when hot and under pressure. Tighten clips, renew hoses and repair radiator as necessary.
	Defective radiator pressure	Examine cap for defective seal and renew if necessary.
	Overheating causing too rapid evaporation due to excessive pressure in system	Check reasons for overheating.
	Blown cylinder head gasket causing excess pressure in cooling system forcing coolant out	Remove cylinder head for examination.
	Cracked block or head due to freezing	Strip engine and examine. Repair as required.
Overheating	Insufficient coolant in system	Top up.
	Water pump not turning properly due to slack fan belt	Tighten fan belt.
	Kinked or collapsed water hoses causing restriction to circulation of coolant	Renew hose as required.
	Faulty thermostat (not opening properly)	Check and renew as necessary.
	Engine out of tune	Check ignition setting and carburetor adjustments.
	Blocked radiator either internally or externally	Flush out cooling system and clean out cooling fins.
	Cylinder head gaskets blown forcing coolant out of system	Remove head and renew gasket.
	New engine not run-in	Adjust engine speed until run-in.
Engine running too cool	Missing or faulty thermostat	Check and renew as necessary.

Chapter 3 Carburetion; fuel, exhaust and emission control systems

For information relating to Series 8 thru 12 models, including 4WD, refer to Chapter 13.

Contents

Air cleaner (Series 1 to 4) - removal and installation ... 2	Crankcase emission control - description and maintenance ... 25
Air cleaner (Series 5 and 6) - removal and installation ... 3	Electric automatic choke - adjustment ... 17
Automatic temperature controlled air cleaner - testing ... 4	Electric automatic choke - general ... 16
Carburetor (DCP-340 Series 4, 5 and 6 models) - dismantling and reassembly ... 12	Emission control system - description ... 23
Carburetor (DCP-340 Series 4 models) - idle adjustment ... 14	Evaporative emission control system - description and maintenance ... 26
Carburetor (DCP-340 Series 5 and 6 models) - idle adjustment ... 15	Exhaust emission control system - component description and maintenance ... 24
Carburetor (DCP-340 Series 4, 5 and 6 models) - linkage adjustment 13	Exhaust system - removal and installation ... 27
Carburetor (DRJ-340 Series 1 and 2 models) - dismantling and reassembly ... 6	Fuel filter element - removal and installation ... 22
Carburetor (DRJ-340 Series 1 and 2 models) - idle adjustment ... 8	Fuel pump (electro magnetic) - removal and installation ... 20
Carburetor (DRJ-340 Series 1 and 2 models) - linkage adjustment... 7	Fuel pump (mechanical) - dismantling, inspection and reassembly ... 19
Carburetor - Types DRJ-340 (Series 1 and 2), DRJ-340 4C (Series 3) and DCP-340 (Series 4) - removal and installation ... 5	Fuel pump (mechanical) Series 1 to 3 models - removal and installation ... 18
Carburetor (DRJ-340 4C Series 3 models) - dismantling and reassembly ... 9	Fuel tank - removal and installation ... 21
Carburetor (DRJ-340 4C Series 3 models) - idle adjustment ... 11	General description ... 1
Carburetor (DRJ-340 4C Series 3 models) - linkage adjustment ... 10	

Specifications

Carburetor

Type
	Two barrel downdraft
Series 1 and 2 models	DRJ-340
Series 3 model	DRJ-340 4C
Series 4, 5 and 6 models	DCP-340

	Series 1	Series 2	Series 3	Series 4	Series 5 and 6
Inner air horn diameter	2.95 in	2.95 in	2.95 in	2.95 in	2.95 in
Fuel pressure	3.13 lb/in^2	4.55 lb/in^2	4.55 lb/in^2	4.55 lb/in^2	3.56 lb/in^2
Outlet diameter					
Primary	1.18 in	1.18 in	1.18 in	1.18 in	1.18 in
Secondary	1.35 in	1.35 in	1.34 in	1.34 in	1.34 in
Venturi diameter					
Large					
Primary	0.91 in	0.91 in	0.91 in	0.91 in	0.91 in
Secondary	1.14 in	1.14 in	1.14 in	1.14 in	1.14 in
Small					
Primary	0.32 in	0.32 in	—	—	—
Secondary	0.35 in	0.35 in	—	—	—
Main jet					
Primary	0.043 in	0.045 in	No. 115		
Secondary	0.061 in	0.061 in	No. 115		
Series 4					
Primary				No. 100 (f)/No. 112 (C)	
Secondary				No. 155	

Chapter 3/Carburetion; fuel, exhaust and emission control systems

Series 5 and 6
  Primary ... No. 103 (f)/No. 114 (C)
  Secondary ... No. 170 (f)/No. 160 (C)

Slow jet
Series 1 and 2
  Primary ... 0.020 in
  Secondary ... 0.032 in
Series 3 ... No. 50
Series 4
  Primary ... No. 70
  Secondary ... No. 80
Series 5 and 6
  Primary ... No. 50 (f)/No. 45 (C)
  Secondary ... No. 100

Power jet
Series 1 and 2 ... 0.016 in
Series 3 ... No. 40
Series 4 ... No. 45 (f)/ No. 40 (C)
Series 5 and 6 ... No. 40 (f)

Main air bleed

	Primary	Secondary
Series 1	0.022 in	0.020 in
Series 2	0.026 in	0.020 in
Series 3	No. 65	No. 70
Series 4	No. 65	No. 60
Series 5 and 6	No. 50 (f)/No. 80 (C)	No. 50

Slow air bleed

	Primary	Secondary
Series 1	0.063 in	0.039 in
Series 2	0.059 in	0.039 in
Series 3	No. 150	No. 100
Series 4	No. 120 (f)/No. 130 (C)	No. 100
Series 5 and 6	No. 160 (f)/No. 140 (C)	No. 70

	Series 1	Series 2	Series 3	Series 4	Series 5 and 6
Coasting jet	0.017 in	0.017 in	No. 43	No. 43	No. 60 (f; MT)/No. 67 (C)
Coasting air bleed	0.098 in	0.098 in	No. 290	No. 290	No. 290
Vacuum jet					1.5 (Series 6)
Idle port	0.07 in	0.07 in	0.07 in	0.07 in	0.07 in
Accelerator - pump nozzle diameter	0.02 in	0.02 in	0.02 in	0.02 in	0.02 in
Accelerator - pump delivery (1 stroke)	0.026 in^3	0.026 in^3	0.026 in^3	0.026 in^3	0.026 in^3
Power valve opening pressure	5.12 in Hg	5.12 in Hg	5.12 in Hg	6.3 in Hg	6.3 in Hg
Throttle valve opening angle (check valve closed)	17°	18.5°	17.5°	17°	17°
Diaphragm spring (set force)	0.64 lbs	0.64 lbs	0.64 lbs	0.64 lbs	0.80 lbs
Engine speed at which diaphragm opens (engine fully loaded)	50°	50°	50°	50°	47°
Main nozzle inside diameter					
Primary	0.091 in	0.091 in	0.091 in	0.091 in	0.091 in
Secondary	0.098 in	0.098 in	0.098 in	0.098 in	0.098 in
Throttle valve closing angle					
Primary	10°	10°	10°	10°	10°
Secondary	20°	20°	20°	20°	20°

Emission control
Air pump type ... ECP200 - 12 (Series 1 and 2)
               ECP200 - 12A (Series 1 and 4)

Construction ... Positive displacement type

Number of vanes ... 2

Inlet diameter	0.748 in
Outlet diameter	0.748 in
Theoretical delivery	11.29 cu in
Maximum speed	6000 rpm Series 1 to 4
	7000 rpm Series 5
Relief valve opening pressure	6.12 lb/in^2 Series 1 to 3
	5.3 lb/in^2 Series 4
	2.84 - 4.98 lb/in^2 Series 5
Check valve type	CV27 - 12
Mixture control valve type	AV54 - 11

Automatic temperature controlled air cleaner
Sensor
1 118.4°F ... 2.756 in Hg
2 100.4°F ... 6.693 in Hg

Vacuum motor
1 Starting negative pressure ... 2.756 in Hg
2 With rod lifted 0.236 in ... 6.693 in Hg

Idle valve opening and compensator: closing temperature	122° - 140°F - Series 1 to 3
	122° - 149°F - Series 4
Series 5 and 6	Oxidizing catalytic converter system
Converter size	16.98 x 8.98 x 3.5 in
Converter volume	160 cu in
Converter (with catalyst) weight	18 lbs
Catalyst dimension	Pellet 1/8 in min - 1/4 in max

Over temperature control system
Warning system operating temperature	Over 1830°F (at converter)
Engine speed sensor	'On' over engine speed of 1600 rpm

Exhaust gas recirculation (EGR) system
EGR valve operating negative pressure
Starting ... 3.5 in Hg
Fully opening ... 8 in Hg

Thermal vacuum valve opening temperature	Over 180° - 126°F

Controlled combustion system
Thermo sensor operating temperature
Starting ... 100°F
Fully opening ... 111°F

Idle compensator operating temperature	115° - 126°F

Evaporating control system
Check and relief valve operating pressure
Check valve open ... 1.0 - 1.4 in Hg
Relief valve open ... 0.2 - 0.6 in Hg

Chapter 3/Carburetion; fuel, exhaust and emission control systems

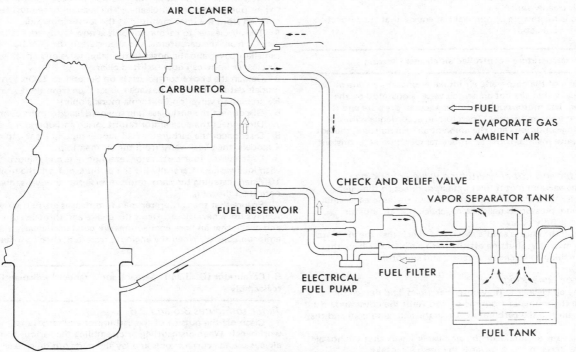

Fig. 3.1. The fuel system layout (Series V)

1 General description

All models are equipped with a rear mounted fuel tank, a fuel filter and, on Series 1 to 3 models, a mechanically operated fuel pump. On Series 4 and 5 the fuel pump is of an electric magnetic type.

A two-barrel downdraft carburetor is fitted to all models although the type depends on the year and model.

The carburetors incorporate emission control features and are fitted with a renewable element type of fuel filter, and on some models the air cleaner has an automatic temperature-control unit fitted.

The general information on the emission control equipment is detailed in the appropriate Section in this Chapter.

Fig. 3.2. Lift the air cleaner unit from the carburetor then disconnect the remaining hoses

2 Air cleaner (Series 1 to 4) - removal and installation

1 The air cleaner fitted to all Series 1 to 4 models is a wet paper element type. The air cleaner unit is fitted with an automatic intake air temperature control system and a hot idle compensator.
2 To remove the cleaner unit, first unscrew the wing nut and detach the rubber hoses on the cover. Also remove the rubber hose from the vacuum motor.
3 Unscrew the air cleaner cover bracket bolts.
4 Detach the hot air hose from the hot air duct, and the air pump hose from the cleaner. Also detach the vacuum hose from the joint nipple side of the inlet manifold.
5 Slacken the air cleaner to carburetor retaining bolt and separate the cleaner unit body from the carburetor (photo). Now disconnect the PCV hose from the camshaft cover, and the rubber hoses from the check and relief valves. The air cleaner unit can now be removed.
6 Installation is a reversal of removal but ensure that all hose connections are good.

3 Air cleaner (Series 5 and 6 models) - removal and installation

1 Refer to Fig. 3.3 and disconnect the PCV hose from the cylinder head cover and the ECS hose from the air cleaner unit.

2.5 Lift the air cleaner from the carburetor and disconnect the hoses

2 Detach the air hose from the air pump and unscrew the retaining bolts from the air cleaner. Loosen the clamp bolts.
3 Lift the air cleaner sufficiently to disconnect the CCS vacuum hose. Remove the air cleaner unit.
4 Installation is the reverse of removal but ensure that the respective hose connections are good.

4 Automatic temperature - controlled air cleaner - testing

1 The principle of the automatic air intake temperature control system is basically that the hot air switch valve is operated by the vacuum motor. The motor operates under vacuum pressure and is controlled by the thermo-sensor. Therefore in accordance with the engine operating conditions and the ambient air temperature, the hot air switching valve maintains the air intake temperature at a constant level.

Vacuum motor and hot air switching valve - checking

2 Connect the vacuum motor inlet pipe hose direct to the inlet manifold nipple. View thru the air intake side of the air cleaner and when the vacuum pressure is nearly zero, check that the hot air switching valve is fully closed.
   Start and run the engine at idle speed and check that the air switch valve operates instantly, shutting off the fresh air intake. Check that the hot air switching valve closes instantly and smoothly.

Thermo-sensor - checking

3 Remove the air cleaner as previously described, and detach the thermo couple thermometer from the sensor. Refit the cover and refit the air cleaner unit to the carburetor. Allow the engine to cool and then restart it.
   Viewing thru the air intake of the air cleaner check that the hot air switching valve starts to move towards the fresh air intake.
   When the valve starts to operate, note the thermo couple thermometer reading. Check again several times and note the mean reading. Any deviation between 100°F to 111°F indicates that the thermo sensor unit is faulty and must be renewed.

Hot idle compensator

4 Connect a thermocouple thermometer as when testing the thermo-sensor. Run the engine at idle speed with the air cleaner cover detached so that the idle compensator is supplied with hot air.
   As soon as the hot idle compensator starts to operate, note the thermometer reading. The hot idle compensator starts to operate when the sound of outside air intake is audible. If the hot idle compensator actuates between 115°F and 126°F it is in good order. If the reading is outside of these figures then it must be renewed.

5 Carburetor - Types DRJ-340 (Series 1 and 2) DRJ-340 4C (Series 3) and DCP-340 (Series 4) - removal and installation

1 Unscrew the air cleaner wing nut and detach the rubber hoses from the air cleaner cover. The rubber hose must also be disconnected from the vacuum motor.
2 Unscrew and remove the air cleaner bracket bolts. Detach the air cleaner cover and lift out the air cleaner filter element.
3 Detach the hot air hose from the hot air duct, and the air hose to the air pump from the air cleaner. The vacuum hose must be disconnected from the inlet manifold at the joint nipple side.
4 The air cleaner to carburetor bolt is now loosened, lift the air cleaner from the carburetor, and then detach the PCV hose to the camshaft cover and the hoses to the relief valve and check valve. The air cleaner body can now be lifted clear.
5 Detach the choke control cable on Series 1 to 2. On Series 3 and 4 models detach the signal passage rubber pipe from the Exhaust Gas Recirculation valve (on California models only).
6 Disconnect the anti-dieselling solenoid lead from its connection.
7 Disconnect the accelerator return spring and operating cable.
8 Disconnect the carburetor fuel supply pipe (photo) and on Series 4 models, the check valve from the air manifold.
9 Unscrew the four carburetor retaining nuts and remove with lockwashers. Carefully lift the carburetor from the manifold.
10 When installing the carburetor, reverse the removal sequence but note the following.
   Ensure that the carburetor manifold flanges are perfectly clean. Always use a new gasket. Reset the choke and throttle adjustments and check that all hose connections are good particularly those of the emission control. When the engine is restarted, check for fuel leaks.

6 Carburetor (DRJ-340 - Series 1 and 2 models) - dismantling and reassembly

Refer to Figures 3.5 and 3.6

1 Clean off the outside of the carburetor and remove to the workbench. When dismantling or assembling the various components always use the correct tools and lay the parts out on a clean area in the sequence of removal.
2 Commence dismantling by disconnecting the accelerator pump lever as in Fig. 3.7.
3 Undo the retaining screws and remove the throttle switch and bracket unit in Fig. 3.8.
4 Disconnect and remove the throttle return spring. Then undo the retaining screws and remove the diaphragm return spring (photo). Detach the choke chamber unit complete with choke cable bracket and disconnect the connection rod (Fig. 3.9).
5 Unscrew the float chamber to throttle chamber retaining bolts - one bolt/screw to the upper section and three to the lower.
They must be removed with care as one of them extracts the negative pressure developed in the venturi. With the bolts removed lift the float chamber from the throttle chamber.
6 Unscrew the accelerator pump plunger unit retaining screws and then invert the float chamber to extract the plunger assembly (Fig. 3.10).
7 Disconnect the fuel pipe nipple joint, strainer and float needle valve unit as in Fig. 3.11 and then very carefully remove the strainer to ensure that it does not get damaged or distorted.
8 Unscrew the level gauge cover retaining screws and disconnect the cover and float. Take care not to lose the float collar, and if

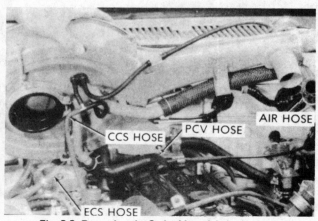

Fig. 3.3. Removing the Series V model air cleaner unit

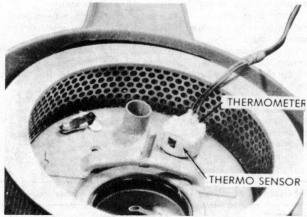

Fig. 3.4. The thermosensor in the air cleaner

Fig. 3.5. Side elevation view of the DRJ-340 type carburetor

1 Float
2 Needle valve
3 Fuel nipple
4 Sec. slow air bleed
5 Sec. slow jet
7 Sec. main air bleed
8 Sec. small venturi
9 Sec. main nozzle
10 Sec. air vent
11 Choke valve
12 Pri. air vent
13 Pri. main nozzle
14 Vapor hole
15 Pri. main air bleed
16 Accelerator air bleed
17 Pri. slow air bleed: 1st
18 Plug
19 Pri. slow air bleed: 2nd
20 Vacuum piston
21 Diaphragm spring
22 Diaphragm
23 Sec. main jet
24 Sec. emulsion tube
25 Coasting hole
26 Sec. throttle valve
27 Sec. vacuum jet
28 Pri. vacuum jet
29 Pri. throttle valve
30 Pri. small venturi
31 Bypass hole
32 Idle hole
33 Inner emulsion tube
34 Outer emulsion tube
35 Pri. main nozzle
36 Pri. slow jet
37 Power valve

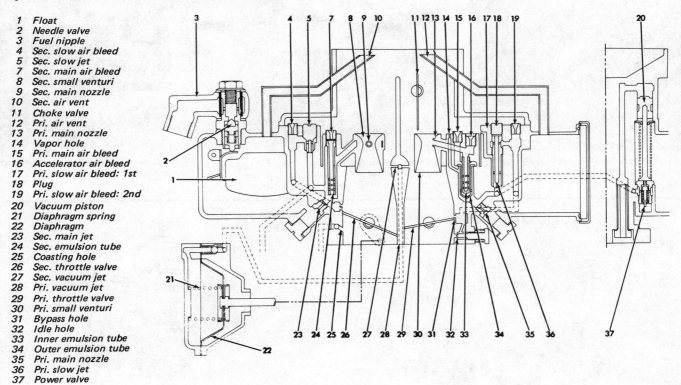

Fig. 3.6. Exploded view of the DRJ-340 type carburetor

1 Choke chamber assembly
2 Float chamber assembly
3 Throttle chamber assembly
4 Throttle adjusting screw
5 Idle adjusting screw
6 Adjusting spring screw
7 Adjusting screw spring
8 Diaphragm, chamber assembly
9 Diaphragm assembly
10 Diaphragm spring
12 (23) Screw
24 (26) Washer
28 Gasket
29 Washer
30 Pin
31 Gasket
32 Washer
33 Gasket
34 Screw
35 Nut
36 (37) Washer
38 Inlet valve
40 (48) Gasket
49 Choke wire bracket
50 Spring hanger
51 Choke connecting rod
52 Choke connecting rod
53 Accelerator switch holder
54 Plate
55 Throttle lever
56 Accelerator switch lever
57 Return spring
58 Accelerator switch bracket
59 Accelerator pump piston
60 Pump return spring
61 Injector weight
62 Pri. small venturi
63 Outer emulsion tube
64 Pump lever
65 Pump connecting lever
66 Return spring
67 Piston plate
68 Sec. Small venturi
69 Float
70 Dust cover
71 Choke connecting lever
72 Sleeve (B)
73 Spring
74 Rubber seal
75 Level gauge
76 Level gauge cover
77 Adjust lever
78 Return plate
79 Sleeve (A)
80 Set screw
81 Joint nipple
82 Spring
83 Float collar
84 Sec. emulsion tube
85 Injector weight plug
86 Main jet plug
87 Strainer
88 Pri. slow jet plug
89 Float needle valve assembly
90 Collar
91 Pri. main air bleed
92 Accelerator air bleed
93 Sec. slow air bleed
94 Sec. main air bleed
95 Pri. slow air bleed
96 Coasting jet
97 Vacuum jet
98 Coasting air bleed
99 Pri. main jet
100 Sec. main jet
101 Pri. slow jet
102 Sec. slow jet
103 Power valve
104 Anti dieseling solenoid
105 Coasting valve solenoid
106 Accelerator switch

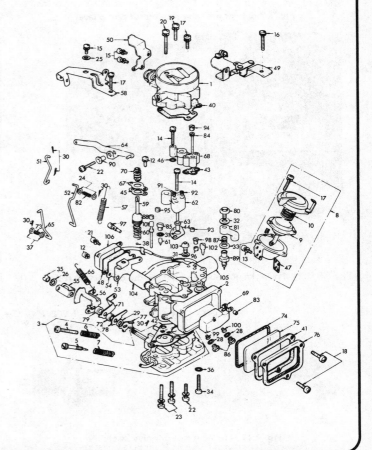

5.8 Disconnect the fuel supply pipe from the float chamber needle valve connection

6.4 View showing 'A' the diaphragm, 'B' the diaphragm return spring, 'C' the throttle return spring and 'D' the throttle switch

6.19d The float chamber with cover removed, showing the needle valve

Fig. 3.7. Disconnect the accelerator pump lever

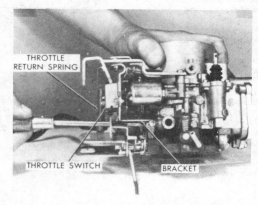

Fig. 3.8. Undo the retaining screws

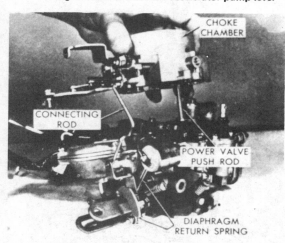

Fig. 3.9. Lift the choke chamber and detach the connecting rod

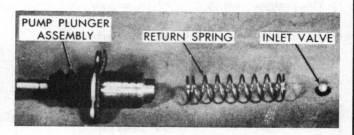

Fig. 3.10. The pump plunger assembly

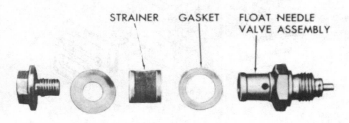

Fig. 3.11. The float needle valve assembly

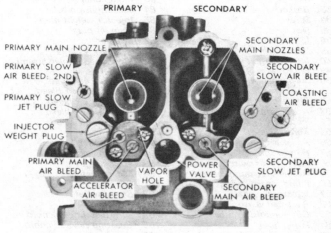

Fig. 3.12. The upper float chamber and jet locations

Chapter 3/Carburetion; fuel, exhaust and emission control systems

possible avoid damaging the gaskets.
9   The coasting valve unit, throttle switch and solenoid units are now removed, taking care not to damage the coasting valve shaft and solenoid valve.
10 Remove the diaphragm chamber and unscrew the diaphragm cover retaining screws. Lift off the cover and separate the spring and diaphragm, but do not lose the ball and spring (Fig. 3.30).
11 From the upper part of the float chamber, the various jets can now be unscrewed and removed as required. If they are all to be removed, mark out their respective locations as in Fig. 3.12 on a piece of paper, and as they are extracted place them in position on the paper accordingly.
12 From the primary and secondary barrels, remove the small venturi then invert the float chamber to remove the outer emulsion tube (Fig. 3.13).
13 Unscrew the accelerator pump plug and invert the chamber to extract the balance weight as in Fig. 3.14.
14 Unscrew and remove the power jet taking care to correctly locate the screwdriver in the slot to avoid damaging the valve rod (Fig. 3.15).
15 Unscrew the two jet plugs (see Fig. 3.16) to remove the primary and secondary main jets.
16 Unscrew the primary vacuum jet and remove it (Fig. 3.17).
17 Apart from the primary and secondary throttle valve and the choke valve, the carburetor is now dismantled for cleaning and inspection. Do not remove the two throttle valves or choke valve as they are preset and sealed to prevent air leaks.
18 Clean the various components in petrol and blow dry. Do not probe the jets with wire. Inspect the parts carefully for signs of wear, damage and cracks in the carburetor body. Renew any parts that are suspect.
19 Reassembly is the reverse of dismantling but the following points should be observed.

General
a) Ensure that the jets are fitted into their correct locations.
b) If for any reason the throttle and choke valves (butterflies) have been removed refit them and on adjustment, to ensure that they close correctly, apply some adhesive solution to prevent air leaks.
c) When the accelerator pump has been reassembled, fill the cylinder with gasoline and check that the fuel is injecting smoothly.

Float level adjustment
d) The correct fuel level is when it is in line with the float chamber window glass line with the engine stationary. Should the fuel level be incorrect, adjustment can be made by bending the float seat accordingly. The effective needle valve stroke should be approximately 0.059 in. This can be adjusted by bending the float stopper.

e) Primary throttle valve opening adjustment
The primary throttle valve is opened by the choke connecting rod. When the choke valve is fully closed the throttle valve should be at an angle of 15° on Series 1 models or 18.5° on Series 2. To check this completely close the choke valve and measure the distance between the wall of the throttle valve chamber in the center of the valve and the throttle valve. The standard clearance is 0.05 to 0.06 inch on Series 1 and 3 or 0.06 to 0.07 in for Series 2 models. See G1 in Fig. 3.18. Fully screw in the throttle stop screw before measuring.
To adjust, bend the connecting rod accordingly.

7 Carburetor (DRJ-340 - Series 1 and 2 models) - linkage adjustment

Refer to Fig. 3.19
1   With the primary throttle valve open at an angle of 50°, the adjust plate and interlocking primary throttle valve are in contact with point 'A' of the return plate. If the primary throttle valve is further opened, the return plate is pulled from the stopper 'B' which allows the secondary throttle valve to open.
2   Measurement of the opening point of the secondary throttle valve is made by measuring the distance between the primary throttle valve and the center of the valve at the throttle chamber wall (G2). The normal clearance is 0.26 to 0.32 in. Adjustment if necessary is made by bending the return plate at point 'A' as required.

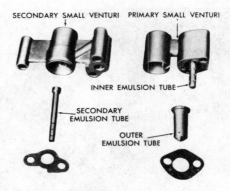

Fig. 3.13. The primary and secondary venturi

Fig. 3.14. Extract the injector weight

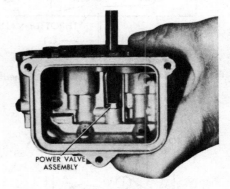

Fig. 3.15. The power jet location

Fig. 3.16. Extract the main jets

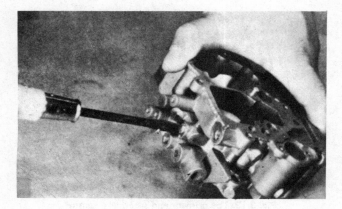

Fig. 3.17. Unscrew the primary vacuum jet

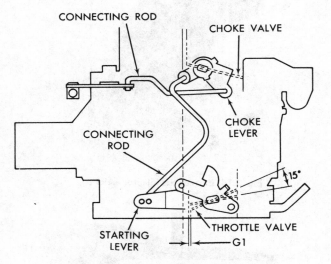

Fig. 3.18. The primary throttle valve adjustment

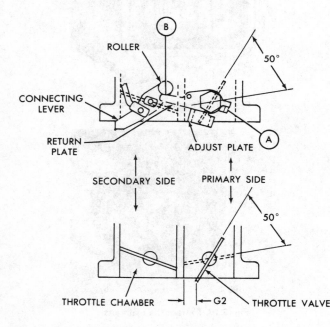

Fig. 3.19. The linkage adjustment

8 Carburetor (DRJ-340 Series 1 and 2 models) - idle adjustment

1 With the engine running at its normal operating temperature, tighten fully the idle adjustment screw and then unscrew it 3½ turns.
2 Now adjust the throttle adjustment screw to give an idle speed of 700 rpm.
3 The idle adjustment screw is then further adjusted so that the engine runs smooth on a fast idle and the throttle screw is then readjusted to enable the engine to run at the correct idle speed.

9 Carburetor (DRJ-340 4C - Series 3 models) - dismantling and reassembly

1 Clean off the outside of the carburetor and remove to the workbench. When dismantling and assembling the various components always use the correct type of tools and lay the parts out on a clean area in sequence of removal.
2 Refer to Fig. 3.21 and detach the accelerator pump lever. Unclip the return spring and remove with bracket and connecting rods. Disconnect the wire arm and choke spring hanger.
3 Withdraw the cotter pin located between the diaphragm rod and secondary throttle lever. Detach the lever and diaphragm unit, disconnect the solenoid valve harness clips and loosen the three bolts retaining the float chamber. Take out the diaphragm (Fig. 3.20).
4 Lift the choke chamber from the float chamber unit, noting that the choke spring hanger is retained by one of the fixing bolts.
5 Remove the float chamber assembly from the throttle unit as in Fig. 3.22 after unscrewing the three lower retaining bolts and one upper bolt. Remove the bolts carefully - one of them relieves the negative pressure from the venturi.
6 Under no circumstances remove the slow cut solenoid and coasting richer solenoid from the float chamber.
7 Unscrew the accelerator pump retaining screws, invert the float chamber and remove the plunger unit.
8 Unscrew the fuel pipe nipple and remove with strainer and float needle valve unit. Do not distort the strainer.
9 Undo the three screws securing the level gauge cover, and remove with cover, baffle plate and float (Fig. 3.24).
10 To dismantle the diaphragm chamber follow the instructions in paragraph 10 in Section 6 and then follow paragraph 11 to remove the jets.
11 From the primary and secondary carburetors remove the small venturi, invert the float chamber to remove the accelerator air bleed, main air bleeds and the secondary emulsion tube (Fig. 3.25), but do not remove the primary emulsion tube.
12 Unscrew the injector weight plug, invert the float chamber and extract the injector weight (Fig. 3.26).
13 Now follow paragraphs 14 to 18 in Section 6. In addition to those items mentioned in paragraph 18, check solenoid and valve wires are in good condition and their connections are firm. If the solenoid

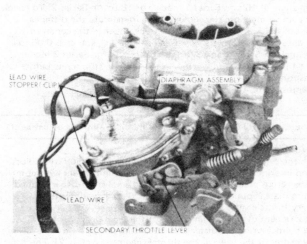

Fig. 3.20. The diaphragm assembly

Chapter 3/Carburetion; fuel, exhaust and emission control systems

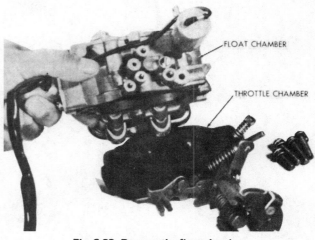

Fig. 3.22. Remove the float chamber

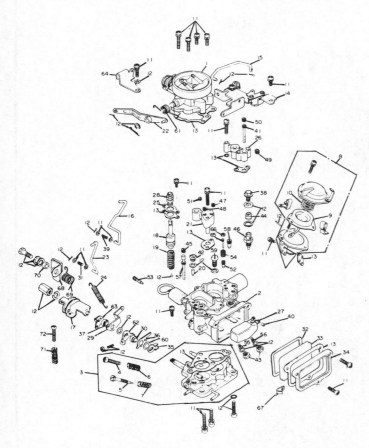

Fig. 3.21. The DRJ-340 4C model carburetor. Illustration numbers 68 to 72 are only fitted to those models having air conditioning

1	Choke chamber assembly	37	Sleeve
2	Float chamber assembly	38	Filter set screw
3	Throttle chamber assembly	39	Spring
4	Throttle adjust screw	40	Collar
5	Idle adjust screw	41	Secondary emulsion tube
6	Throttle adjust screw spring	42	Plug
7	Idle adjust screw spring	43	Drain plug
8	Diaphragm chamber assembly	44	Filter
9	Diaphragm	45	Slow jet plug (primary)
10	Diaphragm spring	46	Needle valve
11	Screw and washer kit, A	47	Accelerator air bleed
12	Screw and washer kit, B	48	Main air bleed (primary)
13	Gasket	49	Slow air bleed (secondary)
14	Choke control arm	50	Main air bleed (secondary)
15	Choke connecting rod	51	Slow air bleed (primary)
16	Choke connecting rod	52	Coasting jet
17	Throttle lever (primary)	53	Vacuum jet
18	Accelerator pump piston	54	Coasting air bleed
19	Piston spring	55	Main jet (primary)
20	Injector weight	56	Main jet (secondary)
21	small venturi (primary)	57	Slow jet (primary)
22	Accelerator pump lever	58	Slow jet (secondary)
23	Connecting rod	59	Power valve
24	Throttle return spring	60	Thrust washer
25	Plate	61	Pump lever return spring
26	Small venturi (secondary)	62	Kick lever
27	Float	63	Crank
28	Dust cover	64	Choke control cable hanger
29	Starting lever	65	Coasting adjust screw
30	Sleeve	66	Locknut
31	Spring	67	EGR vacuum pipe clip
32	Rubber seal	68	Fast idle lever
33	Fuel level gauge	69	Spring
34	Cover	70	Collar
35	Adjust lever, B	71	Fast idle adjust spring
36	Return plate	72	Fast idle adjust screw

and valve are defective they must be renewed as a unit for they cannot be removed separately
14 Reassembly of the carburetor is a direct reversal of dismantling, but note instructions given in paragraph 19 of Section 6, and check the float level, primary throttle linkage and idle adjustments, as described in the following paragraphs.

Float level adjustment
15 Refer to paragraph 19 in Section 6, sub-paragraph 'd'.

Primary throttle valve adjustment
16 Refer to paragraph 19 in Section 6, sub-paragraph 'e'. The instructions are identical but the throttle valve should be at an angle of 17.5° when the choke valve is closed.

10 Carburetor (DRJ-340 4C - Series 3 models) - linkage adjustment

To set the linkage adjustment, refer to Section 7 - the instructions are identical.

11 Carburetor (DRJ-340 4C - Series 3 models) - idle adjustment

1 Run the engine at its normal operating temperature and adjust the idle to 700 rpm.
2 Detach the vacuum line from the air cleaner hot idle compensator and plug at the inlet manifold.
3 Tighten fully the idle adjustment screw and then unscrew it 3½ turns.
4 Adjust the throttle adjustment screw to give an idle speed of 700 rpm.
5 The idle adjustment screw is then further adjusted so that the engine runs smoothly on a fast idle and the throttle screw is then readjusted to enable the engine to run at the correct idle speed.
6 If air conditioning equipment is fitted turn AC on maximum cold and high blower, then open the throttle about a 1/3 rd and allow it to close. The speed up adjustment screw is now set so that the idle speed is 900 rpm. Re-open the throttle 1/3 rd and allow it to close and check that the engine idles at 900 rpm. If not readjust the speed up control adjustment screw.
7 Reconnect the hot air vacuum line from the air cleaner.

12 Carburetor (DCP-340 - Series 4 to 6 models) - dismantling and reassembly)

1 Clean off the outside of the carburetor and remove it to the workbench. When dismantling and assembling the various components always use the correct type of tools and lay the parts out on a clean work area in the sequence of removal.

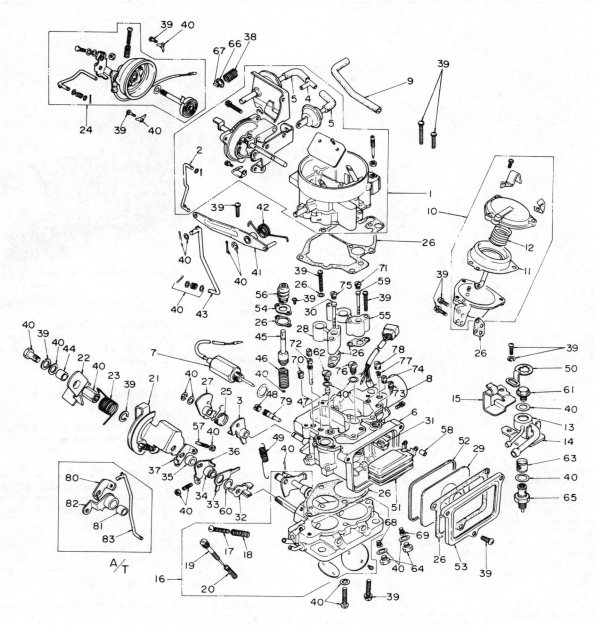

Fig. 3.23. The DCP 340 carburetor component parts

1 Choke chamber assembly	22 Fast idle lever	43 Accelerator pump rod	64 Drain plug
2 Choke connecting rod	23 Fast idle lever spring	44 Fast idle lever collar	65 Needle valve
3 Counter lever	24 Thermostat cover assembly	45 Accelerator pump piston	66 Piston spring carrier
4 Nipple	25 Fast idle cam spring	46 Piston return spring	67 Piston spring stop pin
5 Hose	26 Gasket kit	47 Injector weight	68 Main jet (primary)
6 Float chamber assembly	27 Fast idle cam	48 Vacuum jet plug	69 Main jet (secondary)
7 Anti-dieseling solenoid	28 Small venturi (primary)	49 Throttle return spring	70 Slow air bleed (primary)
8 Coasting richer solenoid	29 Fuel level gauge	50 Lock lever	71 Main air bleed (secondary)
9 Hose	30 Primary emulsion tube	51 Float	72 Slow jet (primary)
10 Diaphragm chamber assembly	31 Baffle plate	52 Rubber seal	73 Slow jet (secondary)
11 Diaphragm	32 Throttle adjusting lever	53 Cover	74 Slow air bleed (Secondary)
12 Diaphragm spring	33 Throttle return plate	54 Plate	75 Main air bleed (primary)
13 Nipple	34 Kick lever sleeve	55 Small venturi (secondary)	76 Power valve
14 Nipple stop plate	35 Throttle lever sleeve	56 Dust cover	77 Coasting jet
15 EGR vacuum hose clip	36 Kick lever	57 Fast idle adjusting screw	78 Coasting air bleed
16 Throttle chamber assembly	37 Fast idle lever	58 Collar	79 Vacuum jet
17 Throttle adjusting screw	38 Auto choke piston spring	59 Secondary emulsion tube	80 Connecting lever
18 Adjusting screw spring	39 Screw and washer kit, A	60 Thrust washer	81 Collar A
19 Idle adjusting screw	40 Screw and washer kit, B	61 Filter set screw	82 Down shift lever
20 Adjusting screw spring	41 Accelerator pump lever	62 Injector weight plug	83 Pump rod
21 Throttle lever (primary)	42 Pump lever return spring	63 Filter	

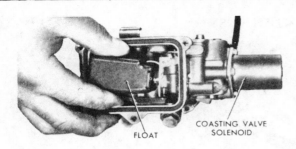

Fig. 3.24. The float chamber and coasting richer solenoid

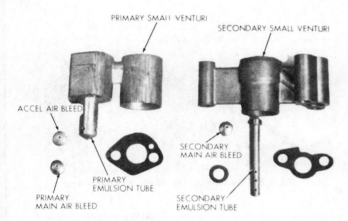

Fig. 3.25. The primary venturi and main air bleed, the secondary venturi and air bleed and the accelerator air bleed

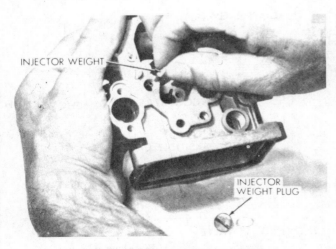

Fig. 3.26. Extracting the injector weight

2 Detach the accelerator pump lever and throttle return spring (Fig. 3.27).
3 Remove the automatic choke lead and unscrew the three thermostat cover to housing screws. Unscrew the fuel pipe nipple and remove with strainer.
4 Unscrew the choke lever to choke shaft retaining screw and move the choke lever fully in towards the choke chamber. Detach the choke connection rod from the counter lever, then remove the choke chamber unit by unscrewing the four securing screws.
5 Withdraw the cotter pin from the diaphragm rod and secondary lever, and detach the lever and diaphragm unit. Unclip the solenoid valve harness clips, slacken the diaphragm chamber attachment bolts to remove the diaphragm unit.
6 Carefully remove the three lower and one upper float chamber to throttle chamber retaining bolts and separate the two assemblies. Note that one of the bolts removes the negative pressure developed in the venturi and must be handled carefully.
7 Unscrew the two screws attaching the accelerator pump plunger unit and invert the float chamber to extract the unit.
8 Unscrew and remove the float needle valve assembly and then unscrew the gauge level cover (Fig. 3.23) retaining screws. Detach the cover and remove the baffle plate and float and note the float collar.
9 To dismantle the diaphragm chamber follow the instructions given in paragraphs 10 to 13 in Section 9.
10 Reassembly of the carburetor is a direct reversal of dismantling but note paragraph 19 of Section 6, and check the float level, primary throttle valve linkage and idle adjustments as described below.

Float level adjustment
11 Refer to paragraph 19 in Section 6, sub-paragraph 'd'.

Primary throttle valve adjustment
12 Refer to paragraph 19 in Section 6, sub-paragraph 'e'. The instructions are identical but the throttle valve should be at an angle of 17° when the choke valve is closed and the standard clearance between the throttle valve chamber and valve is 0.04 to 0.05 in.

13 Carburetor (DCP-340 - Series 4, 5 and 6 models) - linkage adjustment

Refer to Section 7. The instructions are identical to those of the DRJ-340 carburetor but with the exception of the angle of the primary throttle valve, which should be open at 47° and the normal clearance between the primary throttle valve and the center of the valve at the chamber wall is 0.24 to 0.30.

14 Carburetor DCP-340 - Series 4 models - idle adjustment

1 Run the engine to its normal operating temperature, and plug the air pipe at the vacuum pipe connection of the idle compensator.
2 Tighten fully the idle adjustment screw and then unscrew it exactly 2½ turns.
3 Adjust the throttle adjustment screw to give an engine speed of 700 rpm, and then reset the idle adjustment screw to obtain the maximum inlet manifold vacuum.
4 Tighten the idle adjustment screw by 1/8 turn and reset the throttle adjustment to give 700 rpm.
5 If air conditioning is fitted, turn AC to maximum cold and high blower, open the throttle by about 1/3 and allow it to close. Now adjust the speed up controller setting screw to idle at 900 rpm. Open the throttle (1/3) and allow it to close and recheck the rpm. If it does not idle at 900 rpm, repeat the above procedure until correct.
6 Refit the air cleaner hot idle compensation line.
7 If a CO meter is available, follow the instructions in paragraphs 1 and 2 above and then set the throttle adjustment screw so that the engine idles at 700 rpm.
8 Adjust the idle adjustment screw to idle at CO 1%. Then reset the throttle screw to give the 700 rpm engine speed, and check that the idle CO is within $1 \pm 0.5\%$.
9 The procedure for checking with a CO meter is otherwise identical including paragraph 5 concerning air conditioning if fitted.

Fig. 3.27. The accelerator pump lever and connecting rod

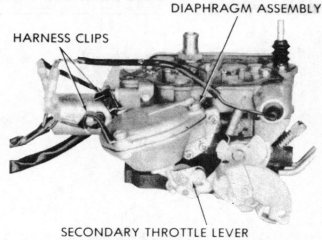

Fig. 3.28. The diaphragm unit showing the harness clip position

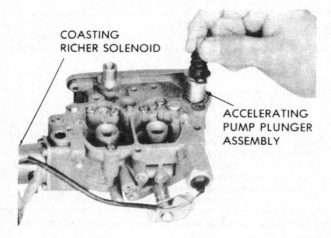

Fig. 3.29. Withdraw the accelerator pump plunger assembly

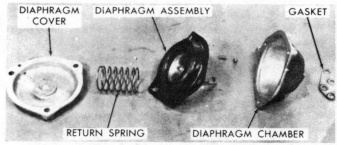

Fig. 3.30. The components of the diaphragm unit

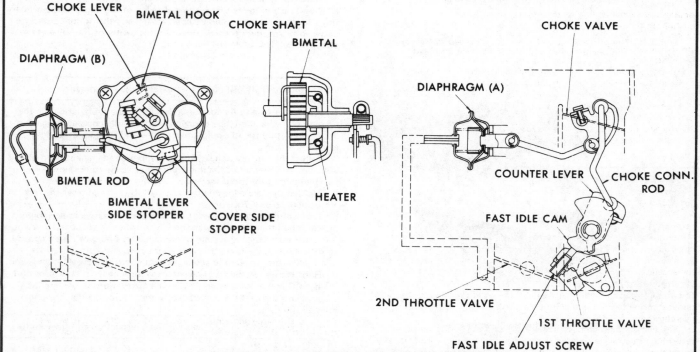

Fig. 3.31. The principal components of the automatic choke

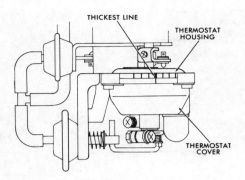

Fig. 3.32. The thick line mark for alignment of the thermostat housing and cover

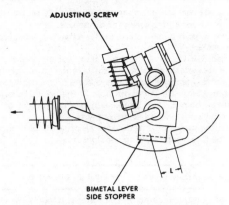

Fig. 3.33. The adjustment screw - check clearance at point 'L'

Fig. 3.34. The fuel pump lid removed showing cap, seal and strainer

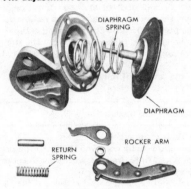

Fig. 3.35. The dismantled pump unit

15 Carburetor (DCP-340 - Series 5 and 6) - idling adjustment

1 Ensure that the parking brake is on and place chocks under the drive wheels, Check that the transmission is in neutral.
2 Run the engine to its normal operating temperature, and if fitted, turn off the air conditioner.
3 Disconnect the distributor vacuum line and also the idle compensator vacuum line which should be plugged on removal.
4 Tighten the mixture idle screw fully and then unscrew it 3 full turns.
5 Set the throttle adjusting screw so that the engine idles at 900 rpm, and then reset the idle adjusting screw to its fastest setting.
6 Re-adjust the throttle screw to 900 rpm, and then turn the idle mixture screw clockwise to reduce the engine speed to 850 rpm.
7 Re-adjust the idle screw turning it counterclockwise by a ½ turn and then reset the throttle screw to give 900 rpm engine speed.
8 If an air conditioning unit is fitted, set the idle as above and then turn the AC to maximum cold and high blower, open the throttle by about a 1/3 and allow the throttle to close. Now set the speed up controller adjustment screw so that the engine idles at 900 rpm.

16 Electric automatic choke - general

Refer to Fig. 3.31

1 An automatic choke assembly is fitted in place of the manually operated choke on some models. The automatic choke incorporates a thermostatic spring which operates the choke valve automatically according to the variation in ambient temperature.
2 The choke valve position is preset by the thermostatic spring when the accelerator pedal is fully depressed and released prior to starting. On starting the engine the negative pressure from the carburetor venturi actuator the choke diaphragm (A) and the choke valve is opened against bimetal spring tension, which in turn prevents the engine from being overchoked.
3 The choke diaphragm 'B' actuates to turn the spring slightly in the returning direction to shorten the choke valve opening. Additionally, on starting the engine, the current flow thru the heater relay unit and heater element flexes the bimetal and opens the choke valve.

4 Since the throttle valve does not automatically return to its position for idling, its opening angle is reduced by the fast idle cam during depressing and releasing of the accelerator pedal with the engine at fast idle and this causes the throttle valve to be repositioned for normal idling. This occurs when the fast idle screw is released from the fast idle cam.

17 Electric automatic choke - adjustment

1 Fit the choke cable lever end to the bimetal hook and align the thick line marking on the housing to that of the cover (Fig. 3.32) and fit the cover.
2 Ensure that the stopper on the bimetal lever side contacts the cover stopper and should a gap exist, adjust the bimetal rod accordingly.
3 Apply finger pressure to the diaphragm and measure the clearance between the cover side stopper and bimetal cover side stopper as in Fig. 3.33. The normal clearance should be 0.28 to 0.29 in.
4 If the measured clearance is incorrect, adjust the screw accordingly.

18 Fuel pump (mechanical) Series 1 to 3 models - removal and installation

1 Clean the outside surfaces of the fuel pump and connecting pipes.
2 Detach the hose from the side of the pump.
3 Unscrew the fuel pipe joint bolt and detach the pipe from the side of the pump.
4 Unscrew the pump unit retaining nuts and withdraw the pump unit.
5 Refitting is the reverse of removal but be sure to fit new joint bolt gaskets if the old ones are worn or broken.

19 Fuel pump (mechanical) - dismantling, inspection and reassembly

1 Unscrew the valve chamber cap retaining bolt and remove with washer. Detach the cap, strainer, seal and screen, Fig. 3.34.

2 To separate the fuel pump upper and lower body sections, unscrew and remove the five retaining screws and washers.
3 Unscrew the valve retainer screws and remove the valve retainers, the valves and seals.
4 With a suitable drift, drive out the rocker arm shaft and disconnect the rocker arm diaphragm, spring and return spring.
5 Clean the components in gasoline and blow dry. Check for cracks in the pump body and cap. If the pin hole in the body is worn the pump will have to be renewed. Similarly, if the diaphragm or valves are defective the pump unit must be renewed.
6 Reassemble the pump unit in the reverse order of dismantling and use new gaskets. Check on refitting for signs of leaks.

20 Fuel pump (electro-magnetic) - removal and installation

1 This type of pump is located on the inner-surface of the third crossmember on the left-hand side.
2 It is of the 'totally-enclosed' variety and cannot be dismantled for repair. If defective, the unit must therefore be renewed.
3 To remove the pump, disconnect the ground cable from the battery and then detach the lead and rubber hoses from the fuel pump.
4 Unscrew the retaining bolts and nut and remove the fuel pump unit.
5 Installation is the reverse of removal.

21 Fuel tank - removal and installation

1 Detach the battery ground cable.
2 Place a suitable clean metal container beneath the drain plug and unscrew the plug. When fully drained, withdraw the container and store in a safe area and place a lid over it.
3 Detach the fuel filler cap and unscrew the fuel filler pipe bracket bolts (Fig. 3.37).
4 From its fuel tank terminal, disconnect the sender unit.
5 Undo the retaining clips of the three rubber hoses on the left side of the fuel tank and detach the hoses.
6 Undo the clips retaining the two hoses from the check and relief valve to the fuel filter at the front of the tank and detach the hoses.

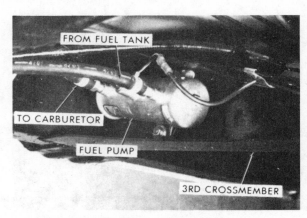

Fig. 3.36. The electro magnetic fuel pump location

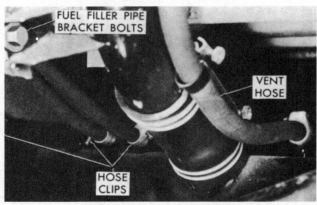

Fig. 3.37. The fuel filler pipe bracket and retaining bolt location

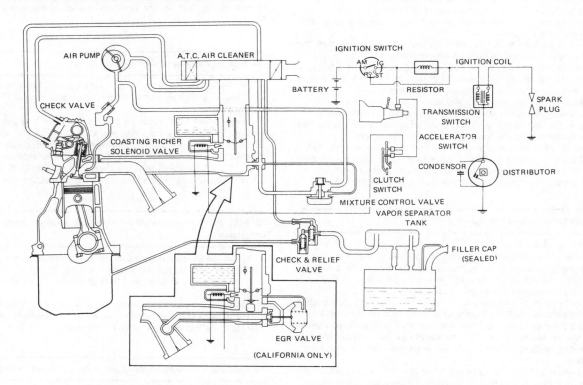

Fig. 3.38. The emission control system - Series 1 to 4 models

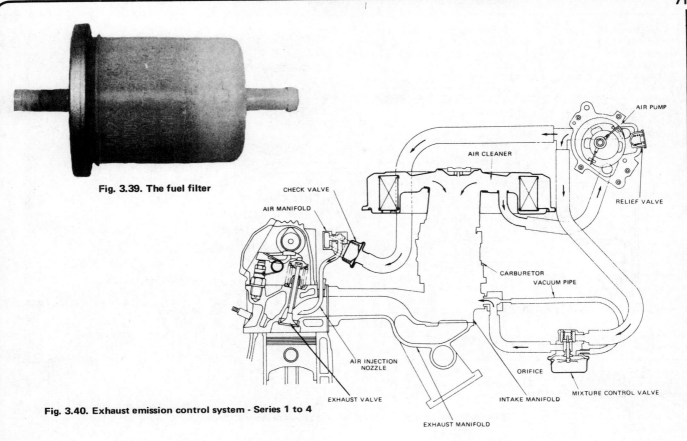

Fig. 3.39. The fuel filter

Fig. 3.40. Exhaust emission control system - Series 1 to 4

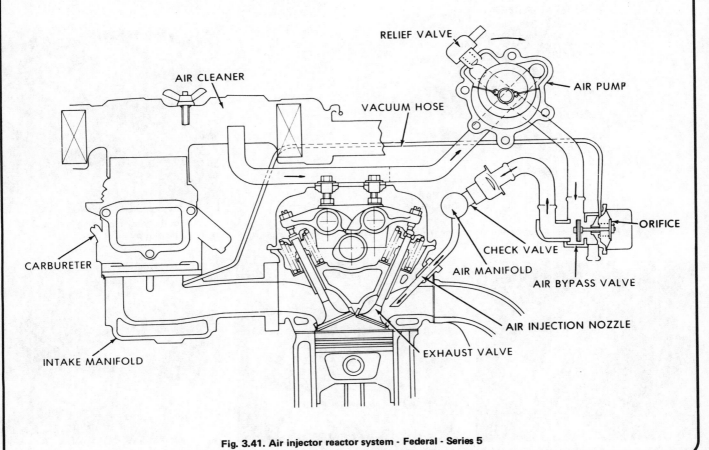

Fig. 3.41. Air injector reactor system - Federal - Series 5

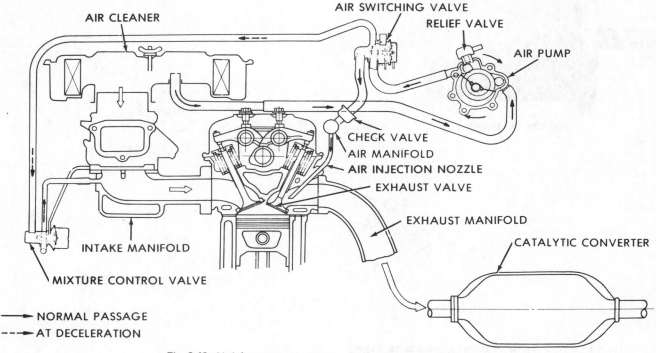

→ NORMAL PASSAGE
--→ AT DECELERATION

Fig. 3.42. Air injector reactor system - California models - Series 5

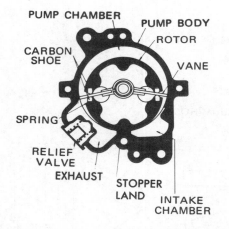

Fig. 3.43. End elevation showing component parts of the air pump

24.11 Withdraw the mounting bolts and remove the pump

24.16 The air manifold check valve

24.17 The mixture control valve

24.22 The coasting richer solenoid

7 Unscrew and remove the fuel tank mounting bolts from the front and rear sides. Withdraw the fuel tank.
8 To install the fuel tank, reverse the above procedure but on refilling, check that there are no signs of fuel leakage from the tank and connection hoses.

22 Fuel filter element - removal and refitting

1 The fuel filter element must be renewed every 12000 miles or sooner if contaminated or damaged.
2 To remove, simply unclip the connecting fuel hoses and withdraw it.
3 Refit in the reverse sequence and ensure that the hose connections are good.

23 Emission control system - description

1 The principle of the emission control system is to prevent the discharge of harmful gases from the vehicle.
2 The emission control system is basically divided into three parts, these being the exhaust emission control system to reduce the exhaust gas; the crankcase emission control which recirculates the blow-by gases from the crankcase, together with fuel vapors, into the combustion chamber via the inlet manifold to be burned; and the evaporative emission control system which stores fuel vapor from the fuel supply system and leads it to the combustion chamber.
3 From Series 3 models, California versions were fitted with an exhaust gas recirculation (EGR) valve and on Series 5 California models, a catalytic converter is incorporated into the exhaust system and a over temperature control system fitted which operate in conjunction with each other.
4 A dual contact points distributor control system is fitted to all models. This allows the ignition timing to be automatically selected to suit the engine operating conditions. The retarded contact breaker points operate when the engine is accelerating or decelerating which momentarily causes the harmful content in exhaust gases to increase. Details of the distributor setting and adjustments are given in Chapter 4.
5 The emission control system can only function correctly when the engine and its associate components are in good condition and correctly adjusted. All interconnecting hoses and electrical connections within the system must be regularly checked and in good condition in order that the system functions efficiently and complies with the strict regulations.

24 Exhaust emission control system - component description and maintenance

Air pump - description
1 The air pump comprises mainly of the pump body, cover, vanes and relief valve which is a press fit. The pump is driven by a connecting belt to the water pump shaft pulley which in turn is driven by the crankshaft pulley.
2 Air enters the air cleaner and then into the pump suction chamber where it is transferred by the vanes and rotor to the outlet chamber which is interconnected to the air manifold. The relief valve is normally closed, but when the outlet air pressure exceeds 2.84 to 4.98 psi the valve opens to release excess air. To operate efficiently the pump drivebelt must be correctly adjusted at all times.

Air pump drive belt - adjustment
3 The tension of the air pump drivebelt should be periodically checked at the same time as the fan belt.
4 If the belt is worn or cracked it must be renewed as soon as possible.
5 To retension the air pump drivebelt, loosen the air pump mounting bolts and hinge the pump unit away from the engine to tighten the tension. Do not overtension the belt. The correct lateral play should be 0.4 in when finger pressure is applied in the middle of the belt run between the pulleys (Fig. 3.46).
6 When pivoting the pump, do not force it with a lever or the pump may distort.

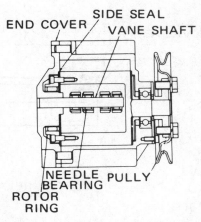

Fig. 3.44. Side elevation showing component parts of the air pump

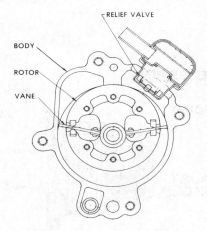

Fig. 3.45. The Series 5 air pump

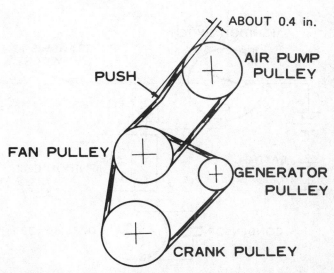

Fig. 3.46. The air pump and generator driving belts layout, showing the adjustment deflection

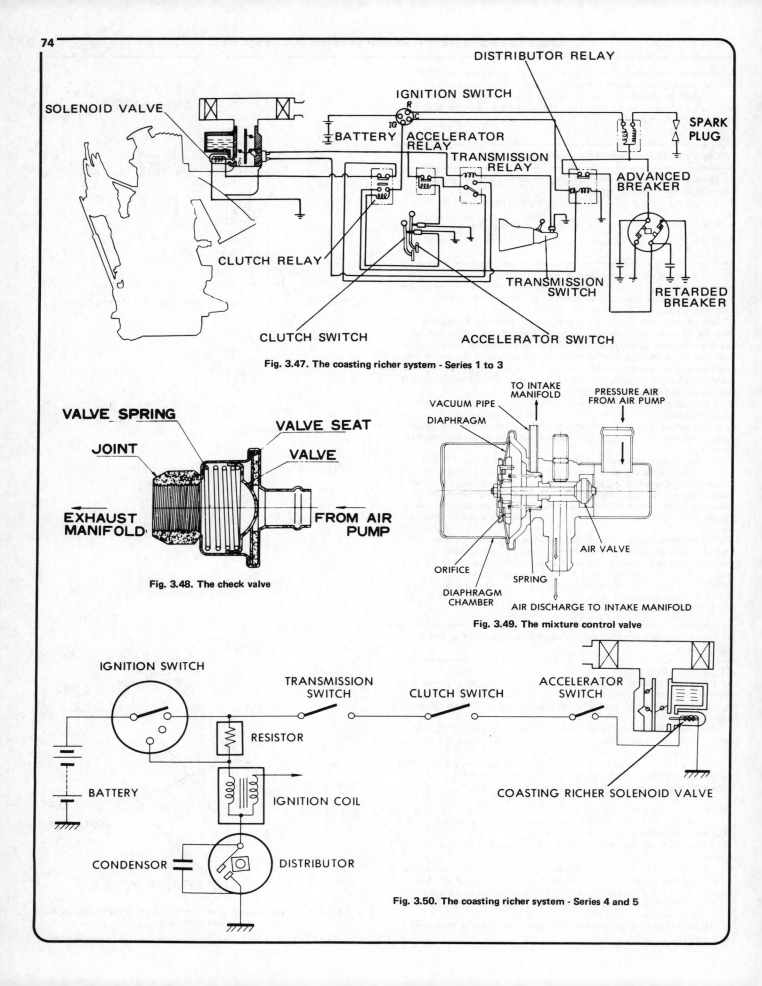

Fig. 3.47. The coasting richer system - Series 1 to 3

Fig. 3.48. The check valve

Fig. 3.49. The mixture control valve

Fig. 3.50. The coasting richer system - Series 4 and 5

Chapter 3/Carburetion; fuel, exhaust and emission control systems

7 On adjustment, retighten the pump bolts and recheck the lateral belt play. If a new belt has been fitted check it again after usage as it may have stretched and need readjusting.

Air pump relief valves - checking
8 Start the engine and increase the speed to 2000 rpm, when at this point the relief valve will open and allow the airflow. If the air blows thru the valve prior to this, it will have to be removed and renewed.

Air pump, Series 1 to 4 - removal, inspection and reassembly
Note: *The air pump on Series 5 and 6 models is non-serviceable and therefore if defective is renewed as a unit.*
9 Slacken the adjustment bolt and the mounting bolts. Detach the hoses from their pump connections.
10 Pivot the pump towards the engine and remove the drivebelt from the pump drive pulley.
11 Withdraw the adjustment and pump mounting bolts (photo) and remove the pump.
12 Clean off the outside of the pulley if the pump is noisier than usual; the bearing may possibly be defective. To check this, turn the pulley ¾ of a turn clockwise and then ¼ of a turn counter-clockwise. If the pulley is binding and not very smooth then this is indicative that the bearing is at fault.
13 Dismantle the pump and lay the respective components on a clean work area. Clean the components and inner pump body.
14 Examine and check the vanes and rotor for wear also the inner pump body. Check the needle roller bearings and, if damaged or badly worn, then the pump unit should be renewed. The rear side seal must be renewed if it is worn or damaged. Renew the carbon shoes and/or vanes if they are badly worn or damaged.
15 Reassemble the pump in the reverse order and install in position in engine. Refit the drivebelt and tension as described in previous Section.

Check valve - description and inspection
16 This is located on the air manifold (photo) and ensures that the air flow under pressure from the air pump can only flow one way. Any counter flow of gases from the exhaust manifold shuts the valve and safeguards the air pump and hoses. To check the valve, disconnect it from the manifold, and blow air into the valve from the pump side and then from the manifold side. Air pressure should only be allowed thru the valve from the pump side. Should air be allowed thru the valve from the manifold side then the check valve is faulty and must be renewed, although a small leakage is allowable.

Mixture control valve
17 The mixture control valve (photo) supplies air to the inlet manifold to prevent the mixture from becoming rich when the throttle valve is closed suddenly. The valve is normally kept closed under pressure from the normal air supply in the intake manifold but when this pressure becomes negative, the valve opens and air is supplied via the air pump. To check the valve, detach the mixture control valve to inlet manifold hose. Plug the inlet manifold side of the valve and start the engine. Depress the accelerator fully and then release it quickly. If the valve is normal, air will continue to blow from the mixture control valve for up to 5 seconds. If the air flow continues after this time, the valve is in need of renewal.

Air injection nozzle manifold - Series 1 to 4 - removal, inspection and installation
18 To remove the air injection nozzles it is first necessary to detach the camshaft carrier as described in Chapter 1. Disconnect the check valve hose and then unscrew each nozzle joint bolt and carefully detach the manifold. Slacken the air injection nozzle nipple and disconnect it by rotating it thru 180°.
19 Wash each nozzle and blow dry with an air line. Inspect them carefully for signs of burning, cracks or distortion. If defective in any way they must be renewed.
20 Installation is the reversal of removal but on restarting the engine, check for leaks.

Series 5 and 6
21 Disconnect the check valve hose and unscrew each of the nozzles from their position near the exhaust valves. Now follow the instructions in paragraphs (19) and (20) above.

Coasting richer system - description
22 This system incorporates a solenoid valve attached to the secondary side of the carburetor (photo). The solenoid is connected in relay to a series of switches which detect engine coasting conditions. These switches are as follows.
Federal manual transmission models: - accelerator switch, clutch switch, transmission 4th/3rd gear switch.
California manual transmission models: - accelerator switch, clutch switch, transmission neutral switch and engine speed sensor.
California automatic transmission models: - accelerator switch, inhibitor switch, engine speed sensor.
23 When these switches are activated, the solenoid valve and magnet are energised and open the secondary valve. On opening the valve, the fuel supply from the float chamber is metered into the coasting jet, mixed with air and metered thru the coasting air bleed.
24 This mixture then enters the lower secondary throttle valve via the coasting valve. This enriches the mixture drawn into the engine to assist the exhaust gas reburning. An engine speed sensor fitted to California models only detects the engine revolutions via the ignition pulse from the coil. When the speed of the engine exceeds 1600 rpm the sensor turns on to operate the coasting richer circuit.
25 When the accelerator or clutch pedal are depressed or the transmission shifted to neutral, the series circuit automatically switches off and the coasting valve shuts.
26 To check the operation of the coasting richer system, listen carefully for the noise that accompanies the operating of the solenoid valve. If a malfunction of the system is suspected, check the respective adjustments of the clutch and accelerator switches and the transmission

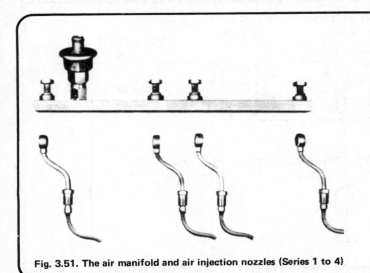

Fig. 3.51. The air manifold and air injection nozzles (Series 1 to 4)

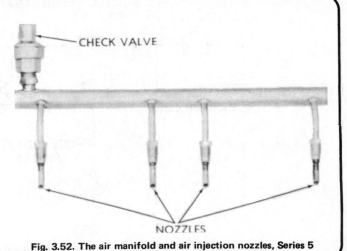

Fig. 3.52. The air manifold and air injection nozzles, Series 5

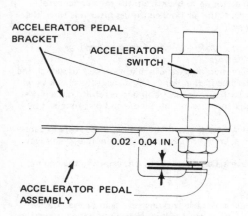

Fig. 3.53. The accelerator switch and adjustment to pedal

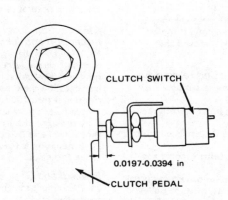

Fig. 3.54. The clutch switch and adjustment to pedal

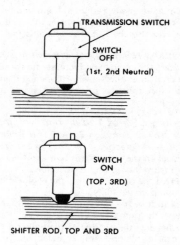

Fig. 3.55. The transmission switch showing on and off positions

switch, (see below). If these are in order, and the system remains inoperative, have it checked by your local dealer.

Accelerator switch and clutch switch - checking

27 Check the respective pedal to switch clearances and if necessary adjust by loosening the locknut and turning the switch accordingly to the correct clearance as in Fig. 3.53 and Fig. 3.54. Test the switches by disconnecting the wires and connect a circuit tester. The switch must be off when the pedal is depressed and on when released. Renew the switch if faulty.

Transmission switch - checking

28 Disconnect the switch wiring and connect a circuit tester. The switch is operating correctly when, with the gearshift moved into 3rd or 4th gear, the switch is turned on. If not, the switch is defective and must be renewed.

Engine speed sensor (California only) - checking

29 Detach the wiring connector from the engine speed sensor and with suitable interconnecting cables connect 'B', 'BR' and 'BY' color coded wiring terminals together. Start the engine and check the continuity between 'LgB' color coded wiring terminals. If continuity exists when the engine speed is in excess of 1500 to 1700 rpm, then the sensor is in order. If not it should be renewed.

Accelerator relay - checking

30 Ensure that the supply voltage is being received at the relay terminals. If the voltage at terminal 3 (Fig. 3.56) equals that measured at terminal 1 with the switch in the off position, then the relay is in order. If the voltage measured at terminal 3 is lower or not being supplied then the contact breaker points are defective and in need of renewal or attention.

31 If the supply of voltage to terminal 3 is interrupted when the accelerator pedal is depressed, then the accelerator relay is in working order. If however, the voltage supply to terminal 3 is not discontinued on depressing the pedal, the relay coil is faulty and should be renewed.

Clutch relay - checking

32 Ensure that the accelerator pedal is released so that the switch is off. Engage 3rd or 4th gear to turn on the transmission switch. Check that the relay terminals 1 and 2 (Fig. 3.56) have a voltage supply and that the clutch switch is operating.

33 If relay terminal 3 has voltage and terminal 4 has not when the clutch pedal is released then the clutch relay is in order. If, however, terminal 3 has no voltage supply or a reduced supply to that of terminal 1, the breaker points are in need of attention or should be renewed.

34 If the supply of voltage to terminal 3 continues when the clutch pedal is depressed or there is no voltage reading at terminal 4 when the pedal is depressed, the relay coil is open or the points seized up, in which case they must be repaired or renewed.

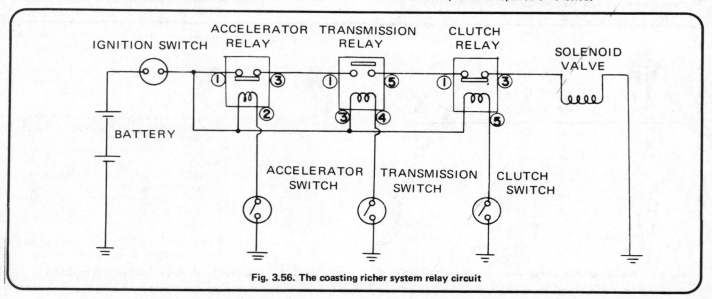

Fig. 3.56. The coasting richer system relay circuit

Chapter 3/Carburetion; fuel, exhaust and emission control systems

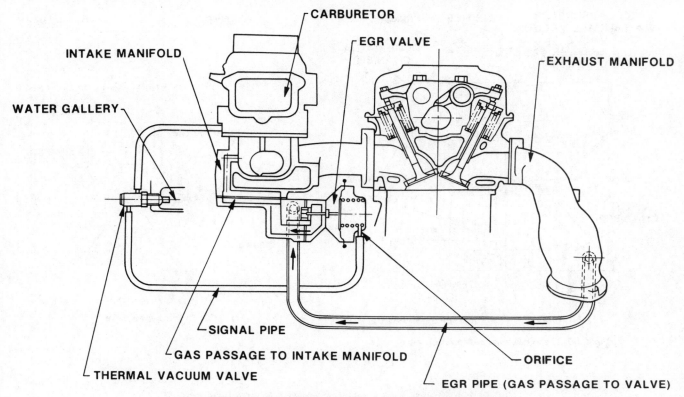

Fig. 3.57. The exhaust gas recirculation system (California models)

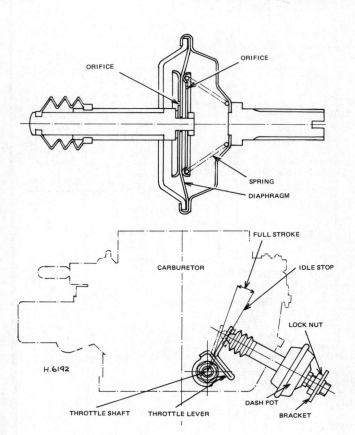

Fig. 3.58. The dash pot showing location in carburetor

Transmission relay - checking
35 Check the voltage supply at terminals 1 and 4 in Fig. 3.56, with the accelerator pedal released. If there is a supply at both terminals this indicates that the relay is operating normally. This is checked further if the voltage supply at terminal 5 is discontinued whilst terminal 6 receives voltage when in neutral (switch off). If there is no voltage supply at terminal 6 or a drop in voltage, the contact points are suspect and should be checked and renewed if necessary. As a further check, shift the transmission into 3rd or 4th gear and check that terminal 5 is receiving voltage but terminal 6 is not. If the voltage supply to these two terminals is not reversed when 3rd or 4th gear are selected then again the contact points are suspect.

Exhaust gas recirculation valve (EGR) - California only - description and checking
36 Mounted beneath the inlet manifold, the exhaust gas recirculation valve controls the recycled gas flow entering the inlet manifold from the exhaust manifold. When the vacuum caused by the throttle valve opening reaches 3.5 in Hg the diaphragm opens the exhaust gas metering valve allowing the exhaust gases into the engine intake. A thermal vacuum valve situated on the inlet manifold is connected between the vacuum port in the carburetor and the EGR valve. When the engine coolant temperature is below 118° to 126°F (46° to 51°C), the valve remains closed, as does the EGR valve. When the coolant temperature rises above these figures the valve opens and operates the EGR system.
37 To check the operation of the EGR valve, start the engine and open the throttle to give an engine speed of 2000 to 2500 rpm and ensure that the diaphragm operating shaft moves upwards, and when the engine speed drops to normal idle, returns downwards.
38 To check the diaphragm, a vacuum gauge is needed and also an alternative source of vacuum supply. If available, connect the vacuum supply to the top of the diaphragm chamber port tube to the carburetor. The diaphragm must not 'leak down' and should be fully raised at about 8 in of vacuum.

Dash pot (California models only) - inspection and adjustment
39 On California models, a throttle dash pot is incorporated into the emission control system and prevents the throttle valve from closing rapidly on deceleration. To check the dash pot, ensure that it operates smoothly when the shaft is moved endwise by hand. If defective or

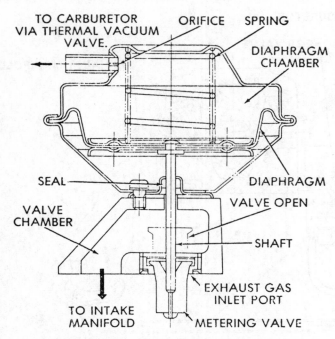

Fig. 3.59. The exhaust gas recirculation valve

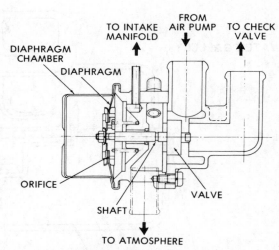

Fig. 3.60. The air bypass valve

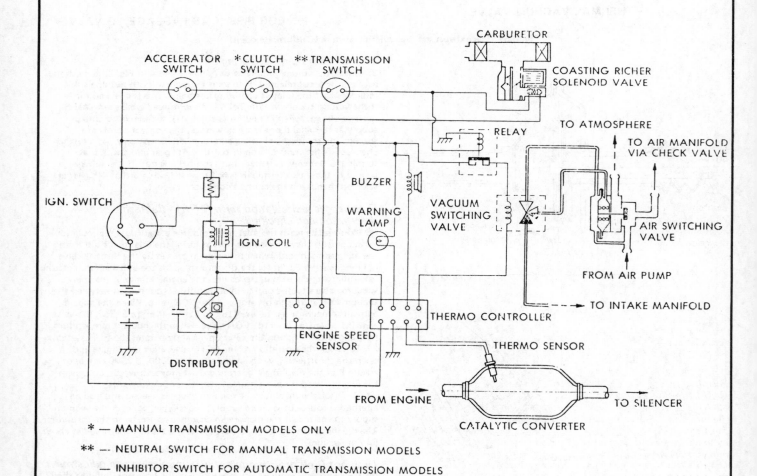

* — MANUAL TRANSMISSION MODELS ONLY
** — NEUTRAL SWITCH FOR MANUAL TRANSMISSION MODELS
— INHIBITOR SWITCH FOR AUTOMATIC TRANSMISSION MODELS

Fig. 3.61A. The over temperature control system

Chapter 3/Carburetion; fuel, exhaust and emission control systems

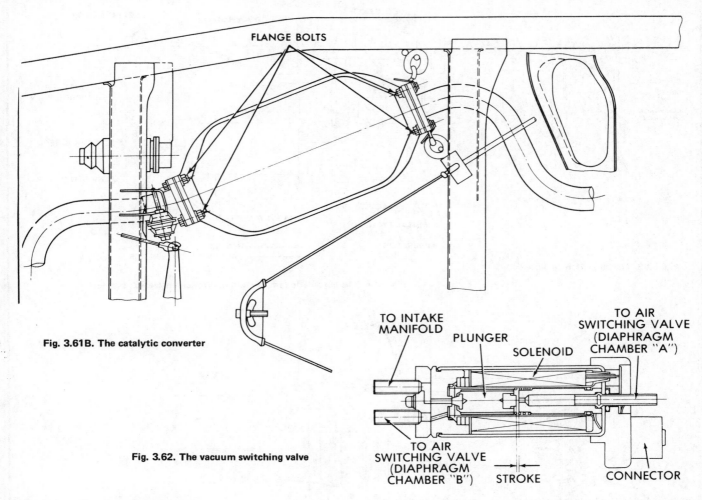

Fig. 3.61B. The catalytic converter

Fig. 3.62. The vacuum switching valve

bending, the dash pot unit must be renewed.
40 To adjust, slacken the dash pot locknut (Fig. 3.59) and unscrew the dash pot fully counter clockwise. The carburetor fast idle cam is now set to the 2nd stop position which should give a throttle valve opening of 13°. Now tighten the dash pot (clockwise) so that the end of the dash pot shaft just contacts the dash pot lever. Tighten the locknut. With the engine running and the fast idle cam in the second stop position the engine speed should be within 2400 to 2800 rpm. Any deviation from this necessitates readjustment of the dash pot.

Air bypass valve (Federal) - Series 5 and 6
41 Located between the air pump outlet and air distribution manifold, the air bypass valve controls the secondary air flow, and prevents after burning in the exhaust system when decelerating. To check that it is operating satisfactorily, depress fully the accelerator pedal and then release it suddenly. The secondary air flow thru the bypass valve should not blow out for more than 5 seconds. If it does then the air bypass valve is faulty and will need renewing.

Oxidizing Catalytic converter system (Series 5 and 6 models - California only) - description
42 A catalytic converter is incorporated into the exhaust system and the thru blow of exhaust gases from the engine are chemically changed. The harmful hydro carbon and carbon monoxide gases are processed by oxidation within the converter to water and carbon dioxide. The catalytic converter operates in conjunction with the over temperature control system, see Fig. 3.61A. To remove the catalytic converter, apply the parking brake, jack up the vehicle and disconnect the catalytic converter flange bolts. Remove the catalytic converter and gaskets. Installation is the reversal of removal but use new gaskets, and check for leaks.

Over temperature control system (California - Series 5 and 6 models)
43 This prevents the catalytic converter overheating and operates in the following manner. When the engine is coasting, the coasting richer system operates to prevent the catalytic converter overheating due to poor combustion. The secondary air injection operates simultaneously. On reaching 1350°F in the catalyst due to continued high speed, hills or high load driving, the secondary air is diverted into the atmosphere to reduce chemical reaction within the catalyst. If the catalyst temperature reaches 1830°F due to engine faults, the warning light and buzzer are actuated by a thermo sensor and thermo controller to indicate a failure.

Vacuum switching valve
44 The vacuum switch valve incorporates three ports. Two of these ports are opened or closed by the electrically controlled solenoid plunger. This operates when the catalyst temperature reaches 1350°F in driving (not coasting conditions). When actuated the switch valve connects the diaphragm chamber 'B' to the inlet manifold, allowing the manifold vacuum to draw off diaphragm chamber 'B'. When deactivating, the solenoid plunger shuts the port and connects chamber 'B' to chamber 'A'. The vacuum switching valve is also actuated by the transmission controlled spark advance system (TCS) as described in Section 1, Chapter 4.

Air switching valve - checking
45 This valve switches the air flow from the air pump, and is actuated by manifold vacuum and air pump pressure, being switched by a vacuum switching valve. Air pump pressure in diaphragm chamber 'B' activates the vacuum switching valve, which flows the air from the pump to check valve. Manifold vacuum switched by the vacuum switching valve acts in diaphragm chamber 'B'. The valve shuts

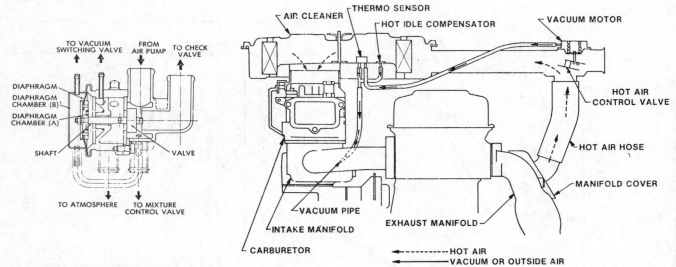

Fig. 3.63. The air switching valve

Fig. 3.64. The automatic temperature controlled air cleaner system

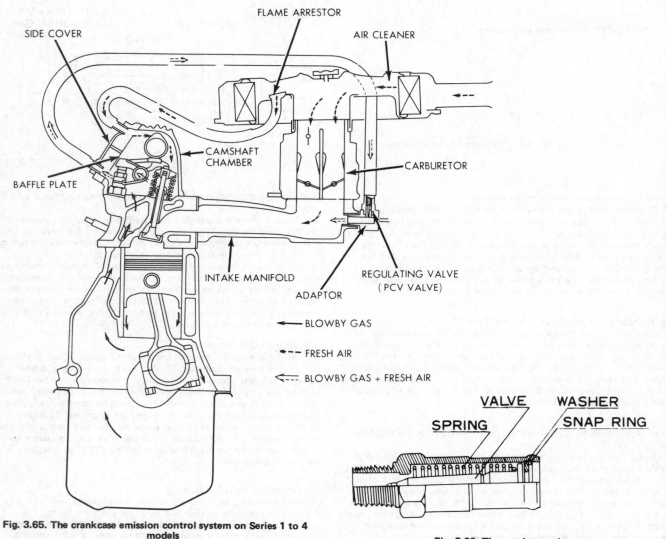

Fig. 3.65. The crankcase emission control system on Series 1 to 4 models

Fig. 3.66. The regulator valve

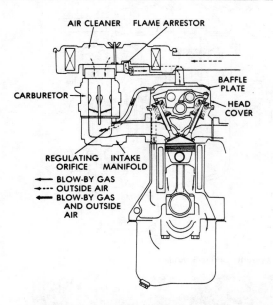

Fig. 3.67. The crankcase emission control system on Series 5 models

the air passage to the check valve and opens the atmosphere port to divert the air flow from the air pump.

46 To check the switching valve, with the engine off electrically operate the switching valve and start the engine. The secondary air should continue to blow out the air switching valve. If it does not, then it is faulty and in need of replacement.

Automatic temperature - controlled air cleaner system - checking

47 The air cleaner contains a viscous type filter element which must be renewed at the specified service intervals. Removal of the element for inspection and/or renewal is described in Section 2 and 3. Additionally check that the vacuum motor, idle compensator and thermo sensor retaining screws are tight. Check the hose connections and condition. Renew where necessary.

25 Crankcase emission control - description and maintenance

1 The crankcase emission control system forces the blowby gases from the engine crankcase into the intake manifold so that they are recirculated into the engine with the fuel air mixture.

2 This system has a baffle plate in the side cover to separate the oil particles from the gases, a regulating valve to control the suction of gases and interconnecting hoses, including a hose to supply fresh air from the air cleaner, (Fig. 3.65). Series 5 models differ in that the baffle plate is located in the head cover as in Fig. 3.67.

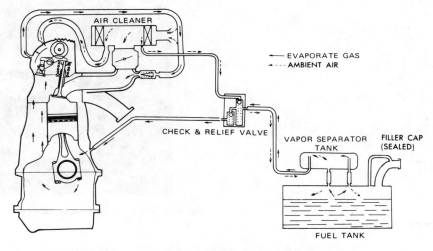

Fig. 3.68. The evaporative emission control system - Series 1 to 3

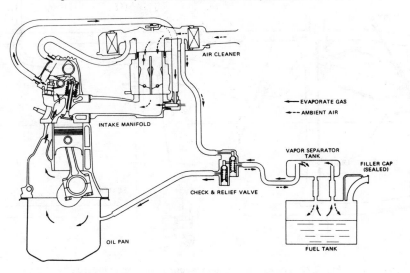

Fig. 3.69. The evaporative emission control system - Series 4 models

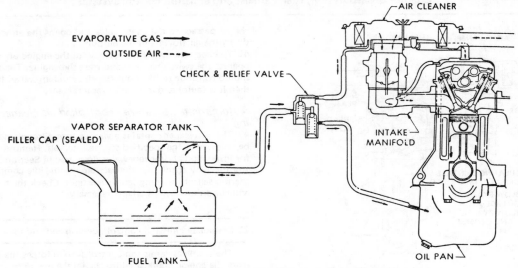
Fig. 3.70. The evaporative emission control system - Series 5 models

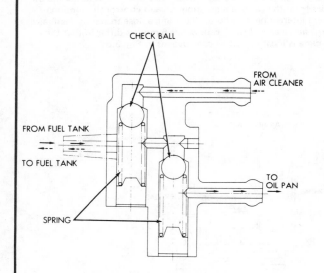
Fig. 3.71. The check and relief valve

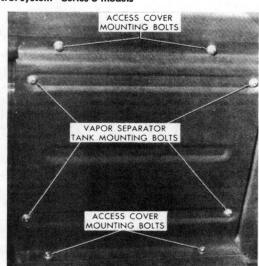
Fig. 3.72. The vapor tank access cover and retaining bolt positions

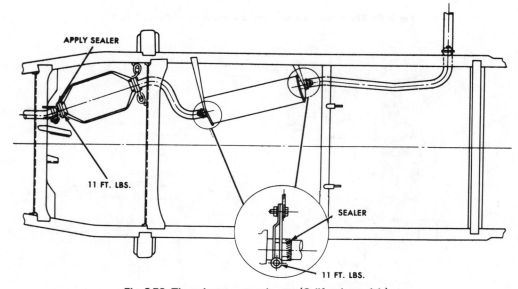
Fig. 3.73. The exhaust system layout (California models)

Chapter 3/Carburetion; fuel, exhaust and emission control systems

Baffle plate - removal inspection and refitting

3 Detach the hoses from the side cover on the camshaft carrier on Series 1 to 4 models or from the head cover in Series 5. Undo and remove the nuts and bolts and washers securing the side cover/head cover.
4 Remove the side cover/head cover. The baffle plate is located on the inner face of the right-hand side cover on Series 1 to 4 or on the underside of Series 5 models. Wash the inner parts of the baffle plate in detergent oil and blow clean with compressed air. If the cover, or baffle plate is damaged in any way, it must be renewed.
5 Refit the cover in the reverse order and fit a new sealing gasket. Ensure the hose connections are good.

Regulating valve - inspection and overhaul

6 Detach the regulating valve hose and carefully inspect the inner valve parts for signs of wear or damage. When applying finger pressure, the valve movement must be light and return easily, if not the valve must be overhauled or renewed. Similarly if the valve is lower in the manifold side and/or the inner valve is clogged or fouled up, it must be renewed or overhauled.
7 To remove the regulator valve from inlet manifold, unscrew it, and then with a pair of snap-ring pliers extract the snap-ring and withdraw the washer valve and spring (Fig. 3.66). Wash the parts in detergent oil and inspect the parts for damage. Renew the unit if any parts are found to be defective. Reassemble in reverse ensuring that the narrow end of the spring is inserted upwards into the body.

26 Evaporative emission control system - description

1 This system enables fuel vapor from the fuel tank to be directed into the engine crankcase and together with the blow by gases is redirected into the inlet manifold.
2 The system consists of a vapor separator tank, a check and relief valve and interconnecting tubes (Figs. 3.68, 3.69, and 3.70).

Check and relief valve - inspections

3 To inspect the check and relief valve, disconnect the hoses and remove the attachment bolts. Remove the valve unit and check for leakage by blowing air into parts. If found to be defective the unit will have to be renewed.
4 The check valve is normal if on applying air from the fuel tank side air passes thru the crankcase side but not the air cleaner side.
5 Apply air from the check side and if the passage of air is restricted then the valve is normal.
6 Apply air from the relief side (from the air cleaner) and air should pass into the fuel tank side but not the check side. Renew the rubber connecting tubes if they are found to be cracked or perished.

Vapor separator tank - removal and installation

7 Unscrew and remove the four access panel mounting bolts on the rear body and also the four vapor tank unit bolts (Fig. 3.72). Remove the panel.
8 Detach the three rubber hoses from the vapor tank. Remove the tank unit.
9 Reassemble in the reverse order and ensure that the hose connections are good.

27 Exhaust system - removal and installation

1 Position the vehicle over a pit or jack it up with the parking brake firmly applied. Secure with chassis stands or blocks.
2 Remove the respective mounting bracket bolts and exhaust pipe to manifold flange retaining bolts.
3 It may be necessary to wire brush the bolt threads clean and apply penetrating oil if they are corroded into position.
4 Remove the exhaust pipe and muffler and on California models, the catalytic converter.
5 Installation is the reversal of removal but semi-tighten the mounting bolts when the system is in position and ensure that the pipes and muffler are clear of surrounding body and fittings, then tighten the bolts to 11 lb f ft (15 Nm).
6 Check for leaks and if necessary apply a sealant to the pipe connections.

28 Fault diagnosis - Carburetion; fuel, exhaust and emission control systems

Symptom	Reason/s
Fuel consumption excessive	Air cleaner choked and dirty giving rich mixture. Fuel leaking from carburetor, fuel pump, or fuel lines. Float chamber flooding. Generally worn carburetor. Distributor condenser faulty. Balance weights or vacuum advance mechanism in distributor faulty. Carburetor incorrectly adjusted, mixture too rich. Idling speed too high. Contact breaker gap incorrect. Valve clearances incorrect. Choke valve incorrectly set. Incorrectly set spark plugs. Tires under-inflated. Wrong spark plugs fitted. Brakes dragging. Emission control system faulty (see later in this Section).
Insufficient fuel delivery or weak mixture due to air leaks	Partially clogged filter in pump, carburetor or fuel line.. Incorrectly seating valves in fuel pump. Fuel pump diaphragm leaking or damaged. Gasket in fuel pump damaged. Fuel pump valves sticking due to fuel gumming. Too little fuel in fuel tank (prevalent when climbing steep hills). Union joints on pipe connections loose. Split in fuel pipe on suction side of the fuel pump. Inlet manifold-to-block or inlet manifold-to-carburetor gaskets leaking. Fuel tank relief valve stuck closed.

Power reduced	Clogged main jets. Accelerator linkage requires adjustment. Fuel filter blocked. Air cleaner blocked.
Erratic idling	Slow jet clogged. Secondary throttle valve operating incorrectly. Worn throttle valve shafts. Broken carburetor flange gasket. Incorrect adjusted carburetor.
Flat spot or hesitation	Clogged jets. Emulsion tube clogged. Secondary throttle valve operating incorrectly.
Engine will not start	Fuel level incorrect. Lack of fuel. Incorrect setting of mixture screw. Incorrect fast idle adjustment.

Emission control system faults

Engine lacks power - difficult to start	Loose vacuum pipes/hose connections. Manifold retaining nuts loose. Faulty EGR valve. Hot air control valve defective. Fuel vapor control valve defective. Coasting richer solenoid defective. Mixture control valve defective. Air pump drivebelt loose/worn. Check valve defective. Air pump/relief valve defective.

Chapter 4 Ignition system

For information relating to Series 8 thru 12 models, including 4WD, refer to Chapter 13.

Contents

Condenser - testing, removal and installation ... 4	Ignition coil ... 10
Distributor contact points - removal, installation and adjustment ... 3	Ignition system (mechanical contact breaker) - fault diagnosis ... 11
Distributor - dismantling, servicing, and installation ... 6	Ignition timing ... 8
Distributor - removal and installation ... 5	Routine maintenance ... 2
Dwell angle - checking ... 7	Sparking plugs and HT leads ... 9
General description ... 1	

Specifications

Spark plugs

Series 1 to 4 models

Type ...	NGK BP6ES
Thread diameter ...	0.55 in
Pitch ...	1.25 in
Reach ...	0.75 in
Spark gap:	
Series 1 models ...	0.028 - 0.032 in
Series 2 and 3 ...	0.030 - 0.040 in
Series 4 ...	0.030 in
Spark plug torque setting ...	18 - 25 ft lbs (2.4 to 3.4 kg fm)

Series 5 and 6 models

Type ...	NGK BPR6ES
Reach ...	0.75 in
Spark gap ...	0.030 in

Firing order
... 1 - 3 - 4 - 2

Ignition coil

Make ...	Nippon Denso
Model - Series 1 to 4 ...	6CR - 200
- Series 5 ...	C6R - 207 029700 - 3500
Voltage ...	12V
Secondary voltage (Series 1 to 4) ...	15,000V @ 1,500 rpm
Primary resistance (Series 1 to 4) ...	1.5 ohms
Secondary resistance (Series 1 to 4) ...	11.5 K ohms
Resistance (Series 5) ...	1.3 to 1.7 ohms

Distributor

Make ...	Nippon Denso
Model - Series 1 ...	D414 - 62
- Series 2 ...	D414 - 64
- Series 3 ...	D417 - 62
- Series 4 ...	D417 - 64 (Fed.) or D417 - 62 (California)
- Series 5 ...	029100 - 3280
- Series 6 ...	029100 - 3281
Rotational direction ...	Counter-clockwise (from top)
Ignition interval (Series 1 to 4) ...	90° ± 1°
(Series 5 and 6) ...	88.5° - 91.5°

Contact breaker point pressure - Series 1 to 4	1.1 - 1.43 lbs
- Series 5 and 6	0.83 - 1.27 lbs

Contact breaker points gap

Series 1 and 2	0.018 - 0.022 in (advance)
	0.016 - 0.024 in (retard)
Series 3 and 4	0.018 - 0.022 in
Series 5 and 6	0.016 - 0.020 in

Dwell angle

Series 1 to 4	49° to 55° with 0.020 in point gap
Series 5 and 6	47° to 57°

Condenser capacity

Series 1 and 2 - advanced	0.20 - 0.24 Micro Farads
- retarded	0.05 Micro Farads ± 15%
Series 3 and 4 - advanced	0.20 - 0.24 Micro Farads
Series 5 and 6	0.25 Micro Farads

1 General description

In order that the engine can run correctly, it is necessary for an electrical spark to ignite the fuel/air mixture in the combustion chamber at exactly the right moment in relation to engine speed and load. The ignition system is based on feeding low tension (LT) voltage from the battery to the coil where it is converted to high tension (HT) voltage. The high tension voltage is powerful enough to jump the spark plug gap in the cylinders many times a second under high compression pressures, providing that the system is in good condition and that all adjustments are correct.

The ignition system is divided into two circuits: the low tension circuit and the high tension circuit.

The low tension (sometimes known as the primary) circuit consists of the battery lead to the ignition switch lead, from the ignition switch to the low tension or primary coil windings (+ terminal) and the lead from the low tension coil windings (— terminal) to the contact breaker points and condenser in the distributor.

The high tension circuit consists of the high tension or secondary coil windings, the heavy ignition lead from the center of the coil to the center of the distributor cap, the rotor arm, and the spark plug leads and spark plugs.

The system functions in the following manner. Low tension voltage is changed in the coil into high tension voltage by the opening and closing of the contact breaker points in the low tension circuit. High tension voltage is then fed via the carbon brush in the center of the distributor cap to the rotor arm of the distributor cap, and each time it comes in line with one of the four metal segments in the cap, which are connected to the spark plug leads, the opening and closing of the contact breaker points causes the high tension voltage to build up, jump the gap from the rotor arm to the appropriate metal segment and so via the spark plug lead to the spark plug, where it finally jumps the spark plug gap before going to earth.

The ignition is advanced and retarded automatically, to ensure the spark occurs at just the right instant for the particular load at the prevailing engine speed.

The ignition advance is controlled both mechanically and by a vacuum operated system. The mechanical governor mechanism comprises two weights, which move out from the center under centrifugal force and are connected to the shaft by two light springs, and it is the tension of the springs which is largely responsible for correct spark advancement.

The vacuum control consists of a diaphragm, one side of which is connected via a small bore tube to the carburetor, and the other side to the contact breaker plate. Depression in the inlet manifold and carburetor, which varies with engine speed and throttle opening, causes the diaphragm to move, so moving the contact breaker plate and advancing or retarding the spark. A fine degree of control is achieved by a spring in the vacuum assembly. A resistor is incorporated in the ignition circuit for Series 5 and 6 models so that during starting, with the engine being cranked by the starter motor, full battery voltage is applied at the coil to maintain a good spark at the plug electrodes which would not be the case should a drop in voltage occur.

On vehicles equipped with a full emission control system a spark timing control system is employed. The system is designed to advance or retard the ignition timing in accordance with the prevailing engine operating conditions in order to reduce the emission of noxious exhaust fumes particularly during periods of deceleration. A dual point distributor was used for this purpose on Series 1 and 2 models.

The contact breaker point sets are connected in parallel and the retard contact breaker circuit incorporates a relay.

The accelerator and transmission relays and throttle and ignition switches are connected to the distributor relay in parallel (Fig. 4.3).

The advance contact breaker points are actuated when the relay is energized by the following:

a) The transmission is not engaged in 3rd or 4th gear, and the throttle valve is not open in excess of 7°.
b) The clutch pedal is depressed.
c) The throttle valve is fully open.
d) When the ignition switch is turned on.

On California Series 5 and 6 models, an engine speed sensor is incorporated into the ignition circuit, which detects the ignition pulse from the coil. At engine speeds in excess of approximately 1,600 rpm, the speed sensor is actuated and this then energizes the coasting richer circuit in the emission control system.

On some California models, a transmission controlled spark advance (TCS) system is fitted. This again works in conjunction with the emission control system, and consists of a transmission switch and a vacuum switching valve, which controls the ignition timing to reduce the nitrogen oxide and hydrocarbon emissions. When this system is fitted, the vacuum advance is eliminated in third gear, and is operated in the following manner.

The vacuum switching valve incorporates a three way port system. Two of the ports are opened or closed by the electrically controlled solenoid plunger. When the vehicle transmission is in 1st, 2nd, 4th or reverse gear, the solenoid leads have continuity and retain the plunger spring in the compressed position, and the carburetor and distributor port sides are in contact. The ignition timing can then be advanced in the normal manner according to the throttle movement.

When the transmission is in third gear the circuit opens and the plunger, actuated by the return spring, closes the carburetor to distributor vacuum port. The air cleaner side port then opens to allow negative pressure in the distributor advance unit to be released, thus preventing the ignition timing being advanced.

2 Routine maintenance

1 The ignition system is one of the most important and most neglected systems in any vehicle, and attention to routine maintenance cannot be over-emphasized. The maintenance intervals are given in the Routine Maintenance Section at the beginning of this Manual.

Spark plugs
2 Remove the plugs and thoroughly clean away all traces of carbon. Examine the porcelain insulation round the central electrode inside the

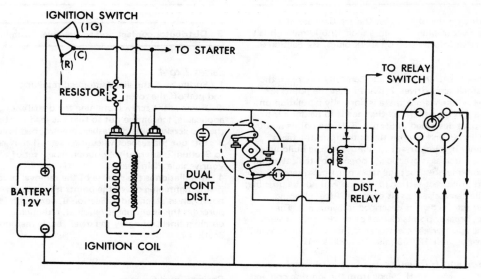

Fig. 4.1. The dual point ignition circuit - Series 1 and 2

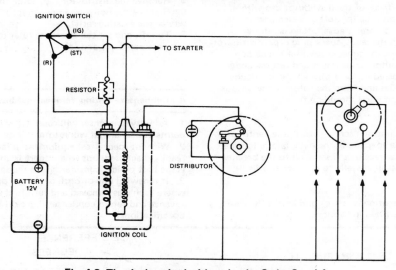

Fig. 4.2. The single point ignition circuit - Series 3 and 4

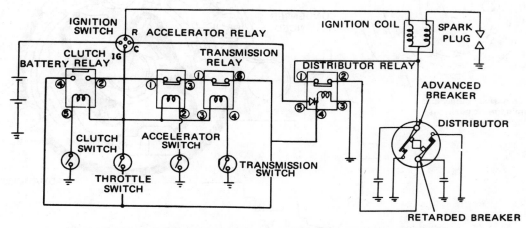

Fig. 4.3. The dual point distributor ignition and associate emission control circuit

plug, and if damaged discard the plug. Reset the gap between the electrodes. Do not use a set of plugs for more than 12,000 miles; it is false economy. For further information on spark plugs, see Section 9.

Distributor contact points
3 Remove the distributor cap and rotor, and carefully pry open the contact points with a small screwdriver. Provided that the points are clean (and this is not very likely since there is normally a build-up on one contact and a cavity in the other), all that needs to be done is to check and reset the gap, as described in Section 3. If the points require attention, refer to Section 3, and remove them; then rub the faces smooth using an oilstone. It is important to remove all the build-up on the one contact, and as much as is practically possible of the cavity on the other contact. Also remember that the faces must be square to each other when re-installed or correct setting of the gap will not be possible; a slightly domed profile can assist for this purpose. After cleaning, install the contacts and set the gap, as described in Section 3. The ultimate life of contact breaker points cannot be predicted, but with correct attention at the appropriate maintenance intervals, they should be capable of lasting for at least 12,000 miles (20,000 km).

Distributor cap, rotor and HT leads
4 Remove the distributor cap and HT leads from the ignition coil and spark plugs. Wipe the end of the coil, the HT leads and spark plug caps and the internal and external surfaces of the distributor cap with a lint-free cloth moistened with gasoline (petrol) or a cleaning solvent. Ensure that all traces of dirt and oil are removed, then carefully inspect for cracked insulation on the leads, spark plug caps, distributor cap and ignition coil end. At the same time, check that the carbon brush in the center of the distributor cap is intact and returns under the action of its spring when pressed in. Carefully scrape any deposits from the distributor cap electrodes and from the rotor; if any serious erosion has occurred, replacement parts should be fitted.

Lubrication
5 Apply two drops of engine oil to the screw head in the center of the distributor cam, and through the aperture in the baseplate to lubricate the advance mechanism. Apply a trace of the same oil to the breaker point pivot(s), and a trace of petroleum jelly to the cam profile.

Ignition timing
6 After completing the aforementioned, check and adjust the dwell angle and ignition timing, as described in Sections 7 and 8.

3 Distributor contact points - removal, installation and adjustment

Series 1 to 4
1 Spring back the distributor cap retaining clips, remove the cap and pull off the rotor.
2 Remove the screws retaining the distributor point set(s) and, where applicable, loosen the nut so that the lead can be detached. Take care that no screws, nuts or washers are dropped inside the distributor.
3 Lift out the respective contact sets. If they are to be cleaned, refer to Section 2; alternatively obtain a new set of the correct type from your vehicle main dealer.
4 Installation is the reverse of the removal procedure, but before finally tightening down the points it is necessary to set the contact point gap as described in Section 6, paragraph 29(e). For most practical purposes this is all that is required, although it is advantageous to obtain a final setting by the dwell angle method, as described in Section 7.

Series 5 and 6
5 Remove the distributor cap, rotor and dust proof cover.
6 Remove the points as in paragraphs 2 and 3 above and renew or service the existing set(s).
7 To refit and set the clearance refer to Section 6, paragraph 29, sub-paragraphs (d) and (e).

4 Condenser - testing, removal and installation

1 A faulty condenser will cause rapid burning of the contact breaker points at best and at worst total failure of the ignition system.
2 Without special test equipment, a faulty condenser cannot be readily diagnosed, but where there is an indication of malfunction it is the best policy, considering the moderate cost, to renew it/them.
3 If renewing one or both of the condensers be sure to note from where each was removed and refit in their original positions as the advance and retard condensers have different microfarad ratings (see Specifications).

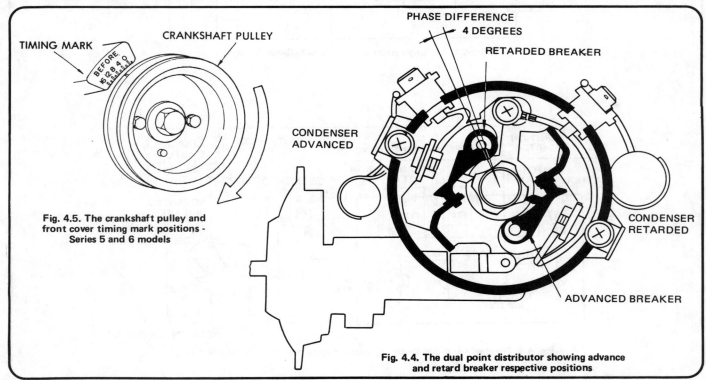

Fig. 4.5. The crankshaft pulley and front cover timing mark positions - Series 5 and 6 models

Fig. 4.4. The dual point distributor showing advance and retard breaker respective positions

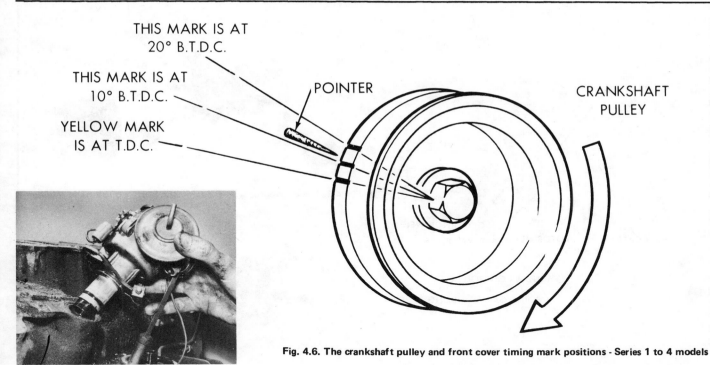

Fig. 4.6. The crankshaft pulley and front cover timing mark positions - Series 1 to 4 models

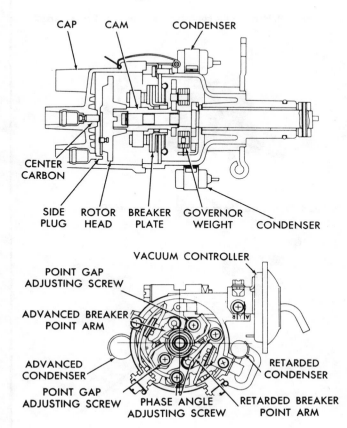

5.6 Removing the distributor

Fig. 4.7a. The Series 1 and 2 distributor component parts

5 Distributor - removal and installation

1 On Series 5 and 6 models, disconnect the battery ground cable.
2 To remove the distributor complete with cap, detach the sparking plug leads from the plugs (note their respective positions). Detach the HT lead from the ignition coil.
3 Disconnect the low tension lead from the coil.
4 Detach the vacuum line from the distributor.
5 To simplify re-installation of the distributor, rotate the crankshaft by turning the crankshaft, using a wrench on the pulley retaining nut clockwise and align the crankshaft pulley and front cover timing pointer at the TDC position (see Fig. 4.5). Index mark the relative position of the distributor body with the fixing plate and cylinder block, and the rotor in relation to the distributor body.
6 Unscrew and remove the distributor fixing plate retaining bolt, and withdraw the distributor from the engine (photo).
7 Installation is a direct reversal of removal, but ensure that when the distributor is fully located, that the relative marks are in alignment with the crankshaft at the TDC position.

6 Distributor - dismantling, servicing and reassembly

Series 1 to 4
1 With the distributor cap removed, lift out the rotor.
2 Remove the vacuum control unit retaining screw and lift the connecting rod from the pivot pin. Withdraw the vacuum controller unit.
3 Unscrew the low tension lead retaining screw and disconnect the lead.

Series 1 and 2
4 Lift out the two sets of contact breaker points, but do not disturb the adjustment plate setting retaining the retarded side breaker points.

Series 3 and 4
5 Undo the retaining screws and withdraw the contact breaker points.
6 Detach the primary side terminal and withdraw the breaker plate.
7 Support the distributor shaft and coupling in a 'V' block and using a suitable drift, remove the roll pin.
8 Undo and remove the screw from the cam head and detach the timing lever.

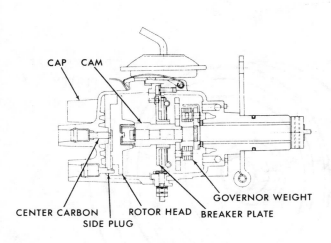

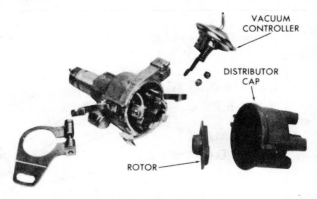

Fig. 4.8. The distributor with vacuum controller, cap rotor and retaining plate removed

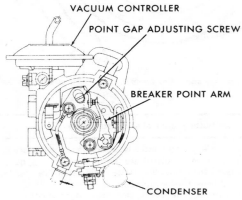

Fig. 4.7b. The Series 3 and 4 distributor component parts

Fig. 4.9. Remove the contact breaker points - Series 1 and 2

Fig. 4.10. Remove the contact breaker points - Series 3 and 4

Fig. 4.11. Removing the roll pin

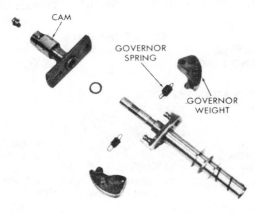

Fig. 4.12. The distributor, shaft, governor weights and springs

Measuring plug gap. A feeler gauge of the correct size (see ignition system specifications) should have a slight 'drag' when slid between the electrodes. Adjust gap if necessary

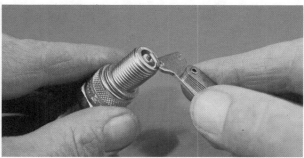

Adjusting plug gap. The plug gap is adjusted by bending the ground electrode inwards, or outwards, as necessary until the correct clearance is obtained. Note the use of the correct tool

Normal. Gray brown deposits, lightly coated core nose. Gap increasing by around 0.001 in (0.025 mm) per 1000 miles (1600 km). Plugs ideally suited to engine, and engine in good condition

Carbon fouling. Dry, black, sooty deposits. Will cause weak spark and eventually misfire. Fault: over-rich fuel mixture. Check: carburetor mixture settings, float level and jet sizes; choke operation and cleanliness of air filter. Plugs can be re-used after cleaning

Oil fouling. Wet, oily deposits. Will cause weak spark and eventually misfire. Fault: worn bores/piston rings or valve guides; sometimes occurs (temporarily) during running-in period. Plugs can be re-used after thorough cleaning

Overheating. Electrodes have glazed appearance, core nose very white – few deposits. Fault: plug overheating. Check: plug value, ignition timing, fuel octane rating (too low) and fuel mixture (too weak). Discard plugs and cure fault immediately.

Electrode damage. Electrodes burned away; core nose has burned, glazed appearance. Fault: pre-ignition. Check: as for 'Overheating' but may be more severe. Discard plugs and remedy fault before piston or valve damage occurs

Split core nose (may appear initially as a crack). Damage is self-evident, but cracks will only show after cleaning. Fault: pre-ignition or wrong gap-setting technique. Check: ignition timing, cooling system, fuel octane rating (too low) and fuel mixture (too weak). Discard plugs, rectify fault immediately

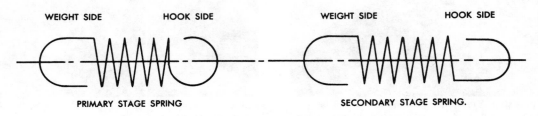

GOVERNOR SPRINGS

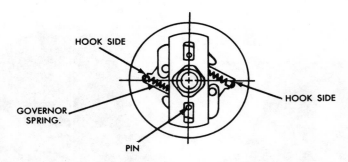

Fig. 4.13. The primary and secondary springs and their respective positions

9 Note their respective positions and unhook the two governor springs and remove with governor weights (Fig. 4.12).
10 Clean thoroughly the respective parts and inspect for wear and damage.
11 Refer to Section 2, paragraph 5, for checks and cleaning details of the cap and rotor. Unless the points are known to be in good condition it is advisable to install new ones. Check the weights for excessive looseness on their pivots; it is advisable to fit new springs in view of their modest cost. Suck (by mouth) on the vacuum diaphragm unit and check that the link is drawn in. When the vacuum is held by placing the tongue or a finger over the tube, the link should remain in; if it fails to do so, the vacuum unit must be renewed. Ensure that the driven gear is securely pinned to the cam spindle, and that the gear is undamaged. Also check the shaft for wear, and the governor weights and springs. Renew any suspect parts.
12 Apply a trace of grease to the bearings and sliding surfaces.
13 Refit the cam and pinion taking care to refit the respective components to their original positions. Two types of governor springs are used in the centrifugal advance.
   The primary stage spring has a round hook on its end.
   The secondary stage spring has a sharp hook on one end as in Fig. 4.13. Do not overstretch these springs when refitting.
14 The primary stage spring is hooked to the governor weight. The governor weight pin locates into the longest hole in the cam plate.
15 The secondary stage spring is hooked to the governor weight and the weight pin fits into the shorter hole.
16 Refer to Fig. 4.14 and note that the coupling is installed with the rotor head tip end facing to the coupling side on which the slot is offset.
17 Apart from the above items, reassembly is basically a reversal of removal procedures but be sure to fit a new roller pin when refitting the distributor coupling.

Series 5 and 6

18 Lift off the distributor cap and remove the rotor and dust proof cover.
19 Unscrew the terminal bolt locknut and remove the terminal complete with insulator (Fig. 4.15).
20 Remove the condenser from the vacuum housing.
21 Extract the circlip securing the vacuum advance rod and withdraw the vacuum advance unit.
22 Remove the dust proof gasket (Fig. 4.17). Unscrew the breaker

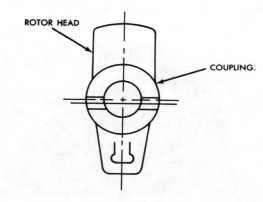

Fig. 4.14. Rotor head to coupling installation position

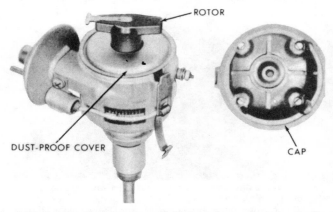

Fig. 4.15. Remove the cap, rotor and dust proof cover for access to the points

arm retaining screws and withdraw the breaker arm, points and damper spring (Fig. 4.18).
23 Detach the distributor cap clamps and the lead wire. Now extract the breaker plate.
24 Undo the screw in the end of the cam assembly and unhook the governor springs and cam. Note the respective spring positions before removing them.
25 Unclip the governor weights retaining circlips and withdraw the governor weights.
26 Remove the caulking from the distributor shaft set pin and then remove the pin from the shaft. Finally remove the collar and distributor shaft (Fig. 4.20)..
27 With the distributor dismantled, wash and clean off the various components but do not wash the inner face of the vacuum advancer.
28 Inspect and service as required the distributor components as described in paragraphs 10 and 11.
29 Reassembly of the distributor is a direct reversal of the removal procedure, but note the following:

   a) Ensure that the governor springs and weights are fully located as in Fig. 4.19 and that they are suitably lubricated.
   b) Fit the cam drive plate with its rear face (marked), turned to the governor plate stop pin.
   c) The cap clip is installed with its projected portion facing the terminal side.
   d) When installing the damper spring, ensure that the clearance between the flat face of the cam and heel is 0.0020 - 0.0040 in (0.05 - 0.10 mm) - see Fig. 4.21.
   e) To fit and adjust the breaker point gap, rotate the cam so that the heel of the breaker arm is against the tip of the cam lobe. The breaker point assembly can now be fitted to the breaker plate and the gap adjusted to the specified clearance (0.015 - 0.019 in) (0.38 to 0.48 mm).

7 Dwell angle - checking

1 For most practical purposes, the breaker points can be set by adjusting the gap, as described in Section 3. However, a more accurate method of setting the points is obtained by the dwell angle method which requires the use of a proprietary dwell angle meter. This type of meter is supplied complete with operating and connecting instructions.
2 Connect the dwell angle meter and set it up in accordance with the manufacturer's instructions for 4-cylinder engines.
3 Start the engine and run it at idle speed.
4 Check the dwell angle and compare it with that specified. If it is satisfactory, all that needs to be done is to disconnect the meter and its connections.
5 If adjustment is required, you will need to increase the breaker point gap, if the angle was too large, or decrease the gap, if the angle

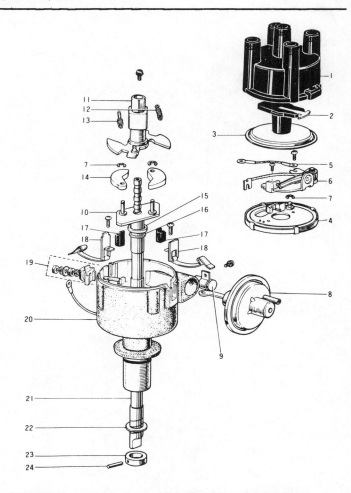

Fig. 4.16. The component parts of the Series 5 model distributor

1 Cap	13 Governor spring (high-speed side)
2 Rotor	14 Governor weight
3 Dust-proof cover	15 Cap washer
4 Breaker plate	16 Thrust washer
5 Lead wire	17 Dust-proof gasket
6 Breaker point	18 Cap clamp
7 Circlip	19 Terminal
8 Vacuum control	20 Housing
9 Condenser	21 Shaft
10 Governor	22 'O' ring
11 Cam assembly	23 Collar
12 Governor spring (high-speed side)	24 Pin

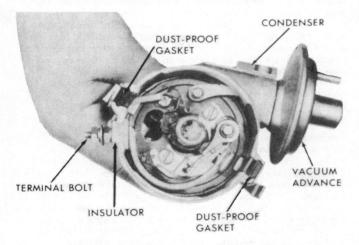

Fig. 4.17. The distributor with dust cover removed

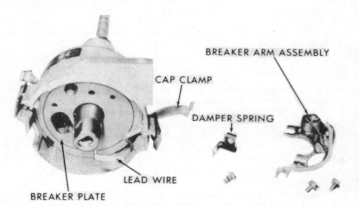

Fig. 4.18. Remove the contact breaker arm

Chapter 4/Ignition system

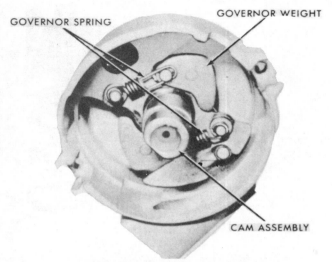

Fig. 4.19. The governor weights and springs

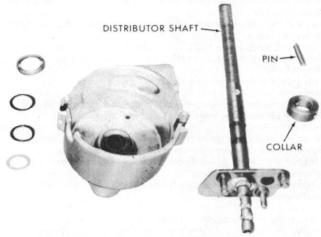

Fig. 4.20. The distributor shaft and collar

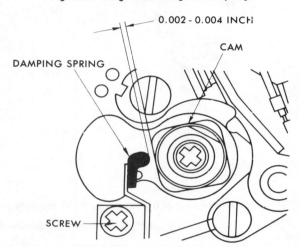

Fig. 4.21. The damper spring to cam position

10.1 The ignition coil and connections - Series 1-4

was too small. This can be done by readjustment, as described in Section 3. However, it is possible to do it whilst the engine is being cranked by an assistant while you carry out the adjustment. First remove the coil HT lead from the distributor and ground it, then remove the distributor cap and rotor. Loosen the breaker point attaching screws then, whilst the engine is being cranked, alter the contact position until the correct dwell angle is obtained. Stop the engine cranking, tighten the attaching screw(s) then recheck the angle. On completion, remove the meter, remake the HT connection, then install the rotor and distributor cap.

8 Ignition timing

Test bulb method

1 Check that the distributor points gap are set correctly as in Section 3, and that the advance contact breaker side is operational.
2 Slacken the distributor location clamp plate bolt, so that the distributor can be rotated for adjustment.
3 Using a suitable wrench, turn the crankshaft pulley to align the pulley mark with the timing cover pointer as in Figs. 4.5 or 4.6 accordingly (No.1 piston on compression stroke).
4 For accurate setting of the timing refer to the Specifications for your model.
5 The instant of time at which the distributor contact breaker points open is the time at which the spark is delivered. Therefore, if the appropriate timing mark is aligned with the timing pointer, the breaker points should just be opening. To be able to judge this more accurately,

connect test bulb leads across the breaker points and switch on the ignition. Rotate the crankshaft at the pulley about 20° counter-clockwise from top-dead-center (TDC) at which time the bulb should be extinguished. Now rotate the crankshaft clockwise until the bulb illuminates (ie. the breaker points are just opening) and check the position of the pulley with respect to the pointer. If the marks are correctly aligned, all is well and good. If the marks are not aligned, it will mean slackening the distributor clamp bolt, turning the distributor slightly, tightening the clamp bolt and rechecking. If the timing is too far advanced (ie. the timing mark is too far from the pointer when the points open) you will need to rotate the distributor clockwise; if it is too far retarded (ie. the timing mark is too near the pointer or has gone past it), you will need to rotate the distributor counter-clockwise.
6 With the ignition timing set, tighten the distributor clamp plate and refit the distributor cap and leads.

Stroboscopic method

7 A more accurate method of timing the ignition is by using a stroboscope connected in accordance with the makers' instructions.
8 Have the engine at normal operating temperature, idling and with the vacuum hose disconnected from the distributor and plugged.
9 Aim the stroboscopic lamp at the timing marks on the front of the engine when the pulley and cover marks should appear to be in alignment. If they are not, release the distributor clamp screw and turn the distributor body until they do align. Re-tighten the screw.
10 Whiten the timing marks if necessary to improve clarity.
11 For best results, use the 6° BTDC mark with the engine idling at 900 rpm.

9 Spark plugs and HT leads

See page 91 for illustrations

1 The correct functioning of the spark plugs is vital for the proper running and efficient operation of the engine.
2 At the intervals specified, the plugs should be removed, examined, cleaned, and if worn excessively, renewed. The condition of the spark plug can also tell much about the general condition of the engine.
3 If the insulator nose of the spark plug is clean and white, with no deposits, this is indicative of a weak mixture, or too hot a plug (a hot plug transfers heat away from the electrode slowly - a cold plug transfers heat away quickly).
4 If the insulator nose is covered with hard black looking deposits, then this is indicative that the mixture is too rich. Should the plug be black and oily then it is likely that the engine is fairly worn, as well as the mixture being too rich.
5 If the insulator nose is covered with light tan to greyish brown deposits, then the mixture is correct, and it is likely that the engine is in good condition.
6 If there are any traces of long brown tapering stains on the outside of the white portion of the plug, then the plug will have to be renewed, as this shows that there is a faulty joint between the plug body and the insulator, and compression is being allowed to leak away.
7 Plugs should be cleaned by a sand blasting machine, which will free them from carbon more than cleaning by hand. The machine will also test the condition of the plugs under compression. Any plug that fails to spark at the recommended pressure should be renewed.
8 The spark plug gap is of considerable importance, as, if it is too large or too small the size of the spark and its efficiency will be seriously impaired. The spark plug gap is given in the Specifications Section.
9 To set it, measure the gap with a feeler gauge, and then bend open, or close, the outer plug electrode until the correct gap is achieved. The centre electrode should never be bent as this may crack the insulation and cause plug failure, if nothing worse.
10 When installing the plugs, remember to connect the leads from the distributor cap in the correct firing order which is 1, 3, 4, 2, No. 1 cylinder being the one nearest the radiator.
11 The plug leads require no maintenance other than being kept clean and wiped over regularly. The leads used are of the carbon cored type which are used to suppress high frequency radio interference from the ignition system. Although these leads can give trouble-free performance for many years, they can sometimes cause starting problems after a considerable period of usage. If they are found to be faulty, consideration should be given to replacing them with the copper cored conductor type and using suppressor-type plug caps. Your automobile electrical specialist will be able to help with the supply of approved types.

10 Ignition coil - removal, checking and installation

1 On Series 1 to 4 models remove the primary and secondary leads, noting their positions.
2 On Series 5 and 6 models, disconnect the battery ground cable, then detach the positive and negative wires from the coil. In addition, remove the resistor lead.
3 Unscrew the coil retaining screws/bolts and nuts/washers and remove the coil.
4 To test the coil requires specialised equipment and this should, therefore, be entrusted to your GM dealer or local auto electrician.
5 Installation is a direct reversal of removal.

11 Ignition system (mechanical contact breaker) - fault diagnosis

Engine fails to start

1 If the engine fails to start and the car was running normally when it was last used, first check there is fuel in the fuel tank. If the engine turns over normally on the starter motor and the battery is evidently well charged, then the fault may be in either the high or low tension circuits. First check the HT circuit. **Note:** If the battery is known to be fully charged; the ignition light comes on, and the starter motor fails to turn the engine check the tightness of the leads on the battery terminals and also the security of the ground lead to its connection on the body. It is quite common for the leads to have worked loose, even if they look and feel secure. If one of the battery terminal posts gets very hot when trying to work the starter motor this is a sure indication of a faulty connection to that terminal.
2 One of the commonest reasons for bad starting is wet or damp spark plug leads and distributor. Remove the distributor cap. If condensation is visible internally, dry the cap with a rag and also wipe over the leads. Refit the cap.
3 If the engine still fails to start, check that current is reaching the plugs by disconnecting each plug lead in turn at the spark plug end, and hold the end of the cable about 3/16th (5 mm) away from the cylinder block. Spin the engine on the starter motor.
4 Sparking between the end of the cable and the block should be fairly strong with a regular blue spark. (Hold the lead with rubber to avoid electric shock). If current is reaching the plugs, then remove them and clean and regap them. The engine should now start.
5 If there is no spark at the plug leads take off the HT lead from the center of the distributor cap and hold it to the block as before. Spin the engine on the starter once more. A rapid succession of blue sparks between the end of the lead and the block indicate that the coil is in order and that the distributor cap is cracked, the rotor arm faulty, or the carbon brush in the top of the distributor cap is not making good contact with the spring on the rotor arm. Possibly the points are in bad condition. Clean and reset them as described in this Chapter.
6 If there are no sparks from the end of the lead from the coil check the connections at the coil end of the lead. If it is in order start checking the low tension circuit.
7 Use a 12v voltmeter or a 12v bulb and two lengths of wire. With the ignition switch on and the points open test between the low tension wire to the coil (it is marked SW or +) and earth. No reading indicates a break in the supply from the ignition switch. Check the connections at the switch and resistor to see if any are loose. Reinstall them and the engine should run. A reading shows a faulty coil or condenser, or broken lead between the coil and the distributor.
8 Take the condenser wire off the points assembly and with the points open, test between the moving points and ground. If there is now a reading, then the fault is in the condenser. Install a new one and the fault is cleared.
9 With no reading from the moving point to ground take a reading between ground and the negative terminal of the coil. A reading here shows a broken wire which needs to be renewed between the coil and distributor. No reading confirms that the coil has failed and must be renewed, after which the engine will run once more. Remember to reinstall the condenser wire to the points assembly. For these tests it is sufficient to separate the points with a piece of dry paper while testing with the points open.

Engine misfires

10 If the engine misfires regularly, run it at a fast idling speed. Pull off each of the plug caps in turn and listen to the note of the engine. Hold the plug cap in a dry cloth or with a rubber glove as additional protection against a shock from the HT supply.
11 No difference in engine running will be noticed when the lead from the defective circuit is removed. Removing the lead from one of the good cylinders will accentuate the misfire.
12 Remove the plug lead from the end of the defective plug and hold it about 3/16th inch (5 mm) away from the block. Restart the engine. If the sparking is fairly strong and regular the fault must lie in the spark plug.
13 The plug may be loose, the insulation may be cracked, or the points may have burnt away giving too wide a gap for the spark to jump. Worse still, one of the points may have broken off. Either renew the plug, or clean it, reset the gap, and then test it.
14 If there is no spark at the end of the plug lead, or if it is weak and intermittent, check the ignition lead from the distributor to the plug. If the insulation is cracked or perished, renew the lead. Check the connections at the distributor cap.

15 If there is still no spark, examine the distributor cap carefully for tracking. This can be recognised by a very thin black line running between two or more electrodes, or between an electrode and some other part of the distributor. These lines are paths which now conduct electricity across the cap thus letting it run to ground. The only answer is a new distributor cap.

16 Apart from the ignition timing being incorrect, other causes of misfiring have already been dealt with under the Section dealing with the failure of the engine to start. To recap - these are that:

   a) *The coil may be faulty giving an intermittent misfire;*
   b) *There may be a damaged wire or loose connection in the low tension circuit;*
   c) *The condenser may be short circuiting;*
   d) *There may be a mechanical fault in the distributor (broken driving spindle or contact breaker spring).*

17 If the ignition timing is too far retarded, it should be noted that the engine will tend to overheat, and there will be a quite noticeable drop in power. If the engine is overheating and the power is down, and the ignition timing is correct, then the carburetor should be checked, as it is likely that this is where the fault lies.

Chapter 5 Clutch

For information relating to Series 8 thru 12 models, including 4WD, refer to Chapter 13.

Contents

Clutch adjustments ... 2	Clutch - removal ... 8
Clutch hydraulic system - bleeding ... 3	Clutch slave cylinder - dismantling and reassembly ... 7
Clutch - inspection and renovation ... 9	Clutch slave cylinder - removal and installation ... 6
Clutch installation ... 11	Fault diagnosis ... 12
Clutch master cylinder - removal and installation ... 4	General description ... 1
Clutch master cylinder - servicing ... 5	Release bearing - removal, inspection and reassembly ... 10

Specifications

Series 1 models
Type ... Single dry plate, diaphragm spring, hydraulically operated

Clutch driven plate
Clutch face material ... Woven SF 105 M
Clutch face diameter (outside diameter x inside diameter x thickness) ... 8 in x 5¾ in x 9/64 in (203 x 127 mm 0.75 x 0.140 mm)
Total friction area ... 24 sq in x 2 (154.8 cm^2)
Driven plate thickness
    Free ... 11/32 in (8.73 mm)
    Compressed ... 5/16 in (7.93 mm)
Number of damper springs ... 6
Clutch surface to rivet head minimum thickness ... 0.008 in (0.203 mm)
Driven plate run out (maximum) ... 0.039 in (1.016 mm)

Diaphragm spring to flywheel distance ... 1.496 in (37.9 mm)
Maximum spline play in direction of rotation ... 0.079 in (2.032 mm)

Clutch pedal height ... 6.3 to 6.7 in (160 to 170 mm)

Clutch pedal play ... Approximately 25/32 in (19.84 mm)

Clutch shift fork play ... 0.118 (3.0 mm)

Clutch master cylinder bore ... 5/8 in (15.87 mm)

Slave cylinder bore ... 3/4 in (19.05 mm)

Minimum diaphragm spring load ... 655 lbs (297 kgs)

Series 2 to 6 models
As per Series 1 but with the following differences

Diaphragm spring to flywheel distance ... 1.299 in (33 mm)
Clutch face diameter (outside diameter x inside diameter x thickness) ... 7.87 x 5.12 x 9.64 in (197.6 x 130 x 3.5 mm)
Total friction area ... 28 sq in x 2 (180.6 cm^2)
Driven plate thickness - free ... 21/64 in (8.33 mm)
Number of damper springs ... 4
Maximum spline play in direction of rotation ... 0.113 in (2.9 mm)
Clutch pedal height (Series 2/4/5 and 6 models) ... 5.9 to 6.3 in (150 to 160 mm)
Series 3 ... 6.3 to 6.7 in (160 to 170 mm)

Torque wrench settings

	lb f ft	kg f m
Shift fork ball stud		
Series 1 to 4 models	70	9.6
Series 5 and 6 models	30	4.1
Clutch cover/flywheel bolts		
Series 1 models	50	6.9
Series 2 to 6 models	13	1.7
Clutch push rod to pedal arm nut		
Series 1 models	25	3.4

1 General description

1 The clutch unit fitted is of the single dry plate (disc) type. The unit comprises the disc assembly, the pressure plate assembly, the clutch cover, and the shift fork assembly.

2 The clutch housing is integral with the transmission casing and contains the clutch shaft (input shaft) bearing retainer and seals.

3 The clutch is hydraulically operated by means of a master cylinder and a clutch slave cylinder. The master cylinder is located in the engine compartment and is attached to the rear panel. The slave cylinder is mounted to the clutch housing.

4 The respective clutch components are riveted together to prevent dismantling for repair and it therefore follows that if the clutch disc or clutch cover unit is worn or damaged, then they must be replaced as a unit. The master cylinder and slave cylinder repairs are confined to the renewal of the rubber seals. The shift fork (operating lever) is adjustable with the pushrod. Periodic checking of this clearance and also topping up the master cylinder is all the maintenance normally required on the clutch system.

2 Clutch - adjustment

Clutch pedal

1 The clutch pedal height is adjusted in the same manner as the brake pedal adjustment. The clutch pedal pushrod and switch correspond to the brake pedal pushrod and stoplight switch. Refer therefore to Chapter 9, Section 3.

Clutch shift fork

2 Disconnect the shift fork return spring and ease the shift fork rearwards slightly. Slacken the adjustment nut (Fig. 5.2) and adjust the pushrod so that it is in contact with the shift fork.

3 Unscrew the pushrod approximately 1¾ turns and lock in position by tightening the locknut. The correct clearance should be 5/64 in (1.98 mm).

4 Any excess clearance between the diaphragm spring fingers and the release bearing will cause the clutch to drag. Too little clearance will cause it to slip.

3 Clutch hydraulic system - bleeding

1 Bleeding the cylinders and fluid lines is necessary when air gets into the system. This is caused by a joint or seal leakage usually, but the system must also be bled when a component in the system is removed and/or replaced. Bleeding entails the process of removing air from the hydraulic circuit.

2 Hydraulic fluid is harmful to paintwork so be very careful not to spill any on the surrounding areas when bleeding or topping up the reservoir.

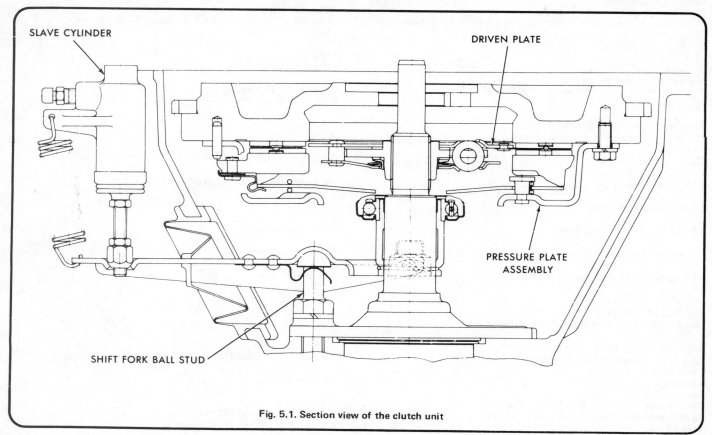

Fig. 5.1. Section view of the clutch unit

3   Always commence bleeding by ensuring that the reservoir is topped up to the level required with the specified type of hydraulic fluid. This level must be kept topped up during the bleeding operation and must not be allowed to drop below ½ of the specified level.
4   Bleeder screws are fitted to both the master cylinder and slave cylinder. Both are bled in the same manner. To bleed the master cylinder, have an assistant pump the clutch pedal several times and then whilst holding it down, slacken the bleeder screw slowly. Now tighten the bleeder screw whilst the assistant gradually releases the clutch pedal. Repeat this operation until the air bubbles disappear from the fluid being pump out.
5   Detach the clutch slave cylinder bleed screw dust cap and connect a length of suitable rubber or vinyl tube to the nipple. Insert the other end of the tube into a clean jar and put about 1 inch (2.5 cm) of fluid into the jar.
6   Slacken the bleed nipple and get the assistant in the cab to depress the clutch pedal slowly and fully down. Tighten the bleed valve.
7   The pedal is now released and the bleed valve loosened again and the procedure repeated until the air bubbles in the circuit are fully expelled. Check that the reservoir is topped up during the operations and that the end of the tube is kept submerged in the fluid within the jar.
8   Remove the tube on completion and refit the bleed valve caps. Always discard the old fluid as it is a false economy to use it again, since it will undoubtedly be contaminated with dirt and moisture.
9   Check the clutch operation on a test drive and readjust the pedal or linkage as described in Section 2.

Fig. 5.2. Adjustment of the clutch push-rod - Series 5

4 Clutch master cylinder - removal and installation

1   Detach the clutch pedal arm from the pushrod.
2   Disconnect the fluid line from the master cylinder and drain the fluid into a suitable container, taking care not to spill any fluid on the paintwork.
3   Remove the master cylinder flange retaining bolts and withdraw the unit from the engine compartment.
4   Installation is the reverse of removal but check the pedal height and linkage adjustment as previously described and bleed the system (Sec. 3)

5 Clutch master cylinder - servicing

1   Drain off the hydraulic fluid from the master cylinder into a container and commence dismantling by removing the rubber boot and circlip to remove the pushrod.

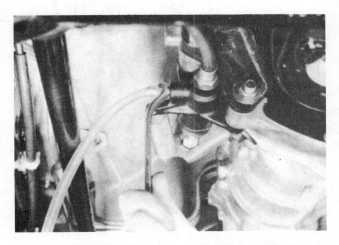

Fig. 5.3. Slacken the bleed nipple on the slave cylinder

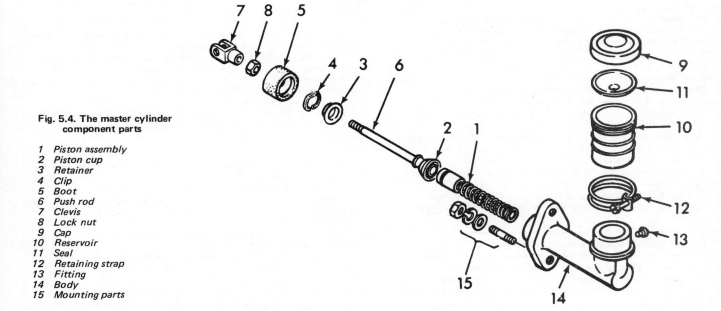

Fig. 5.4. The master cylinder component parts

1   Piston assembly
2   Piston cup
3   Retainer
4   Clip
5   Boot
6   Push rod
7   Clevis
8   Lock nut
9   Cap
10 Reservoir
11 Seal
12 Retaining strap
13 Fitting
14 Body
15 Mounting parts

Chapter 5/Clutch

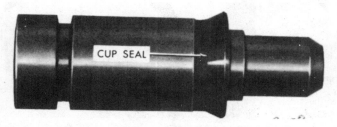

Fig. 5.5. Take care when fitting the cup seal and position as shown

8.4 Remove the clutch pressure plate and disc (driven plate)

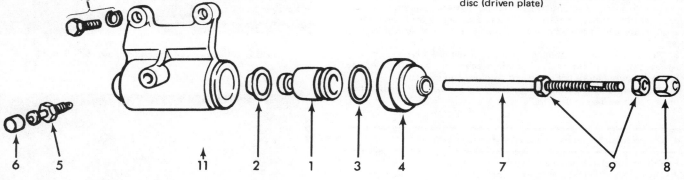

Fig. 5.6 The slave cylinder component parts fitted to Series 1 to 4, Series 5 and 6 models have a different mounting flange, otherwise they are identical

1 Piston
2 Piston cup
3 'O' ring
4 Boot
5 Bleeder screw
6 Cap
7 Push rod
8 Nut
9 Nut
10 Bolt
11 Spring

2 Withdraw from the body of the master cylinder the stopper, piston, cup and return spring, noting the order of sequence and fitting.
3 Lay the respective parts out in order on a sheet of clean paper. Wash all the components in clean hydraulic fluid, isopropyl alcohol or methylated spirit. Examine carefully the surface of the piston and the bore of the cylinder for scoring or bright wear areas. Where these are evident, renew the master cylinder complete.
4 Where components are in good condition, discard the rubber seals and obtain a repair kit.
5 Reassembly of the master cylinder is the reversal of removal procedure but lubricate all parts with clean hydraulic fluid before fitting. When fitting the cup seal take care not to damage the lipped portions.

6 Clutch slave cylinder - removal and installation

1 Undo the slave cylinder securing bolts and remove the pushrod from the shift fork.
2 Disconnect the flexible hydraulic hose from the slave cylinder and plug the hose to prevent further loss of fluid and also the ingress of dirt. Remove the cylinder.
3 To reinstall the slave cylinder, reverse the removal instructions and top up the master cylinder fluid level. Bleed the system as described in Section 3 and if necessary, readjust the pedal and pushrod to shift fork clearance to 5/64 inch (1.98 mm) as described in Section 2.

7 Slave cylinder - dismantling and reassembly

1 Although not normally requiring attention, if the slave cylinder is to be dismantled be sure to obtain a repair kit containing the necessary parts before dismantling. It is recommended that all rubber parts are renewed.

2 Having removed the slave cylinder from the vehicle, clean the outside using a lint-free cloth. Do not attempt to clean the slave cylinder or its components with gasoline (petrol), kerosene or cleaning solvents. It is normally adequate to use new hydraulic fluid for cleaning but isopropyl alcohol or methylated spirit is also suitable.
3 Remove the dust boot from the cylinder and withdraw the pushrod.
4 Remove the piston by tapping the pushrod end of the cylinder sharply on a wooden block. Alternatively, compressed air applied at the inlet connection can be used.
5 Clean all the components as described in paragraph 2, but discard the rubber cups. Examine the bore for wear, scoring or roughness; check the clearance of the piston in the bore and replace any parts which are unsatisfactory. Check the bleed valve and inlet valves are clean and that the bleed valve is in good working condition.
6 When reassembling, dip the respective components in clean hydraulic fluid before fitting. Ensure that the piston cup is carefully fitted and positioned as shown in Fig. 5.6.
7 On refitting the slave cylinder, recheck the pushrod to shift fork clearance and adjust if required. Bleed the system and top up the reservoir.

8 Clutch - removal

1 Remove the transmission unit as described in Chapter 6 or alternatively the engine and/or transmission unit as described in Chapter 1.
2 As applicable, separate the engine and transmission by removing the clutch housing to engine securing bolts.
3 Prior to removing the clutch unit from the flywheel, mark their respective positions in relation to each other, so that they may be refitted in the correct position.
4 Unscrew the clutch assembly securing bolts a turn at a time in diametrically opposed sequences until the tension of the diaphragm

Chapter 5/Clutch

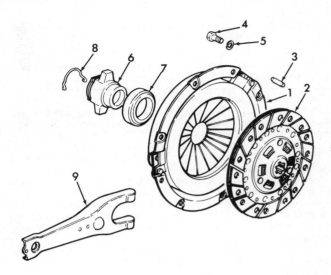

Fig. 5.7. The clutch unit component parts

1 Plate assembly
2 Disc assembly
3 Pin
4 Bolt
5 Washer
6 Clutch shift block
7 Bearing
8 Spring
9 Fork assembly

spring is released. Withdraw the bolts and lift the pressure plate assembly clear.
5 The driven plate (clutch disc) will fall from the location between the flywheel and pressure plate.

9 Clutch - inspection and renovation

1 Clean off the clutch components and wipe dry with a clean cloth.
2 Commence inspection by examining the pressure plate and flywheel surfaces for signs of scoring. Slight score marks may be left but if deep marks are apparent, then the pressure plate will have to be renewed. If the flywheel is badly scored then it must be dismantled (as described in Chapter 1) and taken to your GM dealer or auto engineering specialist for further inspection. It may prove possible to machine across the face to resurface it, without upsetting the balance of the engine and flywheel. If a new flywheel is the only answer, then it must be balanced to match the original.
3 Check the driven plate (friction disc) linings for wear, loose rivets or damaged torsion springs. Also check the driven plate hub to input shaft splines, which must be an easy sliding fit. The driven plate must be renewed if the lining surface is worn to within 0.008 in (0.203 mm) of the rivet head. If the lining surfaces show signs of breaking up or black areas where oil contamination has occurred, it should also be renewed. If facilities are readily available for obtaining and fitting new friction linings to the existing disc this may be done but the saving is relatively small compared with obtaining a complete new disc assembly which ensures that the shock absorbing springs and the splined hub are renewed also. The same applies to the pressure plate assembly which cannot be readily dismantled and put back together without specialised riveting tools and balancing equipment. An allowance is usually given for exchange units.
4 If on inspection oil is present on the driven plate, then check the pilot bush, the transmission input shaft seal and the engine rear main bearing oil seal. Also inspect the front bearing support gaskets and renew defective items as necessary before reassembling the clutch unit.

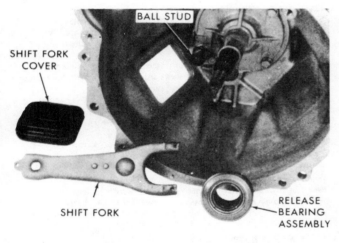

Fig. 5.8. The clutch release bearing and shift fork

10.3 The shift fork showing spring clip in position over the ball stud

10.6 Check the clutch release (throwout) bearing for excessive wear or damage

10.9 Refit the shift fork

11.5 Centralizing the driven plate using an old input shaft

10 Clutch release bearing - removal, inspection and reassembly

1 To inspect or renew the clutch release bearing it is necessary to remove the engine and/or transmission unit from the vehicle as described in Chapters 1 and 6.
2 Detach the shift fork cover from the transmission casing.
3 The shift fork spring clip is now disconnected from the ball stud (photo) and the shift fork and release bearing removed. If required the shift fork ball stud can be unscrewed and removed from its location in the front cover.
4 Clean the respective parts with a clean rag - do not wash the release bearing in gasoline, oil or any other cleaning solvent.
5 Check the clutch fork ball socket and fingers for signs of wear and inspect the retaining spring for any damage. The spring must be able to retain the fork to the ball stud firmly.
6 To check the clutch release bearing, spin it and listen for noisy operation or if it vibrates, renew it. Renew any other suspect clutch release fork components and reassemble as follows.
7 Screw the shift fork ball stud into position in the front cover and tighten to a torque figure of 30 lb f ft (4.147 kg f m).
8 Lubricate the inner collar groove of the shift block and the shift fork ball seat and fingers with graphite grease and refit the shift fork and clutch release bearing unit together with the shift fork spring clip. Ensure that the fork is fully engaged with the ball stud and that the release bearing can operate smoothly.
9 Finally refit the shift fork (photo) cover to the transmission casing.

11 Clutch - installation

1 Clean the flywheel and pressure plate surfaces.
2 Apply a little Lubriplate or a similar lubricant to the splines of the driven plate, and also smear the pilot bearing in the center of the flywheel with a high melting point grease.
3 Locate the driven plate against the flywheel.
4 Position the pressure plate assembly on the flywheel and ensure that the relative marks are in alignment.
5 Insert the retaining bolts, but do not tighten. The disc must now be centralized on the flywheel and this can be best achieved using the transmission input shaft (if the transmission has also been dismantled) as alignment tool. An old input shaft, if available, would also align the disc (photo). Alternatively a suitable round shaft of wood or steel may be utilized.
6 When the disc is centralized, tighten the respective bolts evenly in a diametrically opposed manner to a specified torque of 13 lb f ft (1.79 kg f m).
7 Remove the alignment shaft, and check that the clutch release lever operates freely before installation of the transmission/engine.

12 Fault diagnosis - Clutch

Symptom	Reason/s
Judder when taking up drive	Loose engine transmission mountings. Badly worn friction surfaces or contaminated with oil. Worn splines on transmission input shaft or driven plate hub. Worn input shaft spigot bush (pilot bearing) in flywheel.
Clutch spin (failure to disengage) so that gears cannot be meshed	Incorrect release bearing to diaphragm spring finger clearance. Driven plate sticking on input shaft splines due to rust. May occur after vehicle standing idle for long period. Damaged or misaligned pressure plate assembly.
Clutch slip (increase in engine speed does not result in increase in vehicle road speed - particularly on gradients)	Incorrect release bearing to diaphragm spring finger clearance. Friction surfaces worn out or oil contaminated.
Noise evident on depressing clutch pedal	Dry, worn or damaged release bearing. Insufficient pedal free travel. Weak or broken pedal return spring. Weak or broken clutch release lever return spring. Excessive play between driven plate hub splines and input shaft splines.
Noise evident as clutch pedal released	Distorted driven plate. Broken or weak driven plate cushion coil springs. Insufficient pedal free travel. Weak or broken clutch pedal return spring. Weak or broken release lever return spring. Distorted or worn input shaft. Release bearing loose on retainer hub.

Chapter 6 Part A : Manual transmission

For information relating to Series 8 thru 12 models, including 4WD, refer to Chapter 13.

Contents

Clutch gear shaft unit - dismantling ... 5	Transmission - Series 5 and 6 - dismantling ... 4
General description ... 1	Transmission - Series 1 to 4 - reassembly ... 8
Mainshaft dismantling ... 6	Transmission - Series 5 and 6 - reassembly ... 9
Transmission - cleaning and inspection ... 7	Transmission - Series 1 to 6 - removal and installation ... 2
Transmission - Series 1 to 4 models - dismantling ... 3	Fault diagnosis ... 10

Specifications

Series 1 to 4

Type	4 forward and 1 reverse gear with synchromesh on all forward gears
Model	MSE
Control system	Direct with floor mounted shift lever

Gear ratios

1st	3.51 : 1
2nd	2.18 : 1
3rd	1.42 : 1
4th	1.00 : 1
Reverse	3.93 : 1

Speedometer gear ratios

Series 1 and 2	17.5
Series 3 and 4	5.18

Lubricant capacity	1.3 qt
Lubricant type	SAE 30
Synchromesh blocker ring to gear face maximum clearance allowance	0.040 to 0.080 in (1.01 to 2.03 mm)

Series 5 and 6

Type	4 forward and 1 reverse gear with synchromesh on all forward gears
Model	MSG
Control system	Direct with floor mounted shift lever

Gear ratios

1st	3.51 : 1
2nd	2.18 : 1
3rd	1.42 : 1
4th	1.00 : 1
Reverse	3.83 : 1

Speedometer gear ratio	6.20

Lubricant type

Below 50°F (10.0°C)	SAE 10W - 30
Above 50°F (10.0°C)	SAE 40
Between 0°F and 90°F (17.8°C and 32.2°C)	SAE 30

Chapter 6 Part A: Manual transmission

Lubricant capacity ... 1.35 qt

Dimensions and backlash specifications
Reverse idle shaft gear to bush clearance - maximum allowance ... 0.006 in (0.152 mm)
Reverse idler shaft standard outside diameter
    Series 5 ... 0.709 in (18 mm)
    Series 6 ... 0.866 in (22 mm)
Maximum allowance for 1st gear width to collar width ... 0.019 in (0.5 mm)
Mainshaft run out limit ... 0.002 in (0.05 mm)
Synchromesh blocker ring to gear clearance allowance ... 0.032 in (0.8 mm)
Synchromesh hub spline to mainshaft spline clearance allowance ... 0.008 in (0.20 mm)

Minimum shift arm thickness
3rd/4th ... 0.256 in (6.5 mm)
1st/2nd and reverse ... 0.276 in (7.1 mm)
Shift rod detent spring length - minimum allowance ... 1.084 in (27.5 mm)

Standard gear backlash
1st ... 0.0039 to 0.0071 in (0.07 to 0.17 mm)
2nd ... 0.0039 to 0.0067 in (0.07 to 0.15 mm)
3rd ... 0.0039 to 0.0067 in (0.07 to 0.15 mm)
4th ... 0.0039 to 0.0067 in (0.07 to 0.15 mm)
Reverse gear - main ... 0.0039 to 0.0071 in (0.07 to 0.17 mm)
Reverse idle gear ... 0.0039 to 0.0071 in (0.07 to 0.17 mm)
Maximum limit allowance ... 0.016 in (0.40 mm)

Inside diameter for gears
1st ... 1.773 in (45.03 mm)
2nd ... 1.615 in (41.02 mm)
3rd ... 1.615 in (41.02 mm)
Limit allowance ... 0.004 in (0.101 mm)

Torque wrench settings

Series 1 to 4
	lb f ft	kg f m
Mainshaft rear bearing nut	80	11.0
Rear extension retaining bolts	10	1.3
Top cover bolts	36 - 48	4.9 - 6.6
Front cover bolts	10	1.3
Clutch shift fork ball stud	65 - 70	8.9 - 9.6

Series 5 and 6
	lb f ft	kg f m
Idler gear shaft lock plate bolt	14	1.9
Countershaft reverse gear nut	100	13.8
Mainshaft locknut	94	12.9
Rear extension bolts	27	3.7
Shifter cover bolts - Series 5	10	1.3
Shifter cover bolts - Series 6	14	1.9
Front bearing retainer bolts	14	1.9
Clutch fork ball stud	30	4.1

1 General description

The transmission fitted to all Series 1 to 4 models is a four forward speed and one reverse type with synchromesh on all forward gears. Gear selection is via a floor mounted shift lever.

The aluminium transmission casing incorporates the clutch housing at the front and there is a detachable rear cover, the top part of which is a quadrant box containing the transmission control mechanism.

All forward gears are helically cut to ensure quiet operation. The reverse gear and reverse idler gear are straight cut spur gears.

The Series 5 and 6 models gearbox is similar in construction except that the mainshaft and counter gear shafts are located in a detachable centre support.

The overhaul procedures differ on the Series 5 and 6 models from those of the earlier types and reference should therefore be made to the relevant Section when major work is undertaken.

Automatic transmission is also available on the Series 5 and 6 models and for information on this refer to Part B of this Chapter.

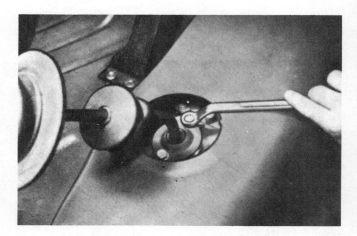

Fig. 6.1. Undo the gearlever retaining bolts

Chapter 6 Part A: Manual transmission

2 Transmission - Series 1 to 6 - removal and installation

To remove the engine and transmission units complete refer to Chapter 1, Section 6.
1 Detach the ground cable from the battery.
2 Remove the air cleaner unit from the carburetor and detach the accelerator linkage from the carburetor throttle lever.
3 Remove the starter motor attachment bolts and connecting cable. Withdraw the starter motor.
4 From inside the cab, lift the gearshift lever boot up the lever and with a suitable wrench, unscrew and remove the gearshift lever attachment bolts. Lift the lever clear.
5 Unless vehicle is located over a work pit, jack up and block up sufficiently to enable work to be carried out underneath the vehicle.
6 Unscrew and remove the exhaust pipe to transmission hanger.
7 Detach the speedometer cable from the transmission unit (photo).
8 Refer to Chapter 7 and remove the propeller shaft.
9 Drain the oil from the transmission, and plug the rear extension to prevent spillage of oil during removal.
10 Unscrew the clutch slave cylinder securing bolts and detach the slave cylinder. Tie it to the side frame with a piece of wire or cord to keep it out of the way.
11 Disconnect the flywheel stone shield and undo the three frame bracket to transmission rear mounting bolts.
12 Place jacks under the engine and transmission and raise the units sufficiently to undo and remove the crossmember to frame bracket bolts.
13 Lower the engine and gearbox and support the rear of the engine with blocks or jack.
14 From the top of the gearbox, detach the TCS switch (photo), and reverse lamp switch where fitted. On California models disconnect the neutral switch wires on Series 5 and 6 models.
15 Support the transmission and remove the engine bellhousing bolts. Withdraw the gearbox from the engine by pulling to the rear and when clear of the clutch, tilt the gearbox downwards and remove.
16 Installation of the transmission unit is direct reversal of removal but note the following.

a) When the slave cylinder and pushrod are reconnected adjust the shift fork as given in Chapter 5.
b) Adjust the clutch pedal height - see Chapter 5
c) Refit the propeller shaft and align the rear flange marks as described in Chapter 7.
d) Tighten all retaining bolts/nuts to the specified torque where given.
e) Refit the drain plug and refill the transmission with new oil of the correct grade and quantity.

3 Transmission - Series 1 to 4 - dismantling

1 Having removed the transmission unit from the vehicle, clean the exterior housing and fitting and drain any remaining oil. Remove to clean work area.
2 From the clutchhousing remove the clutch fork cover and disconnect the clutch fork and release bearing (Fig. 6.2).
3 Unscrew and remove the front bearing retainer bolts and withdraw the retainer (photo).
4 Unscrew and remove the eight bolts securing the gearbox top cover. Prise the top cover from the gearbox and note the gasket and baffler plate (photo).
5 Extract the three detent springs and balls.
6 Remove the four retaining bolts from the gear shift quadrant and remove the quadrant (photo) and gasket.
7 Remove the TCS and reversing light switches if still fitted, taking care not to lose the reverse light switch detent ball and actuating pin.
8 Refer to photo and drive out the retaining pin from the shift control fulcrum and remove the fulcrum bracket, reverse idler gear control lever and shift block (Fig. 6.5).
9 Remove the selector rod shift fork roll pins (photo).
10 Withdraw the reverse shifter shaft from the front and lift out the shift fork. Do not lose the detent balls at the front of the casing when removing the shifter shafts.
11 From the rear extension disconnect the speedometer adaptor

2.7 Unscrew the speedo cable connection

2.14 Remove the TCS switch from the top of the gearbox

3.3 Withdraw the bearing retainer

3.4 Remove the top cover

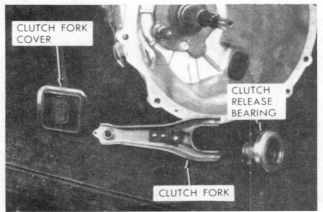
Fig. 6.2. Remove the clutch fork, cover and release bearing

3.6 Remove the gearshift quadrant

3.8 Remove the shift control fulcrum roll pin

3.9 The selector rod shift forks and locating roll pins

3.11 Remove the speedo motor adaptor from the rear extension

3.13 Remove the lockplate

3.14 Use a dummy shaft to remove the counter gearshaft

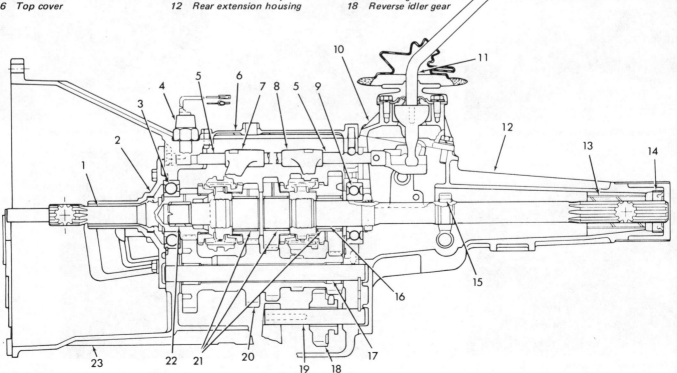

Fig. 6.3. Cross section view of the four speed manual transmission - Series 1 to 4 models

1 Clutch drive gear
2 Front bearing retainer
3 Front bearing
4 TCS switch
5 Shift rails
6 Top cover
7 3-4 shift fork
8 1-2 shift fork
9 Rear bearing
10 Shift quadrant
11 Shift lever assembly
12 Rear extension housing
13 Extension housing bushing
14 Extension seal
15 Speedometer drive gear
16 Mainshaft
17 Countergear shaft
18 Reverse idler gear
19 Reverse idler gear shaft
20 Countergear
21 Caged roller bearings
22 Clutch gear pilot bearing
23 Transmission case

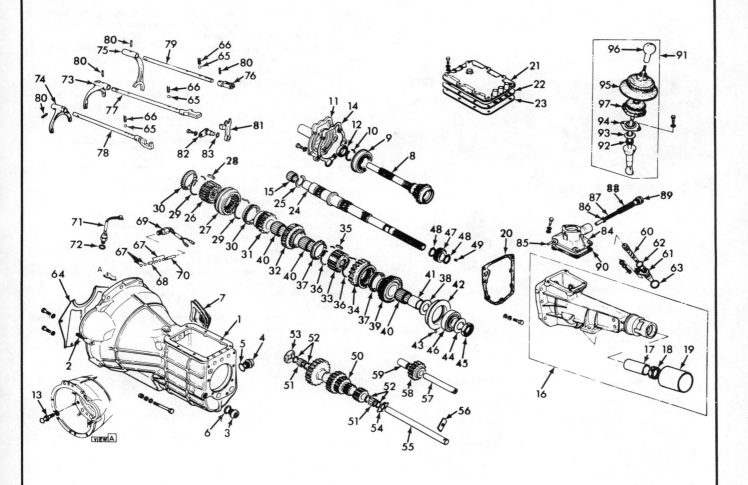

Fig. 6.4. The gearbox component parts - Series 1 to 4 models

1 Transmission case
2 Stud; starter to case
3 Plug; oil filler
4 Plug; magnetic
5 O-ring; drain plug to case
6 Packing; filler plug
7 Cover; clutch shift fork
8 Shaft; clutch gear
9 Bearing; front
10 Snap ring; bearing top gear
11 Cover assembly; front
12 Oil seal; front cover
13 Support; clutch fork
14 Gasket; front cover
15 Bearing; main shaft
16 Extension housing assembly
17 Bushing; prop shaft yoke
18 Oil seal; rear cover
19 Cover; dust rear cover
20 Gasket, rear cover
21 Cover assembly
22 Gasket, top cover
23 Plate
24 Main shaft
25 Snap ring; main shaft
26 Hub; 3-4
27 Sleeve; 3-4
28 Insert; clutch hub
29 Spring; insert
30 Blocker ring; 3-4
31 3rd gear assembly
32 2nd gear assembly
33 Hub; 1-2

34 Reverse gear
35 Insert; clutch hub
36 Spring; insert
37 Blocker ring; 1-2
38 Thrust washer
39 Low gear
40 Bearing; low, 2nd, 3rd
41 Collar, low
42 Adapter assembly; mainshaft
43 Pin; adapter
44 Spacer
45 Nut; main shaft
46 Bearing; main shaft
47 Speedometer drive gear
48 Snap ring; speedometer gear
49 Key; speedometer gear
50 Counter gear
51 Needle roller
52 Spacer; counter gear
53 Thrust washer
54 Thrust washer
55 Counter gear shaft
56 Lock plate
57 Shaft; reverse idle
58 Gear; reverse idle
59 Bushing; reverse idle gear
60 Speedometer driven gear
61 W/O-ring bushing
62 O-ring; speedometer gear bush
63 O-ring; speedometer gear
64 Cover; front transmission case
65 71 ball, detent, gear shift

66 Spring, detent ball, gear
67 Ball, inner lock, gear shift
68 Pin, lock, inner lock
69 Switch assembly, reverse lamp
70 Plunger, reverse lamp switch
71 Switch assembly, top-3rd
72 Gasket, switch
73 Arm, shift, top-3rd
74 Arm, shift low-2nd
75 Arm, shift, reverse
76 Block, shift w/pin, reverse
77 Rod, gear shift, top-3rd
78 Rod, gear shift, low-2nd
79 Rod, gear shift, reverse
80 Pin, spring, shift arm
81 Lever, reverse, idle gear control
82 Bracket, fulcrum, lever control
83 O-ring, fulcrum bracket
84 Box, quadrant
85 Pin, quadrant box
86 Plunger, reverse stop
87 Spring, reverse stop, inner
88 Spring, reverse stop, outer
89 Cap, reverse stop, spring
90 Gasket, quadrant box
91 Lever assembly, gear shift control
92 Spring, control lever
93 Cage, control lever
94 Cover, control lever
95 Grommet, lever, gear shift
96 Knob, lever, gear shift control
97 Cover, dust, control lever

Chapter 6 Part A: Manual transmission

(photo) then undo the retaining bolts and remove the rear housing.
12 The 3rd/4th and 1st/2nd shifter shafts can now be withdrawn thru the rear of the gearcase, and the shift forks extracted.
13 Detach the backplate from its location between the reverse idler shaft and the countershaft on the rear of the gearcase (photo).
14 Drive the countershaft thru the rear of the gearcase using a dummy shaft (photo).
15 The mainshaft assembly can now be removed from the rear (photo).
16 Withdraw the clutch shaft (primary shaft) assembly from the front of the gearcase (photo) taking care not to lose any of the needle roller bearings from the mainshaft aperture.
17 Lift the countergear and reverse collar out of the gearcase.

4 Transmission - Series 5 and 6 - dismantling

1 Drain any remaining oil from the casing.
2 Detach the clutch fork, rubber boot and throwout bearing.
3 Undo and remove the bearing retainer bolts and withdraw the retainer gasket and spring washer (Fig. 6.9).
4 Unscrew the speedometer gear bush retaining bolt and withdraw the speedometer driven gear assembly.
5 Undo the retaining bolts and remove the gear shift cover and gasket.
6 Remove the reverse light switch and CRS switches.
7 Remove the rear extension retaining bolts and detach the rear extension and gasket.
8 Remove the reverse idler gear and thrust washers from the gear shaft (Fig. 6.10).

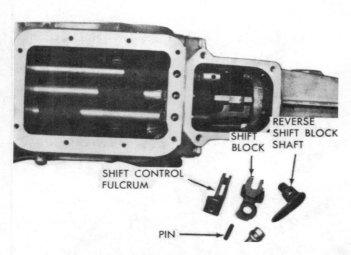

Fig. 6.5. Remove the shift blocks and control fulcrum

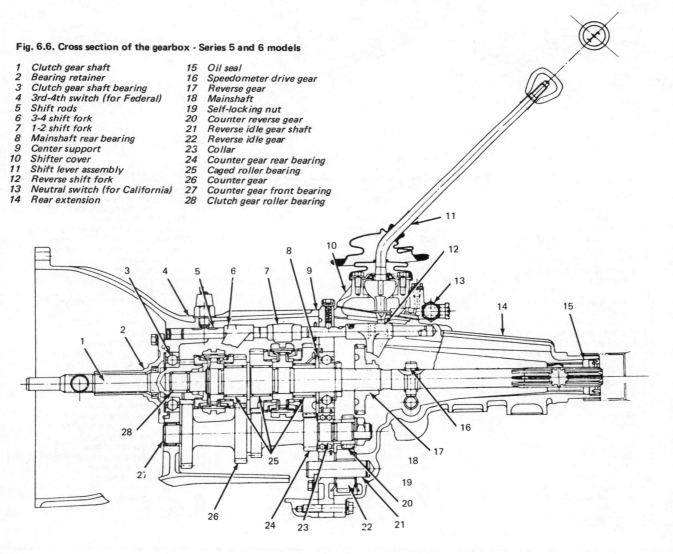

Fig. 6.6. Cross section of the gearbox - Series 5 and 6 models

1 Clutch gear shaft
2 Bearing retainer
3 Clutch gear shaft bearing
4 3rd-4th switch (for Federal)
5 Shift rods
6 3-4 shift fork
7 1-2 shift fork
8 Mainshaft rear bearing
9 Center support
10 Shifter cover
11 Shift lever assembly
12 Reverse shift fork
13 Neutral switch (for California)
14 Rear extension
15 Oil seal
16 Speedometer drive gear
17 Reverse gear
18 Mainshaft
19 Self-locking nut
20 Counter reverse gear
21 Reverse idle gear shaft
22 Reverse idle gear
23 Collar
24 Counter gear rear bearing
25 Caged roller bearing
26 Counter gear
27 Counter gear front bearing
28 Clutch gear roller bearing

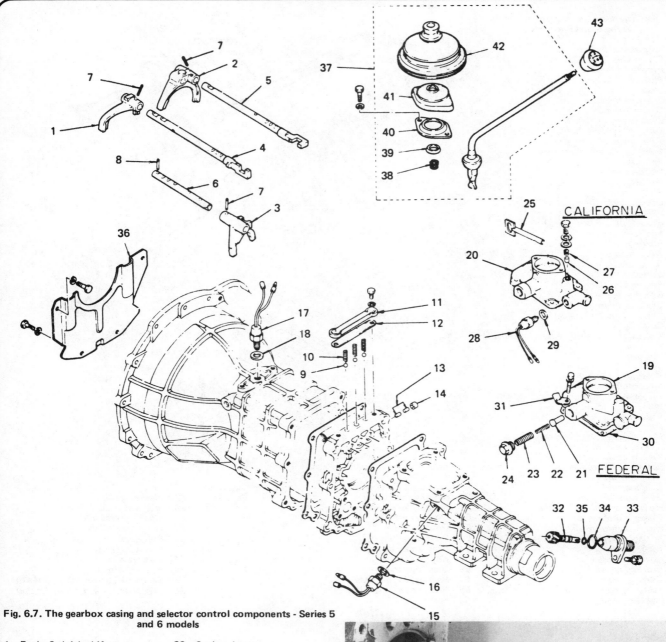

Fig. 6.7. The gearbox casing and selector control components - Series 5 and 6 models

1 Fork, 3rd-4th shifter
2 Fork, 1st-2nd shifter
3 Fork, reverse shifter
4 Rod, gear shift, 3rd-4th
5 Rod, gear shift, 1st-2nd
6 Rod, gear shift, reverse
7 Pin, spring, shift fork
8 Pin, spring, reverse rod
9 Ball, detent, gear shift
10 Spring, detent ball
11 Plate, detent spring
12 Gasket
13 Pin, interlock
14 Plug, interlock
15 Switch assembly, reverse lamp
16 Gasket
17 Switch assembly, 3rd-4th (Federal spec.)
18 Gasket (Federal spec.)
19 Shifter cover (Federal spec.)
20 Shifter cover (California spec.)
21 Plunger, reverse, stop
22 Spring, inner
23 Spring, outer
24 Cap, reverse, stop
25 Rod, neutral switch
26 Damper pad
27 Spring
28 Switch assembly, neutral
29 Gasket
30 Gasket, shifter cover
31 Bracket
32 Gear, speed, driven
33 Bush, W/O-ring
34 O-ring, bush outer
35 O-ring bush inner
36 Cover, front, transmission case
37 Lever assembly, gear shift control
38 Spring, control lever
39 Seat, spring
40 Cover, control lever
41 Cover, dust, control lever
42 Grommet, lever, gear shift
43 Knob, control lever

3.15 Remove the mainshaft assembly

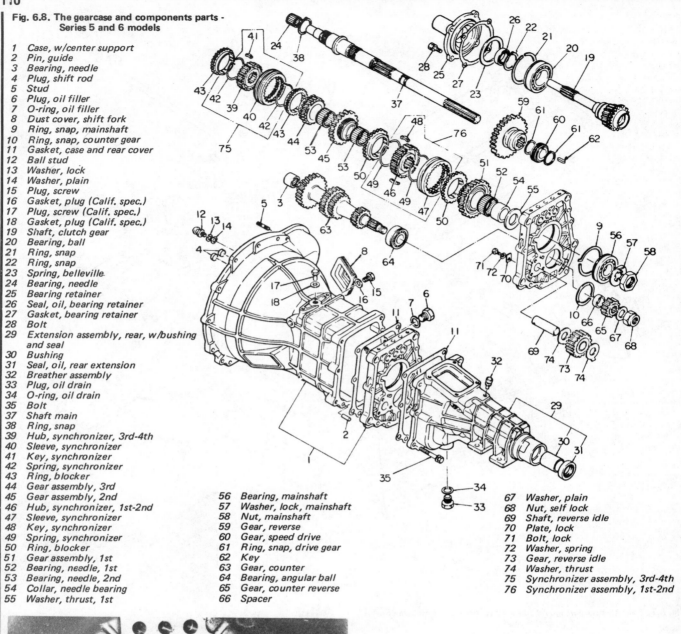

Fig. 6.8. The gearcase and components parts - Series 5 and 6 models

1 Case, w/center support
2 Pin, guide
3 Bearing, needle
4 Plug, shift rod
5 Stud
6 Plug, oil filler
7 O-ring, oil filler
8 Dust cover, shift fork
9 Ring, snap, mainshaft
10 Ring, snap, counter gear
11 Gasket, case and rear cover
12 Ball stud
13 Washer, lock
14 Washer, plain
15 Plug, screw
16 Gasket, plug (Calif. spec.)
17 Plug, screw (Calif. spec.)
18 Gasket, plug (Calif. spec.)
19 Shaft, clutch gear
20 Bearing, ball
21 Ring, snap
22 Ring, snap
23 Spring, belleville
24 Bearing, needle
25 Bearing retainer
26 Seal, oil, bearing retainer
27 Gasket, bearing retainer
28 Bolt
29 Extension assembly, rear, w/bushing and seal
30 Bushing
31 Seal, oil, rear extension
32 Breather assembly
33 Plug, oil drain
34 O-ring, oil drain
35 Bolt
37 Shaft main
38 Ring, snap
39 Hub, synchronizer, 3rd-4th
40 Sleeve, synchronizer
41 Key, synchronizer
42 Spring, synchronizer
43 Ring, blocker
44 Gear assembly, 3rd
45 Gear assembly, 2nd
46 Hub, synchronizer, 1st-2nd
47 Sleeve, synchronizer
48 Key, synchronizer
49 Spring, synchronizer
50 Ring, blocker
51 Gear assembly, 1st
52 Bearing, needle, 1st
53 Bearing, needle, 2nd
54 Collar, needle bearing
55 Washer, thrust, 1st
56 Bearing, mainshaft
57 Washer, lock, mainshaft
58 Nut, mainshaft
59 Gear, reverse
60 Gear, speed drive
61 Ring, snap, drive gear
62 Key
63 Gear, counter
64 Bearing, angular ball
65 Gear, counter reverse
66 Spacer
67 Washer, plain
68 Nut, self lock
69 Shaft, reverse idle
70 Plate, lock
71 Bolt, lock
72 Washer, spring
73 Gear, reverse idle
74 Washer, thrust
75 Synchronizer assembly, 3rd-4th
76 Synchronizer assembly, 1st-2nd

3.16 Withdraw the clutch shaft

Fig. 6.9. Withdraw the bearing container

9 From the mainshaft, unclip the snap-rings, and remove the speedometer drive gear and key.
10 Drive the spring pin from the reverse shifter fork using a suitable drift and, remove the shift fork and reverse gear (Fig. 6.11).
11 Using a pair of snap-ring pliers (Fig. 6.12) remove the snap-ring retaining the clutch gear shaft ball bearing in place.
12 Withdraw the center support unit from the gearcase as shown in Fig. 6.13.
13 Support the center support unit and drive the spring pins from the 1st/2nd, and 3rd/4th shifter forks. Support the shift rods during removal to prevent the rods bending and to ease removal of the pins.
14 Undo the detent spring retaining plate bolts and remove the plate and gasket. Extract the detent springs and balls.
15 Now remove the respective shifter rods, taking care not to lose detent interlock plugs in the rods in the center support. The reverse shifter rod is fitted with a stop pin and must therefore be removed forwards.
16 To prevent the mainshaft rotating, move the synchronizers to the rear. If they are reluctant to move, tap them with a hammer handle as shown in Fig. 6.15.
17 Flatten the mainshaft nut lockwasher tab, and undo the locknut. Remove nut and washer.
18 Undo the locknut from the countershaft and remove the washer, reverse gear and collar (Fig. 6.16).
19 Disengage the snap-ring from the center support using a pair of snap-ring pliers as shown in Fig. 6.17. Tap the front of the center support while expanding the snap-ring.
20 Expand the mainshaft retaining snap-ring and remove the center support.

Fig. 6.10. Remove the reverse idler gear

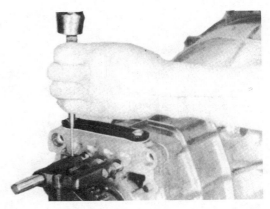

Fig. 6.11. Drift out the spring pin

5 Clutch gear shaft unit - dismantling

1 To remove the front bearing from the clutch gear shaft, extract the snap-rings from the shaft, retaining groove and the bearing outer race groove.
2 Remove the bearing from the shaft using a suitable puller.

6 Mainshaft dismantling

Series 1 to 4

1 Using snap-ring pliers, extract the speedometer drive gear snap-rings and withdraw the speedometer gear and drive key from the shaft.
2 Undo the rear bearing retaining nut and remove with lockwasher. Remove the rear bearing retainer.
3 Withdraw 1st gear thrust washer, caged roller bearing sleeve and blocker ring.
4 Withdraw the 1st/2nd synchronizer hub and gear unit.
5 Withdraw 2nd speed gear, roller bearing and blocker ring.
6 Disengage the 3rd/4th synchronizer hub snap-ring and withdraw the synchronizer unit, blocker ring and roller bearing.
7 Clean and lay the respective parts out ready for inspection.

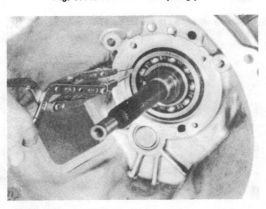

Fig. 6.12. Removing the bearing snap ring

Fig. 6.13. Withdraw the center support

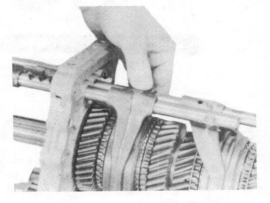

Fig. 6.14. Remove the shifter rods

Fig. 6.15. Move the synchronizers to the rear

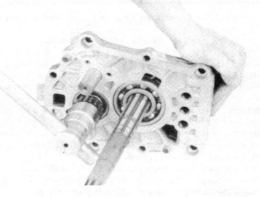

Fig. 6.16. Remove the counter gear self-locking nut

Fig. 6.17. Using snap ring pliers to expand the snap ring for removal of the center support from mainshaft and counter gear

Fig. 6.18 Remove the rear bearing retainer (Series 1 to 4)

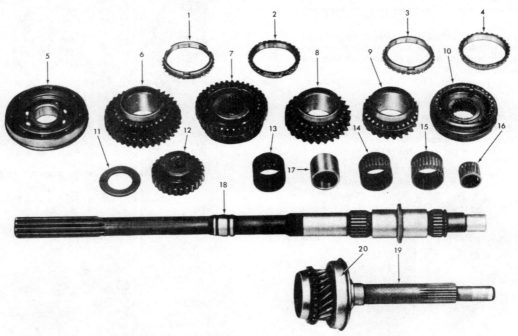

Fig. 6.19. The mainshaft components

1 Blocker ring	6 1st gear	11 1st gear thrust washer	16 Clutch gear pilot roller
2 Blocker ring	7 1st and 2nd synchronizer hub	12 Reverse idler gear	17 Sleeve
3 Blocker ring	8 2nd gear	13 Caged roller bearings	18 Mainshaft
4 Blocker ring	9 3rd gear	14 Caged roller bearings	19 Clutch gear
5 Rear bearing and retainer	10 3rd and 4th synchronizer hub	15 Caged roller bearings	20 Front bearing

Chapter 6 Part A: Manual transmission

Fig. 6.20 Removing the mainshaft rear bearing using special tool J-22912 (Series 5 and 6)

Fig. 6.21. Remove 1st speed gear unit

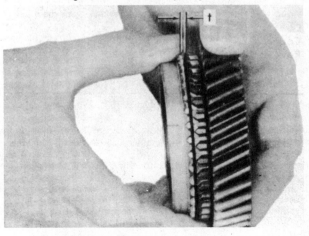

Fig. 6.22. Measure the blocker ring to gear clearance

Series 5 and 6

8 Remove the mainshaft rear ball bearing using a suitable puller or press. If special tool J - 22912 is available, this should be used together with an arbor press to remove the bearing as shown in Fig. 6.20. NB. This tool can also be used to remove the countershaft bearing.
9 Withdraw the 1st speed gear assembly complete with thrust washer, needler roller bearings, collar and blocker ring as in Fig. 6.21.
10 Withdraw the 1st/2nd synchronizer assembly.
11 Withdraw 2nd gear with blocker ring and needle roller bearing.
12 Disengage and remove the snap-ring, then withdraw 3rd/4th synchronizer assembly and blocker ring.
13 Withdraw 3rd speed gear from the shaft with needle roller bearings.

7 Transmission - cleaning and inspection

Cleaning

1 All parts, except seals and ball bearing assemblies, should be soaked in gasoline (petrol) or kerosene (paraffin), brushed or scraped as necessary, and dried with compressed air.
2 Ball bearing assemblies may be carefully dipped into clean gasoline or kerosene, and spun with the fingers. Whilst being prevented from turning, they should be dried with compressed air. After inspection (paragraphs 12 thru 14) they should be lubricated with SAE 90EP transmission oil and stored carefully in clean conditions until ready for use.

Inspection - general

3 Inspect the transmission case and extension housing for cracks, worn or damaged bearings and bores, and damaged threads and machined surfaces. Small nicks and burrs can be locally dressed out using a fine file.
4 Examine the shift levers, forks, shift rods and associated parts for wear and damage.
5 Renew roller bearings which are chipped, corroded or rough running.
6 Examine the countershaft (layshaft) for damage and wear; renew it if this is evident.
7 Examine the reverse idler and sliding gears for damage and wear. Renew them if this is evident.
8 Renew the input shaft and gear if damaged or worn. If the roller bearing surface in the counterbore is damaged or worn, or the cone surface is damaged, renew the gear and gear rollers.
9 Examine all the gears for wear and damage, renewing as necessary.
10 Renew the mainshaft if bent, or if the splines are damaged.
11 Renew the seal in the transmission front bearing cover.

Ball bearing assemblies - inspection

12 Examine the inner and outer raceways for pitting and corrosion, renewing any bearings where this is found.
13 Examine the ball cage and races for signs of cracking, renewing any bearings where this is found.
14 Lubricate the raceways with a small quantity of SAE 90EP transmission oil then rotate the outer race slowly until the balls are lubricated. Spin the bearing by hand in various attitudes, checking for roughness. If any is found after the bearings have been cleaned (paragraph 2) they should be renewed. If they are satisfactory they should be stored carefully whilst awaiting assembly into the transmission.

Synchronizer mechanism

15 Inspect the clutch hub, sliding sleeve, blocker rings and inserts for signs of excessive wear or damage - renew as required.
16 Check the hub taper, the gear teeth and inspect fitting grooves of the blocker rings for excessive wear or damage and renew as required. To inspect the tapered portions locate the blocker ring against the conical section of the gear and, using a feeler gauge, measure the clearance as shown in Fig. 6.22 and renew the blocker ring if the clearance exceeds that specified.
17 Measure the clearance of the clutch hub splines and mainshaft splines and renew if the clearance exceeds that specified.

Mainshaft run-out

18 Place the mainshaft on two 'V' blocks and check for run-out using a dial test indicator. If the run-out exceeds that specified, renew the mainshaft.

Chapter 6 Part A: Manual transmission

8.2 Fit the 3rd gear and roller bearing to the mainshaft ...

8.3 ... followed by the synchronizer hub unit ...

8.4 ... and retain with snap ring

8.5 Locate the needle roller bearing and 2nd gear

8.6 Fit the 1st/2nd synchronizer hub and reverse sliding gear

8.7 Fit the needle roller bearing over the 1st gear sleeve

Reverse idler gear

19 Check the reverse idler gear for wear, damage and the amount of clearance between the gear and bushings. Renew the bushing if clearance is beyond that specified.

8 Transmission - Series 1 to 4 - reassembly

Mainshaft assembly

1 Before reassembling the various components of the transmission ensure that all items are clean and old gaskets and sealant solution are removed from mating surfaces. Lubricate where specified with the recommended transmission oil.
2 Commence reassembly by lubricating the 3rd gear bearing journal on the mainshaft, then fit the caged roller bearing and 3rd gear (photo).
3 Refit the blocker ring and the 3rd/4th synchronizer hub assembly (photo) - the synchronizer hub chamfer faces forwards. Ensure that the blocker rings are assembled with their notches aligned with the synchronizer keys.
4 Retain the gear and synchronizer hub with the snap-ring (photo).
5 Lubricate the 2nd speed journal on the mainshaft and slide the needle roller bearings into position followed by 2nd gear (photo).
6 Fit the blocker ring and 1st/2nd synchronizer hub (see Fig. 6.24) and reverse sliding gear (photo).
7 Lubricate the 1st speed journal and install the needle roller bearing, over the sleeve as shown in the photo.
8 Fit the blocker ring and 1st gear into position (photo).
9 Fit the thrust washer against the 1st gear with its oil groove towards 1st gear.
10 If the rear bearing has been removed from the retainer, this should now be reinstalled with the sealed side towards the front of the retainer (photo).
11 Assemble the bearing and retainer unit to the rear end of the main shaft. Fit the lockwasher and nut (photo) and torque the nut to 80 lb f ft (11 kg f m). Bend over the lock tab on the washer over a flat of the nut to retain it in position.
12 Refit the speedometer drive gear and locate with snap-rings (photo).

13 Install the reverse idler into the casing with the shift fork groove to the rear (photo), and the shaft slot facing the countershaft hole.

Countergear shaft assembly

14 Insert the inner bearing thrust washers into the front and rear of the countergear shaft assembly.
15 Lubricate the needle roller bearings with grease and locate them into the respective housings followed by the outer thrust washers (photo). There are 46 roller bearings in all, 23 to fit in each end.
16 Position the countershaft thrust washers into position in the gear-case (photo) and install the countergear into position in the gearcase with the large diameter gear to the front of the case (photo). Locate temporarily with dummy shaft.

Clutch gear shaft assembly

17 If it has been removed, press the bearing onto the clutch gear and retain with snap-ring (photo), and fit the locating ring to the outer bearing race.
18 Fit the clutch gear shaft unit into position in the front of the gear-case. When locating the bearing into the housing ensure that the gear does not foul the housing wall when pressing or drifting into position.
19 Smear the clutch gear pilot roller bearing with grease and insert into position (photo).
20 Fit the blocker ring to the clutch gear cone.

Transmission final assembly

21 Install the mainshaft assembly into position in the gearcase (photo). Locate the lug on the rear bearing retainer into the recess in the rear face of the gearcase (photo).
22 Rotate the mainshaft assembly to position the countergear for the insertion of the countershaft.
23 Lightly lubricate the countershaft and check that the thrust washers are correctly located. Install the countershaft thru the rear of the casing and tap carefully into position; extract the dummy shaft. The slot in the end of the countershaft is to face the reverse idler shaft (photo).
24 Position the reverse idler gear shaft and countergear shaft and

8.8 Fit the blocker ring and 1st gear

8.10 Install the bearing with the sealed side to the front of the retainer

8.11a Fit the lockwasher and nut

8.11b The assembled mainshaft unit

8.12 Locate the speedometer drivegear with snap rings

8.13 The reverse idler positioned in the gearcase

8.15 The countershaft gear needle bearings and washers in position

8.16a The countershaft thrust washer in position

8.16b Insert the countershaft into the gearcase

8.17 Retain the clutch shaft bearing with snap ring

8.19 Insert the clutch shaft gear pilot roller bearing and fit the blocker ring to the cone

8.21a Install the mainshaft assembly

8.21b The locating lug in the rear face recess

8.23 Fit the countershaft with the slot towards the reverse idler shaft

8.25 The shifter forks located

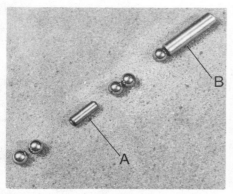

8.27 The front detent balls, pin and plunger positions:
A = Inner lock pin   B = Plunger

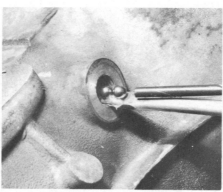

8.31 Installing the detent balls

8.32 Fit the shifter fork to shaft roll pins

8.34 Fit the reverse lampswitch

8.35 Drift the roll pin into position thru the reverse shift block and shaft

8.37a Fit the rear detent balls ...

8.37b ... and springs into position

8.38 Fit the top cover with gaskets and baffle plate

Chapter 6 Part A: Manual transmission 117

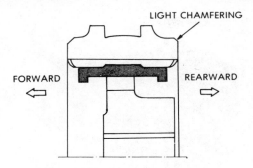

Fig. 6.23. 3rd/4th synchronizer hub directional location

locate the lock plate into position in the key slots of the respective shafts. Carefully tap the lock plate flush to the rear of the case.
25 Position the shift forks into their respective 1/2nd, 3/4th synchronizer sleeve grooves, and the reverse gear shift fork into its groove on the reverse idler gear (photo).
26 Drive the interlocking pin into the 1st/2nd shifter shaft and insert the shaft into the gearcase thru the rear and locate with the 1st/2nd shifter fork.
27 Install two detent balls into the front shifter shaft bosses. See photo for respective ball, pin and reverse plunger positions.
28 Insert the 3/4th shifter shaft from the rear and locate thru the shifter fork and into the front of the casing to retain the detent balls.
29 Fit the rear extension and gasket into position and retain with bolts tightened to the specified torque.
30 If the rear extension oil seal has been removed, carefully lift the new one into position.
31 Install the two detent balls into position (photo) and then insert the reverse shifter shaft thru the front of the casing and reverse shifter fork.
32 Fit the respective roll pins into position thru the shifter forks and shafts (photo).
33 Fit the speedometer driven gear unit into position.
34 Install the detent ball and plunger followed by the reverse lamp switch (photo). Fit the TCS switch.
35 Fit the reverse shift block, fulcrum bracket and reverse idler gear control lever and locate with the roll pin (photo).
36 Refit the top shift quadrant and gasket tighten the retaining bolts to the specified torque.
37 Select neutral position and insert the three detent balls, (photo) and springs (photo) into their location in the top of the rear face of the casing.
38 Fit the gaskets and baffle plate and then the top cover into position (photo). Locate the retaining bolts and tighten to the specified torque.
39 Locate the front cover gasket and fit the front cover. Tighten the retaining bolts to the specified torque.
40 Refit the clutch shift fork and release bearing.

9 Transmission - Series 5 and 6 - reassembly

Mainshaft
1 Smear the needle roller bearing with grease and fit to the mainshaft together with the third speed gear. The gear is fitted with its tapered side to the front.
2 Fit a blocker ring with clutching teeth upwards over the third speed gear synchronising surface.
3 To reassemble the 3rd/4th synchronizer unit, locate the synchronizer hub face with its heavy boss to the sleeve face as in Fig. 6.23. Locate the keys into the groove and insert the springs into the hole in the side face of the synchronizer hub. Check that the hub and sleeve slide smoothly.
4 Fit the 3rd/4th synchronizer into position on the mainshaft with the small chamfer on the sleeve face to the rear of the shaft. Retain with the snap-ring.
5 Invert the mainshaft and fit the 2nd speed gear and needle bearing onto the shaft-grease the bearing before assembly and ensure that the 2nd gear taper face is to the rear of the shaft.
6 Fit the blocker ring with its clutching teeth downwards over the

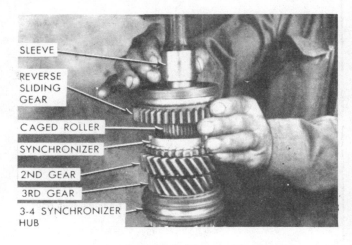

Fig. 6.24. Fit the reverse sliding gear and sleeve

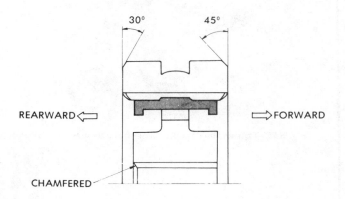

Fig. 6.25. 1st/2nd synchronizer directional location

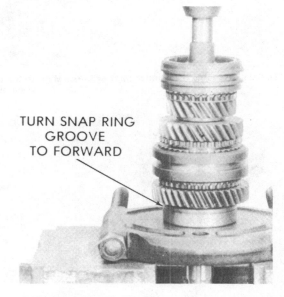

Fig. 6.26. Install the rear bearing with snap ring groove forward

Fig. 6.27. Fit the clutch gear bearing retaining snap ring

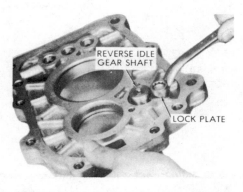
Fig. 6.28. Tighten the lockplate to the specified torque

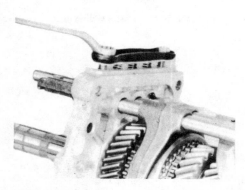
Fig. 6.29. Tighten the detent retainer bolts to specified torques

Fig. 6.30. The detent plug positions in center support

Fig. 6.31. Assemble the counter shaft and main shaft to the center support using special tool no. J-26545-5 if available to keep the shaft assemblies in mesh and parallel

Fig. 6.32. Fit the reverse shifter fork and reverse gear

Fig. 6.33. Fit the reverse idler shaft

Chapter 6 Part A: Manual transmission

Fig. 6.34. Thrust washer location in extension

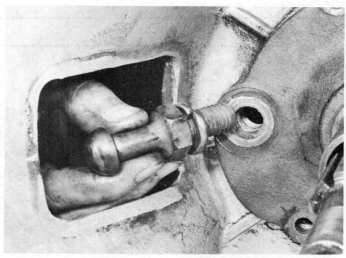

9.45 Refit the ball stud and tighten to specified torque

2nd gear synchronizer surface.

7 To reassemble the 1st/2nd synchronizer, position the hub with the chamfer on the inner edge facing to the heavy chamfer on the outer edge of the sleeve as in Fig. 6.25. Locate the keys into their grooves and synchronizer spring in the hole in the synchronizer hub side face. Check that the hub and sleeve slide smoothly.

8 Fit the 1st/2nd synchronizer unit into position with its sleeve chamfer facing forwards.

9 Fit the blocker ring with its clutching teeth rearwards.

10 Smear the 1st speed gear roller bearing with grease and fit it together with its collar and gear onto the mainshaft. The gear is fitted with the tapered side facing forwards.

11 Fit the 1st speed gear thrust washer with its grooved side against the gear.

12 Press the rear bearing into position on the shaft with the snap-ring groove off set to the front of the mainshaft (Fig. 6.26).

13 Fit the bearing to the clutch gear shaft if it has been removed. The snap-ring groove is positioned offset to the front of the shaft. Retain the bearing with the snap-ring as in Fig. 6.27.

14 Smear the clutch gear pilot needle roller bearing with grease and fit it together with the blocker ring and clutch gear to the mainshaft.

15 Press the countergear ball bearing into position with the snap-ring groove offset to the rear.

16 Refit the snap-rings to their grooves in the mainshaft and countergear shaft locations in the center support.

17 Install the idler gear shaft into position in the center support with the lock plate grooved side positioned correctly. Fit the lock plate and tighten the retaining bolt to the specified torque of 14 lb f ft (1.9 kg f m).

18 The mainshaft assembly and countergear shaft unit are meshed together in parallel and fitted to the center support. If special tool No. J-26545-5 is available this should be used to retain the two gear shaft assemblies in parallel mesh when fitting the center support.

19 Expand the countergear bearing and mainshaft bearing snap-rings in the center support and get an assistant to press the center support over the bearings so that the snap-rings are fitted in their respective grooves.

20 Push the synchronizers to the rear to prevent the mainshaft rotating. If the synchronizers are reluctant to move, tap them with a wooden hammer handle.

21 Now locate on the rear of the countergear the collar, countershaft reverse gear, washer and self-locking nut. Always use a new self-locking nut, and tighten to a torque of 100 lb f ft (13.8 kg f m).

22 Fit the mainshaft locknut and washer. Tighten the nut, fitted with its chamfered side facing the washer, to 94 lb f ft (12.9 kg f m).

23 Grease the detent plugs and install them into their location holes from the middle hole of the center support as in Fig. 6.30.

24 Position the 1st/2nd and 3rd/4th shifter forks into their location grooves in the synchronizers.

25 Insert the 3rd/4th shifter rod thru the rear of the center support, the middle hole, and into the respective shifter forks. The 3rd/4th shifter rod has two detent grooves on the side of the rod.

26 Insert the 1st/2nd shifter rod thru the rear of the center support and into the 1st/2nd shifter fork.

27 Align the shifter forks and shaft spring pin holes.

28 Install the stopper pin in the reverse shifter rod so that the pin is protruding equally from each side of the shaft. Fit the reverse shifter rod from the front of the center support.

29 Support the shifter rods and drift the spring pins into position to locate the shifter forks and shafts.

30 The detent balls, springs and retainer are now installed and the retainer securing bolts tightened to 14 lb f ft (1.9 kg f m).

31 Position the transmission casing upright on wooden blocks and fit the center support assembly and gasket into the casing. Align the dowel pin holes with the dowel pins.

32 Fit the reverse shifter fork to reverse gear and then fit them into position from rear of the mainshaft, and connect to the reverse shifter rod.

33 Drive the spring pin into the reverse shifter fork.

34 Fit the thrust washer and then the reverse idler gear to the idler shaft. The flange on the thrust washer locates in the center support stopper. The undercut teeth on the reverse idler gear face rearwards (Fig. 6.33).

35 Locate the speedometer drive gear snap-ring and key onto the mainshaft.

36 Fit the speedometer drive gear in alignment with the key groove and retain with snap-ring.

37 Carefully fit a new oil seal to the rear extension, then smear the outer thrust washer of the reverse idler shaft with grease and fit to the rear extension (Fig. 6.34).

38 Fit the rear extension and gasket into position aligning the dowel pin and hole. Tighten the retaining bolts to 27 lb f ft (3.6 kg f m).

39 Refit the reverse lamp switch and CRS switch.

40 Fit the shifter cover and gasket and tighten the retaining bolts to 10 lb f ft (1.3 kg f m).

41 Apply some oil to the 'O' ring on the speedometer driven gear unit and fit into position in the rear extension. Tighten the retaining bolt.

42 Fit the new front bearing retainer seal into position.

43 Fit the snap-ring to the outer clutch gear bearing race. Grease the retainer spring washer and position it in the bearing retainer. The dished face of the washer must face the bearing outer race.

44 Fit the bearing and retainer into position in the front of the gearcase and retain with bolts. The shorter bolts are fitted to the countergear front bearing side of the retainer. Tighten the bolts to 14 lb f ft (1.9 kg f m).

45 If it has been removed, refit the ball stud, (photo) and tighten to a torque of 30 lb f ft (4.1 kg f m).

46 Refit the clutch fork, dust boot and throwout bearing, and install the retaining spring. Grease the shift fork support with molybdenum disulphide grease, and ensure that the clutch fork hook is correctly located to the support.

10 Fault diagnosis - manual gearbox

Symptom	Reason/s
Weak or ineffective synchromesh	Synchro. cones worn or damaged Baulk rings worn Defective synchro. unit
Jumps out of gear	Worn interlock plunger Worn detent ball Weak or broken detent spring Worn shift fork or synchro sleeve groove Worn gear
Excessive noise	Incorrect oil grade Oil level too low Worn gear teeth Worn mainshaft bearings Worn thrust washers Worn input or mainshaft splines
Difficult gear changing or selection	Incorrect clutch free movement

Chapter 6 Part B: Automatic transmission

For information relating to Series 8 thru 12 models, including 4WD, refer to Chapter 13.

Contents

Fault diagnosis - Automatic transmission ... 21	Shift lever - dismantling, inspection and reassembly ... 17
Fluid level - checking ... 12	Shift linkage - adjustment ... 14
General description ... 11	Throttle valve control cable - removal, installation and adjustment ... 15
Inhibitor switch - adjustment ... 18	
Rear oil seal - renewal ... 20	Transmission - draining and refilling ... 13
Shift control lever - removal and installation ... 16	Transmission - removal and installation ... 19

Specifications

Type ...	3 speed, fully automatic
Torque ratio ...	2.6

Gear ratios

1st ...	2.74 : 1
2nd ...	1.57 : 1
3rd ...	1.00 : 1
Reverse ...	2.07 : 1
Lubricant capacity ...	6.5 US quarts (5.4 Imp qts)
Lubricant type ...	Dexron 11

Torque wrench settings

	lb f ft	kg f m
Screen to valve body bolts ...	6 to 10	0.8 to 1.3
Oil pan baths ...	10 to 13	1.3 to 1.7
Shift lever to control shaft nut		
Series 5 ...	16	2.2
Series 6 ...	22	3.0
Flexplate to converter bolts ...	30	4.1

11 General description

1 The fully automatic Turbo-Hydromatic 200 Transmission unit basically consists of a three element hydraulic torque converter and a compound planetary gear set.
2 To obtain the required actuation of the compound planetary gear set there are three multiple - disc clutches, a band and a roller clutch. The hydraulic torque is applied via the torque converter which connects the engine to the planetary gears thru oil, and the planetary gears consist of three forward and one reverse gear.
3 The torque converter comprises a pump (driving member), a turbine (driven member) and stator unit. The stator works in conjunction with a one-way roller clutch, which allows the stator to rotate in a clockwise direction only.
4 The engine crankshaft is coupled to the torque converter by a flex plate which only rotates at engine speed. The torque converter housing is filled with oil and the converter pump directs the oil to the turbine. Oil passing thru the turbine causes it to rotate. The stator redirects the oil flow from the turbine so that it does not adversely affect the converter blade action, at higher engine speeds. At lower speeds, the stator roller clutch is actuated and this activates the stator causing it to direct the oil at the converter pump. This has the effect of assisting the converter pump to deliver power or increase engine torque at the lower speeds.
5 The friction components and automatic controls are operated by a hydraulic system, pressurized by a gear pump, which supplies the working pressure required to operate the controls and components.
6 The transmission has six selector positions:
  P Parking position which locks the out shaft to transmission case and incorporates a locking pawl. This is a safety device to prevent the vehicle moving when on an incline. The engine may be started with 'P' selected and this position must always be selected when adjusting the engine while it is running. 'P' must only be engaged when the vehicle is completely stationary.
  R Reverse gear
  N Neutral. Select this position to start the engine or when idling for long periods such as in heavy traffic.
  D Drive, for all normal motoring conditions. The drive range has three gear speeds from starting ratio to direct drive. This range gives maximum economy. Detent down shifts can be made by depressing the accelerator pedal to the floor.
  L2 This range gives the same starting ratio as 'D' range but stops the transmission from engaging a higher gear, and can be used for acceleration or engine braking as required. Once engaged, the transmission will remain in 2nd until the vehicle speed or throttle are changed to obtain 1st gear as in 'D' range.
  L1 The selection of this gear above a road speed of approximately 30 mph (48 kph) will engage 2nd gear and as the speed decreases below this figure the transmission will shift into 1st gear.
7 Due to the complexity of the automatic transmission unit, any internal adjustment or servicing should be left to a GM dealer or automatic transmission specialist. The information given in this Chapter is therefore confined to those operations which are considered within

the scope of the home mechanic. An automatic transmission should give many tens of thousands of miles service provided normal maintenance and adjustment is carried out. When the unit finally requires major overhaul consideration should be given to exchange the old transmission for a factory reconditioned one, the removal and installation being well within the capabilities of the home mechanic as described later in this Chapter.

8 The hydraulic fluid must be periodically drained and refilled as specified in Section 12, and the fluid level regularly checked (see next Section).

9 Periodically the outside of the transmission housing should be cleaned as an accumulation of dirt and oil is liable to cause overheating of the unit under adverse conditions.

10 The engine idle speed must be set as specified.

11 If a new or reconditioned transmission unit has been installed, the maximum speed must not exceed 70 mph for the break-in period of the first 600 miles.

12 Fluid level - checking

1 The transmission fluid level must be checked at every engine oil change period.

2 The oil must be checked at its normal operating temperature (93°C - 200°F) and this is achieved after a minimum of 15 miles highway driving or equivalent city driving.

3 To check the level, park the vehicle on a level surface with the shift lever in 'P' position, and the parking brake applied.

4 Run the engine at its normal idle speed.

5 Withdraw the fluid level indicator and wipe clean. Fully re-insert it until the cap seats and then remove the indicator for a level reading, which should be at the 'F' (full) mark.

6 Top up if necessary, but do not over-fill, with the correct grade of transmission oil as specified.

7 If for any reason the vehicle cannot be driven to warm up the transmission for oil level checking as above, it can be checked at an ambient temperature of 21°C (70°F), as follows.

8 Follow the instruction in paragraphs 3 and 4 above, but do not race the engine under any circumstances!

9 Move the selector shift lever thru each range.

10 Now check immediately the fluid level with the selector engaged in the 'P' position with the engine running. The fluid level on the indicator should be between the two dimples below the 'ADD' mark.

11 If it is necessary to top up the transmission only add sufficient fluid to bring the level up to the point between the two dimples on the dipstick.

12 The need for frequent topping-up indicates a leakage and should be rectified as soon as possible.

13 Transmission - draining and refilling

1 The transmission oil pan must be drained and the screen cleaned every 60,000 miles, and new fluid added. Under extreme working conditions this should be undertaken every 15,000 miles.

2 Drain the fluid immediately after the vehicle has been used so that the oil is at its normal working temperature, but take care - transmission fluid can exceed 177°C (350°F).

3 Raise the vehicle and support the transmission vibration damper with a jack.

4 Place a suitable container under the oil pan and unscrew the oil pan attachment bolts, from the front and side.

5 Slacken the oil pan rear securing bolts approximately four turns and, using a suitable screwdriver, prise the oil pan loose and allow the fluid to drain.

6 When the oil has drained, remove the rear oil pan bolts and lower the pan and gasket.

7 Drain the remaining fluid from the oil pan, and clean the pan using a solvent and blow dry using compressed air.

8 Unscrew and remove the screen to valve body bolts, screen and gasket.

9 Clean thoroughly the screen in solvent and blow dry with compressed air.

10 Fit a new gasket to the screen and refit it, tightening the retaining bolts to 6 - 10 lb f ft (8 - 14 Nm).

11 Fit a new gasket to the oil pan and refit it to the transmission, tightening the retaining bolts to 10 - 13 lb f ft (14 to 18 Nm).

12 Lower the vehicle and top up the transmission with approximately 6 pints (2.8 litres) of the recommended fluid.

13 Now check the transmission oil level as described in paragraphs 7 to 11 in the previous Section.

14 Shift linkage - adjustment

1 Detach the control rod from the shift lever side by withdrawing the cotter pin and removing the washers.

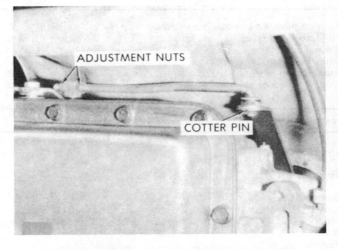

Fig. 6.35. The control rod showing adjustment nuts position

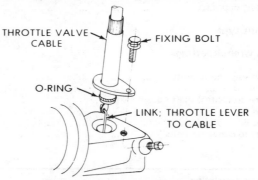

Fig. 6.36. Throttle valve cable location

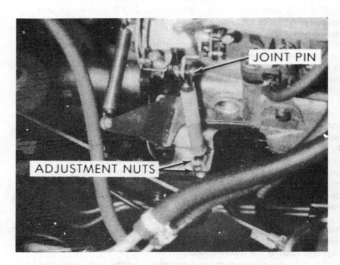

Fig. 6.37. Throttle valve cable adjustment nuts

Chapter 6 Part B: Automatic transmission

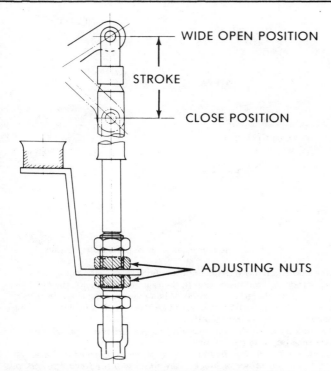

Fig. 6.38. Check the cable stroke between wide open and fully closed

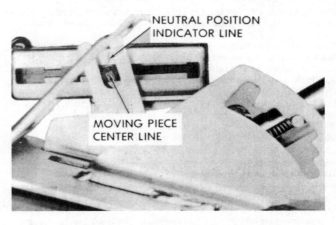

Fig. 6.39. Inhibitor switch showing adjustment positions

2 Turn the transmission manual shaft to stop in the counter clockwise direction viewed from the transmission left side (position '1'), then back the shaft off three stops to the Neutral position.
3 Retain the parts in this position and have an assistant move the shift lever in the cab to Neutral, and push rearward to remove play.
4 Slacken the control rod locknuts and adjust the rod length by turning it so that its end is in alignment with the shift lever hole.
5 Reconnect the control rod and shift lever and retain with washers and cotter pin. Tighten the lock nuts.
6 Finally check the shift lever movement and that the vehicle moves when the shift lever is in 'D' range and reverses when in 'R' range.

15 Throttle valve control cable - removal, installation and adjustment

1 Slacken the throttle valve control cable adjusting nuts and detach the cable from the carburetor lever by removing the cotter pin from the connector pin.
2 Detach the throttle valve cable clip from its location on the right rear side of the cylinder body. Undo the throttle valve control cable flange to transmission retaining bolt, and pull the cable upwards to detach the inner cable end from the throttle lever linkage on the transmission side.
3 Remove the cable assembly.
4 Installation is a direct reversal of removal and when in position must be adjusted as follows.
5 Ensure that the carburetor lever and throttle valve control cable bracket are free from distortion and normal.
6 Slacken the cable adjustment nuts (Fig.6.37) and fully open the carburetor lever. The inner cable adjustment is set by turning the lower adjustment nut on the outer cable by hand to give the inner cable approximately 0.04 in (1.0 mm) play. Tighten the upper lock nut to secure.
7 Check that the inner cable stroke from wide open to closed position is within 1.37 to 1.41 in (34 to 35.8 mm).

16 Shift control lever - removal and installation

1 Locate the shift lever in Neutral position.
2 Withdraw the cotter pin and remove the washers and disconnect the control rod from the shift lever side.
3 Remove the console cover retaining screws and remove the console cover.
4 Disconnect the inhibitor switch and indicator light wiring from the connector.
5 Unscrew the four bolts retaining the shift lever unit and remove it.
6 Installation is a direct reversal of removal but readjust the shift lever setting if required as in Section 13.

17 Shift lever - dismantling, inspection and assembly

1 Unscrew the lever knob setscrew, remove the knob assembly from the lever.
2 Unscrew the four indicator upper cover retaining screws and remove the cover, slider and lower the cover, with indicator light, from the shift lever bracket.
3 Remove the inhibitor switch from the bracket by unscrewing the two securing screws and then prise the rubber boot from the bracket.
4 Undo the nut and remove with the spring washer from the control shaft to shift lever and remove them from the bracket.
5 Remove the shift lever bracket bushings.
6 To disconnect the shaft lever rod, turn it counterclockwise and withdraw it from the upper opening in the shift lever.
7 Now compress the shift lever rod return spring and withdraw the selector pin via the lower shift lever slot by turning it 90° and then remove the spring.
8 Clean and inspect the various components for signs of excessive wear or damage. Test the inhibitor switch for continuity using a circuit tester. The moving piece of the switch must be tested in its various positions of P - R - N - D.
9 Commence reassembly by fitting the shift rod return spring into its location thru the opening in the upper part of the lever.
10 Fit the select pin into its slot in the lower part of the lever by compressing the spring and turning the pin 90°. Smear the pin with a small amount of grease prior to fitting and when fitted its long side end must be positioned to the shift lever bracket.
11 Insert the shift lever rod into the lever with its end projecting from the upper face of the lever by approximately 1.686 in (42.8 mm) on Series 5 models or 1.8 in (46 mm) for Series 6. Screw the select pin into the lever and set the shift lever rod with its taper face to the push button.
12 Fit the two bushings to the control shaft fitting face of the lever bracket. Lightly grease the control shaft, select pin and control shaft fitting face of the bracket and install the control shaft and shift lever. Fit the washer and nut and tighten to a torque of 16 lb f ft (2.2 kg f m) on Series 5 models or 22 lb f ft (3 kg f m) on Series 6 models.
13 Fit the rubber boot to the bracket, and then insert the projected portion of the inhibitor switch in the slot in the shift lever. Fit parts to the bracket and locate the screws but do not fully tighten them. A small amount of grease should be applied to the projected portion of the inhibitor switch but be careful not to grease the points within the switch.

14 Fit the indicator light to the lower cover and assemble to the bracket complete with slider and upper cover. Do not fully tighten the setscrews.
15 Adjust the position of the upper cover by moving the shift lever to each of its positions and the red color mark on the slider is in alignment with the indicator plate window.
16 Fit the knob assembly to the shift lever and retain using a new setscrew. The free play of the push button should be 0.008 to 0.048 in (0.2 to 1.2 mm) for Series 5 or 0.028 to 0.048 in (0.7 to 1.2 mm) with the lever in the N or P position. If adjustment is required refer to paragraph 11. Check that the shift lever operates freely thruout the range, adjust the inhibitor switch as follows.

18 Inhibitor switch - adjustment

1 If the engine can be started in any position other than Neutral or Park adjust the switch as follows.
2 With the two inhibitor switch retaining screws slackened, adjust the position of the switch so that the switch moving piece center is in alignment with the neutral position on the indicator line on the steel case, with the lever set in Neutral position. Tighten screws.

19 Transmission - removal and installation

1 Detach the ground cable from the battery terminal.
2 Slacken the throttle valve cable adjustment nuts, withdraw the cotter pin and joint pin from the carburetor lever and then detach the control lever.
3 Disconnect the oil level gauge from the transmission.
4 Jack up the vehicle and make secure with blocks.
5 Remove the dust cover from the lower part of the converter housing. Drain the transmission oil - see Section 13.
6 Unscrew and remove the starter mounting bolts and withdraw the starter unit.
7 Refer to Chapter 7 and remove the propeller shaft.
8 Detach the control rod from the shift lever.
9 Disconnect the exhaust pipe retaining bracket from the transmission and then detach the speedometer cable from the transmission.
10 Slacken the oil cooler pipe retaining nuts from the right-hand side, remove the pipe securing clips and relocate the pipe towards the body.
11 Refer to Fig 6.40 and remove the dust cover to gain access to the three nuts and bolts coupling the converter to the drive plate. Turn the drive plate accordingly to undo the respective bolts/nuts.
12 Undo and remove the three transmission to frame bracket mounting bolts.
13 Place a jack under the engine and raise the engine and transmission sufficiently to remove the frame bracket to crossmember bolts, and remove the rear mount.
14 Support the transmission with a jack or blocks.

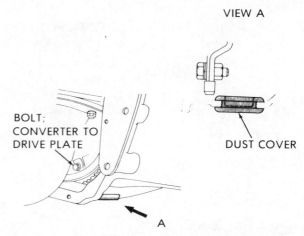

Fig. 6.40. The converter to drive plate bolt position

15 Undo and remove the bellhousing bolts.
16 Withdraw the transmission to the rear complete with the oil filler tube and the throttle valve cable. When manoeuvring the transmission clear of the vehicle take care not to let the torque converter become detached from the transmission.
17 Installation of the transmission unit is a direct reversal of removal sequence but note the following.
   a) Prior to fitting the flexplate to converter bolts ensure that the welded converter brackets are flush with the flexplate, and the converter can be freely rotated by hand. Tighten all of the bolts finger tight before tightening to the specified torque to ensure correct converter alignment.
   b) The throttle valve cable and control rod must be re-adjusted as described in Sections 15 and 14.
   c) Refill the transmission with fluid of the correct type specified and check the fluid level as in Section 12.

20 Rear oil seal - renewal

1 After a considerable mileage, leakage may occur from the seal which surrounds the shaft at the rear end of the automatic transmission extension housing. This leakage will be evident from the state of the underbody and from the reduction in the level of the hydraulic fluid.
2 Remove the propeller shaft as described in Chapter 7.
3 Taking care not to damage the spline output shaft and the alloy housing, pry the old oil seal from its location. Drive in the new one using a tubular drift. Refit the propeller shaft.
4 Check the fluid level and replenish to the correct level as required.

21 Fault diagnosis - Automatic transmission

Symptom	Reason/s
Engine will not start in 'N' or 'P'	Faulty starter or ignition circuit. Incorrect linkage adjustment. Incorrectly installed inhibitor switch.
Engine starts in selector positions other than 'N' or 'P'	Incorrect linkage adjustment. Incorrectly installed inhibitor switch.
Severe bump when selecting 'D' or 'R' and excessive creep when idling	Idling speed too high.
Poor acceleration and low maximum speed	Incorrect oil level. Incorrect linkage adjustment.

The most likely causes of faulty operation are incorrect oil level and linkage adjustment. Any other faults or mal-operation of the automatic transmission unit must be due to internal faults and should be rectified by your GM dealer. An indication of a major internal fault may be gained from the color of the oil which under normal conditions should be transparent red. If it becomes discolored or black then burned clutch bands must be suspected.

Chapter 7 Driveshaft

For information relating to Series 8 thru 12 models, including 4WD, refer to Chapter 13.

Contents

Driveshaft - balance and run-out check 5	General description 1
Driveshaft - dismantling and reassembly 4	Universal joints - inspection 3
Driveshaft - removal and installation 2	

Specifications

Spider pin endplay ... ...   0.004 in

Snap-ring thickness ... ...   0.057 in
0.059 in } as required
0.061 in

Torque wrench setting
Propeller shaft to companion flange ... ...   18 lb f ft (2.4 kg f m)

1 General description

Drive from the transmission to the rear axle is via a driveshaft (propeller shaft) comprising the coupling shaft, universal joints and rear connecting flange yoke.

The front end of the shaft is connected to the transmission unit by means of a splined sliding yoke which permits fore and aft movement of the driveshaft in accordance with axle movement. The rear end of the shaft is connected to the rear axle companion flange by the flanged yoke of the rear end universal joint.

The front and rear universal joints are retained in the yokes by snap-rings (circlips) which are available in varying thicknesses so that the joint spider can be correctly centered.

2 Driveshaft - removal and installation

1 Mark the relative position of the driveshaft rear yoke and rear axle companion flanges to ensure that they are refitted in the same relative position.
2 Unscrew the four bolts connecting the flange yoke with the companion flange. Support the shaft and remove the bolts.
3 Lower the rear of the shaft to clear the axle housing and carefully withdraw the driveshaft assembly from the transmission housing, taking care not to damage the rear oil seal in the transmission.
4 Plug the end of the transmission to prevent loss of lubricant (Fig. 7.2).
5 Installation is the reversal of removal but ensure that the transmission end of the driveshaft is well lubricated with transmission oil and take care not to damage the oil seal as the shaft is inserted.

Also check that the rear yoke and companion flange marks are aligned before fitting the bolts, which must be tightened to the specified torque. Top up the transmission if any lubricant has been lost.

3 Universal joints - inspection

1 Wear in the needle roller bearings is characterised by vibration in the transmission, 'clonks' on taking up the drive, and in extreme cases of lack of lubrication, metallic squeaking and ultimately grating and shrieking sounds as the bearings break up. If a bearing breaks up at high speed it could be lethal, so they should be changed in good time.
2 It is easy to check whether the needle roller bearings are worn with the driveshaft in position, by trying to turn the shaft with one hand, the other hand holding the companion flange when the rear universal is being checked, and the front half coupling when the front universal is being checked. Any movement between the shaft and the front and the rear half couplings is indicative of considerable wear. If, worn, the old bearings and spiders will have to be discarded and a repair kit, comprising new universal joint spiders, bearings, seals and snap-rings purchased. Check also by trying to lift the shaft and noticing any movement in the joints.
3 Examine the driveshaft spline for wear. If worn it will be necessary to purchase a new front half coupling, or if the yokes are badly worn, an exchange shaft. It is not possible to fit oversize bearings and journals to the trunnion bearing holes.

4 Driveshaft - dismantling and reassembly

1 Clean all traces of dirt from the universal joints. If the snap-rings

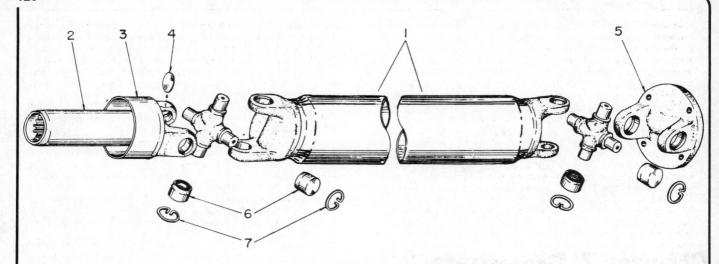

Fig. 7.1. The propeller shaft component parts

| 1 Propeller shaft | 3 Cover | 5 Flange yoke | 7 Snap rings |
| 2 Spline yoke | 4 Plate plug | 6 Bearing caps | |

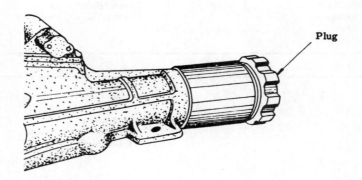

Fig. 7.2. Plug the end of the transmission

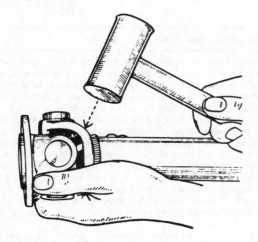

Fig. 7.3. Tap the yoke to withdraw the bearing cup

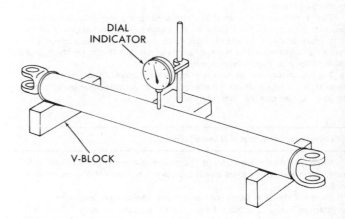

Fig. 7.4. Checking the propeller shaft run-out

have rusted into the yokes, use a wire brush to remove the surface rust and soak the snap-rings in penetrating oil.

2 Remove the snap-rings using suitable pliers or a small screwdriver. New snap-rings must be used on reassembly so it does not matter if they are distorted on removal. If they are difficult to remove, a hammer blow on the end of the joint spider may relieve the pressure on them.

3 To remove the bearing cups, hold the joint in one hand and with a mallet or copper hammer tap the yoke lightly as shown in Fig. 7.3. Alternatively, select two sockets, one being large enough to fit completely over the bearing cup and the other to fit on the end of the cup. By compressing in the jaws of a vice or by hammering on the smaller socket (providing that a degree of caution is exercised), the spiders and cups can be driven out of one side of the yoke. Having removed one cup, the opposite one can then be removed by reversing the direction of applied pressure. The remaining ones are then treated similarly.

4 Wash the respective parts in a solvent and wipe dry. Inspect the various components for wear and other damage. If a joint spider is worn below the limit, it must be renewed regardless of the condition of the bearings.

5 When reassembling the universal joints, ensure that all the parts are free from dirt, then lightly grease the holes in the yokes. Ensure that the bearing cups are approximately 1/3 full of molybdenum-disulphide grease.

6 Insert the spider into the yoke and carefully press in two opposing cups and seals. The vise and sockets can be used for this job as they were when dismantling. Snap-rings are available in selective sizes of (0.057, 0.059 and 0.061 in) so that the universal joint can be positioned centrally and then all endplay can be taken up. Ensure that the opposing snap-rings are of equal size to obtain optimum balance. With the bearing cups in position the endplay of the spider pin must be within 0.004 in (0.1 mm).

5 Driveshaft - balance and run-out check

1 The driveshaft is balanced during manufacture and will not normally require any further balancing provided that the correct snap-rings have been used on the universal joints, and that accident damage has not occurred.

2 If vibration of the driveshaft is experienced, and the universal joints and coupling flange bolts and bolt holes are known to be in good order, then it may be possible that the driveshaft is misaligned and/or out of balance.

3 To check the alignment, remove the driveshaft from the vehicle and lay it with its ends supported in two 'V' blocks as shown in Fig. 7.4. Locate the probe of a dial indicator to locate centrally on the shaft. Turning the shaft slowly, note the dial indicator reading. Any excessive variation indicates that the shaft is misaligned and a replacement must be acquired.

Chapter 8 Rear axle

For information relating to Series 8 thru 12 models, including 4WD, refer to Chapter 13.

Contents

Axle housing - removal and installation ... 5	Fault diagnosis - rear axle ... 6
Axleshaft unit - removal and installation ... 2	General description ... 1
Differential carrier assembly - removal and installation ... 4	Rear wheel bearing and seal - removal and installation ... 3

Specifications

Type	Semi-floating, hypoid
Ratio	4.56 : 1 (manual)
	4.10 : 1 (speed automatic)
Ring gear to pinion backlash	0.005 to 0.008 in (0.127 to 0.203 mm)
Pinion to differential gears	0.001 to 0.003 in (0.02 to 0.07 mm)
Rated capacity	3.500 lbs (1533 kg)
Lubricant type	GL-5
Below 50°F (10°C)	SAE 80 W
0° to 90°F (17° to 32°C)	SAE 90
Consistently above 50°F (10°C)	SAE 140
Lubricant capacity	2.7 pints

Torque wrench settings

	lb f ft	kg f m
Axle to spring 'U' bolt	40	5.5
Shackle pins (Series 1 to 4)	130	17.9
Propshaft to companion flange	18	2.4
Axleshaft locknut	190	26.2
Bearing holder thru bolts	55	7.6
Bearing cap nuts	75	10.3
Pinion nut	190	26.2
Pinion nut initial torque	85	11.7
Ring gear to case bolts (with Loctite)		
Series 1	55	7.6
Series 2	65	8.9
Series 3, 4, 5 and 6	80 to 87	11.0 to 12.0
Carrier to housing bolts		
Series 1	35	4.8
Series 2 and later	18	2.4

Chapter 8/Rear axle

1 General description

1 The conventional rear axle is of the semi-floating type, located by semi-elliptic rear springs and telescopic double acting shockabsorbers. The differential carrier is detachable from the axle casing, permitting the crownwheel and pinion adjustments to be made on the workbench during any major servicing.

2 The final drive crownwheel and pinion are of the hypoid type with the pinion center line below the crownwheel center line. The pinion, which runs in two opposed taper roller bearings, is adjusted for depth of mesh to the crownwheel by means of shims. A collapsible spacer on the pinion shaft allows for the pinion bearing pre-load.

3 The crownwheel ring gear is bolted to the differential case and is located by two opposed taper roller bearings. Crownwheel side play adjustment is made by the use of shims.

4 The semi-floating axleshafts are retained at the outer ends by taper roller bearings located in housings bolted to the axle housing.

5 It is considered beyond the scope of a manual of this type to delve into the complexities of repair to and setting up of the differential since a number of special tools are required. Also, the procedure is rather complicated, and for those people without a good knowledge of this type of gear train there is a possibility of incorrect meshing and preload which will give rise to noise and backlash which will lead to early failure. In this Section you will find details of the removal procedures for the axleshafts and differential carrier. In the event of any failure of the differential assembly, it is recommended that a complete exchange assembly is purchased or the existing faulty assembly is repaired by a vehicle main dealer.

2 Axleshaft unit - removal and installation

1 Chock the front wheels and raise the rear of the vehicle.

2 Remove the rear roadwheel, brake drum, brake shoes and detach the parking brake inner cable. For further information on these procedures refer to Chapter 9.

3 Undo and release the brake pipe from the wheel cylinder, and plug the end of the pipe to prevent the ingress of dirt.

4 Undo and remove the four retaining nuts from the bearing holder.

5 Withdraw the axleshaft unit, and if available use special tool number J-5748. On no account pry against the flange plate.

6 If both axleshaft units have been removed and serviced, commence installation by inserting a 0.079 in (2 mm) shim between the axle tube end flange and the bearing holder. Refit the axleshaft unit into the axle tube and locate with the four flange retaining bolts. Tighten the flange bolts to a torque of 55 lb f ft (6.97 kg f m).

7 If only one axleshaft unit has been removed for servicing or when fitting the second proceed as follows. Refit the opposing axleshaft less shims and when it is fully located against the thrust block in the differential, calculate the bearing holder to flange clearance, add 0.004 in (0.101 mm) to the measured clearance to determine the shim requirement.

8 Withdraw the shaft, and fit the necessary selected shims. Refit the shaft and tighten the flange bolts to 55 lb f ft (6.97 kg f m).

9 Reconnect the brake pipe/s, refit the brake shoes, the parking brake cable, brake drum and road wheels. The hydraulic brake system must now be bled and the parking brake readjusted before lowering the vehicle.

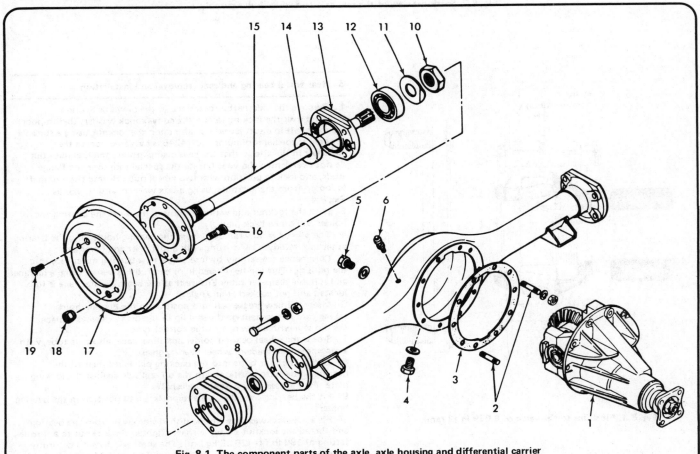

Fig. 8.1. The component parts of the axle, axle housing and differential carrier

1 Differential carrier and case assembly	5 Filler plug	10 Lock nut
2 Mounting bolt	6 Vent	11 Lock washer
3 Gasket	7 Thru bolt	12 Axle shaft bearing
4 Drain plug	8 Oil seal	13 Bearing holder
	9 Shims	14 Grease seal
		15 Axle shaft
		16 Wheel stud
		17 Brake drum
		18 Wheel nut
		19 Drum to flange screw

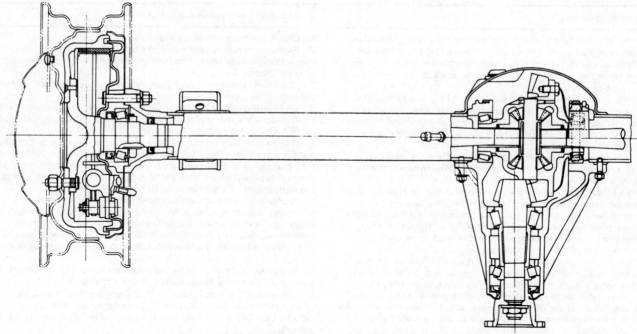

Fig. 8.2. Sectional view of the rear axle and wheel hub assembly

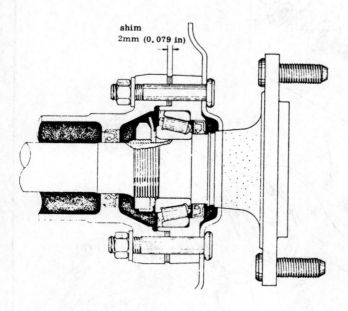

Fig. 8.3. Fit shims to the value of 0.079 in (2 mm)

3 Rear wheel bearing and seal - removal and installation

1 Remove the axleshaft as described in the previous Section.
2 Bend over the locking tab on the convex lock washer, then support the axleshaft in a soft jawed vise and undo the locknut using a suitable wrench. If special tool number J-24246 is available, locate the axleshaft in the vise so that the jaws clamp around the locknut - but do not overtighten the vise. Locate the special tool over the flange studs, and lock in position with two wheel nuts. Rotate the axleshaft to loosen from the locknuts, using a box wrench, and 'L' bar as in Fig. 8.4.
3 With the locknut and washer removed, press off the bearing and holder and brake back plate.
4 The oil seal is now removed from the bearing holder and the bearing outer race removed by drifting with a soft drift and hammer.
5 Commence reassembly by inserting a new bearing outer race into the housing. This can be pressed in or carefully driven in using a hammer and suitable diameter tube. Ensure that the bearing outer race is fully located and not cocked at an angle.
6 Insert the new grease seal into position in the bearing holder taking care not to damage the seal lip or distort it - always replace the old seal with a new one of the correct type.
7 Smear the wheel bearing holder and inner race, also axle tube, with a recommended grade of wheel bearing grease.
8 Locate the four bolts into the backing plate and then fit the bearing holder into position with the oil seal side against the locking plate. Fit the assembly over the axleshaft.
9 Fit the bearing over the axle and press it into position in the bearing holder.
10 Fit a new lockwasher with the dished side away from the bearing and thread the locknut onto the shaft. Tighten the locknut to a torque setting of 190 lb f ft (26.25 kg f m). The shaft will have to be firmly supported to achieve this figure and if special tool J-24246 is available fit the axleshaft to the vise and the tool to the axleshaft as described in paragraph 2, and then tighten the shaft to the nut. Remove the tool.
11 Bend the lockwasher over the nut flat opposite the locating tab as in Fig. 8.5.
12 Refit the axleshaft, brake and wheel assemblies as directed in Section 2.

4 Differential carrier assembly - removal and installation

1 Remove the axleshaft as described in Section 2. It is not necessary to fully remove the shafts unless required, as they only need to be withdrawn sufficiently to allow the differential carrier to be extracted.
2 Remove the axle drain plug and drain the oil into a suitable container.
3 Mark the relative positions of the driveshaft flange to differential companion flange (if not already marked), and disconnect the shaft as described in Chapter 7.
4 Unscrew the ten retaining nuts from the differential carrier and case unit and withdraw it from the axle housing.
5 Commence installation by cleaning down the faces of the rear axle casing and the differential carrier. Also clean inside the axle housing around the differential area to ensure that no particles of metal are present.
6 Smear the gasket of the differential carrier with a coating of non-setting gasket sealant and locate it round the carrier flange. Reposition the carrier to the axle housing and refit the retaining nuts. Tighten them to a torque of 18 lb f ft (2.48 kg f m).
7 Relocate the driveshaft, referring to Chapter 7.
8 Relocate the axleshafts, referring to Section 2 if necessary.
9 Refit the drain plug into the axle housing and top up the axle with lubricant of the specified grade.

5 Axle housing - removal and installation

1 Jack up the rear of the car by placing a jack under the rear axle. Chock the front wheels and place chassis stands or blocks under the body to support.
2 Remove both rear wheels and brake drums. With a 12 mm box wrench compress the retaining fingers and remove the brake springs and shoes.
3 Detach the cables from the leaf spring clip (Fig. 8.6), and then disconnect the brake hoses from the frame bracket (Fig. 8.7). To prevent the ingress of dirt, plug the ends of the brake pipes.
4 Detach the driveshaft from the companion flange of the differential as described in Chapter 7. If removing the driveshaft from the transmission unit, plug the end to prevent oil spillage.
5 Disconnect the shockabsorber lower retaining bolts, and also the 'U' bolts locating the axle to the leaf springs.
6 Get an assistant to steady the axle unit and undo and remove the shackle bolt from each leaf spring at the rear and lower the springs. Lower the trolley jack under the axle and withdraw it to the rear of the vehicle, complete with axle. Take care not to damage the brake pipes.
7 Installation of the axle is a direct reversal of removal but be sure to bleed the brakes and readjust the parking brake if necessary. Also tighten all nuts and bolts to the specified torques, and top up the axle oil level if drained.

Fig. 8.4. Method of undoing the locknut using special tool number J - 24246

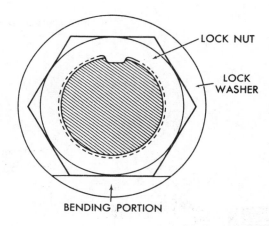

Fig. 8.5. Bend the lock washer opposite the tab

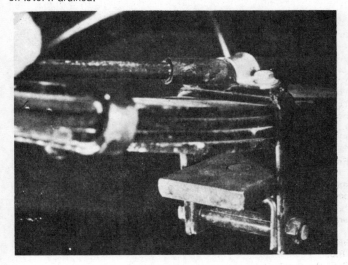

Fig. 8.6. The leaf spring clip and cable

Fig. 8.7. The brake hose and retaining bracket

6 Fault diagnosis - rear axle

Symptom	Reason
Vibration	Worn axle shaft bearing.
	Loose bolts (driveshaft to companion flange).
	Wheels require balancing.
	Driveshaft out of balance.
Noise on turns	Worn differential gear.
Noise on drive or coasting*	Worn or incorrectly adjusted ring and pinion gear.
'Clunk' on acceleration or deceleration	Worn differential gear cross shaft.
	Worn driveshaft universal joints.
	Loose bolts (driveshaft to companion flange).

*It must be appreciated that tire noise, wear in the rear suspension bushes and worn or loose shock absorber mountings can all mislead the mechanic into thinking that components of the rear axle are the source of trouble.

Chapter 9 Braking system

For information relating to Series 8 thru 12 models, including 4WD, refer to Chapter 13.

Contents

Bleeding the hydraulic system ... 5	Fault diagnosis ... 21
Brake adjustments ... 2	Front brake shoes (Series 1 to 4) - inspection, removal and installation ... 10
Brake/clutch pedal - removal and installation ... 18	Front disc brake shoe and lining (Series 5 and 6) - removal, inspection and installation ... 11
Brake disc/hub - removal, inspection and installation ... 13	General description ... 1
Brake hoses - inspection, removal and installation ... 9	Parking brake - adjustment ... 4
Brake master cylinder (single) - dismantling and reassembly ... 8	Parking brake - removal and installation ... 19
Brake master cylinder (single and tandem) - removal and installation ... 6	Parking brake cables - removal and installation ... 20
Brake master cylinder (tandem) - dismantling and reassembly ... 7	Power cylinder - dismantling, inspection and reassembly ... 17
Brake pedal - adjustment ... 3	Power cylinder - removal and installation ... 16
Brake wheel cylinders - removal, servicing and installation ... 15	Rear brake shoes - inspection, removal and installation ... 14
Disc brake caliper - overhaul ... 12	

Specifications

Type ... Hydraulic, duo servo assisted

Front
Series 1 to 4 ... Two leading shoe drum brake
Series 5 and 6 ... Disc type

Rear ... Duo servo type

Parking brake ... Rear wheels only - cable operated by hand

Brake drum inside dia.
Front ... 10.000 in
Rear ... 10.000 in
Maximum refinish diameter ... 10.059 in
Replacement diameter ... 10.079 in
Brake lining replacement thickness ... 0.059 in

Wheel cylinder
Bore diameter front ... 1.06 in
Bore diameter rear ... 0.75 in
Maximum piston to bore - clearance ... 0.006 in

Master cylinder
Bore diameter ... 0.875 in
Maximum piston/bore clearance ... 0.006 in

Series 5 and 6 models specifications are as per above information but for the following items

Front brakes
Maximum disc run out ... 0.005 in
Disc rotor thickness (standard) ... 0.709 in
Minimum refinish thickness ... 0.653 in
Front disc lining replacement thickness ... 0.067 in (Series 5)   0.236 in (Series 6)

Rear brakes
Wheel cylinder bore diameter ... 1.25 in
Maximum piston to bore clearance ... 0.006 in

Chapter 9/Braking system

Torque wrench settings	Series 1 to 4 lb/ft (kg fm)	Series 5 and 6 lb/ft (kg fm)
Flange plate to steering knuckle		
large bolt	55 (7.6)	
small bolt	35 (4.8)	
Flange plate to axle housing	55 (7.6)	55 (7.6)
Wheel cylinder - front		
large bolt	16 (2.2)	10 (1.3)
small bolt	10 (1.3)	10 (1.3)
Brake pipe nuts	12 (1.6)	12 (1.6)
Flexible hose nuts	12 (1.6)	12 (1.6)
Flexible hose bracket nuts	35 (4.8)	35 (4.8)
Master cylinder		
end plug	90 (12.4)	90 (12.4)
two way connector bolt	35 (4.8)	35 (4.8)
fluid reservoir clamp	30 (4.1)	30 (4.1)
mounting nuts	10 (1.3)	10 (1.3)
brake lines to connector	12 (1.6)	12 (1.6)
stopper bolt	14 (1.9)	14 (1.9)
master cylinder bracket bolts	10 (1.3)	10 (1.3)
Power cylinder		
master cylinder flange to power cylinder	10 (1.3)	10 (1.3)
dash panel stud nuts	10 (1.3)	10 (1.3)
Connector to caliper		64 (8.8)
Bleeder to caliper		7 (0.9)
Rotor to hub		36 (4.9)
Support to adaptor		64 (8.8)

1 General description

Series 1 to 4

The braking system employed on all Series 1 to 4 vehicles consist of vacuum-assisted hydraulic self-adjusting type drum brakes. The front drum brakes are of the twin leading shoe type where the front cylinder operates the lower shoe and the rear cylinder the upper shoe.

The rear drum brakes have a single cylinder operating the primary and secondary shoes. All shoe linings are the molded type bonded to the shoes and on the rear brakes the primary shoe lining is smaller than the secondary.

Self-adjusting brakes are fitted all round, the front brakes being adjusted on forward braking movements whilst the rears are taken up on reverse braking.

The master cylinder fitted to Series 1 models is a single acting type and is power assisted by vacuum.

On Series 2 onwards, the master cylinder is of dual (tandem) type having two separate master cylinder reservoirs and incorporates a primary and secondary piston. These operate the front and rear brakes simultaneously, but through separate circuits so that in the event of a failure in one circuit the other will still operate but with reduced braking efficiency.

The parking brake (handbrake) is operated by a hand controlled lever mounted to the dash on the right of the steering column and is locked on by means of a ratchet. The L-handled lever operates the interconnecting cables between the rear brakes and equalizer unit by means of a front cable. The equalizer unit is adjustable if required.

Series 5 and 6 models

Although the braking system on the Series 5 and 6 models is basically the same as the earlier series, the front brakes are now disc type, incorporating a single operating piston which actuates both inner and outer shoes. These are also self-adjusting.

With self-adjusting brakes all round, the maintenance requirements for the brake system is reduced considerably, but should never be neglected. The periodic maintenance and inspection tasks, as described in the maintenance section at the front of the book, must therefore be strictly followed.

2 Brake adjustments

1 Normally, adjustment of the brakes is not necessary as they are self-adjusting. However, after the brake linings have been renewed or the adjuster position altered, an initial adjustment can be made as follows.
2 With the car raised on a hoist and the road wheel removed, use the brake drum as an adjusting gauge, and adjust the upper and lower shoes by an equal number of notches so that the brake drum is slightly dragging against the brake linings.
3 On the front brakes, the upper and lower shoes are then retracted by two notches. On the rear brakes, turn the stave wheel 1¼ turns to retract the shoes.
4 Refit the drum to flange retaining screws.
5 Refit the roadwheels and lower the vehicle and remove the hoist. Refit the flange plate access plugs on the front.
6 The final adjustment is made by driving the vehicle in forward and reverse directions and applying the brakes firmly until the correct brake height is attained.

3 Brake pedal - adjustment

1 Disconnect the ground cable from the battery.

Series 1

2 Withdraw the cotter pin from the pushrod clevis pin and remove the clevis pin. Ensure that the brake pedal and return spring are connected.
3 Measure the distance from the floor pan to the surface of the foot pedal pad as shown in Fig. 9.2. The pedal must be fully released.
4 The correct distance (A) requirement is 6.3 to 6.7 inches. If adjustment is required proceed as follows.
5 Detach the wiring harness from the stop light switch, and slacken the stop lamp switch locking nut. Rotate the switch to obtain the correct pedal height, and then retighten the switch locknut (photo).
6 Slacken the locknut on the pushrod clevis and turn the clevis on the rod until the clevis pin can be refitted. Check that the pedal does not bind and then insert the clevis pin and cotter pin. Note that the pin must not be forced in while depressing the pedal.
7 Securely tighten the clevis locknut, and reconnect the ground cable to the battery.

Series 2 to 6 models

Detach the battery ground cable and then proceed as follows.
8 The correct distance (A) requirement is 5.9 to 6.3 inches. If adjustment is required, detach the stop light switch wiring harness and then remove the switch locknut. Rotate the switch counterclockwise to detach the switch from the bracket.
9 Slacken the pushrod locking nut and adjust the pedal to the

135

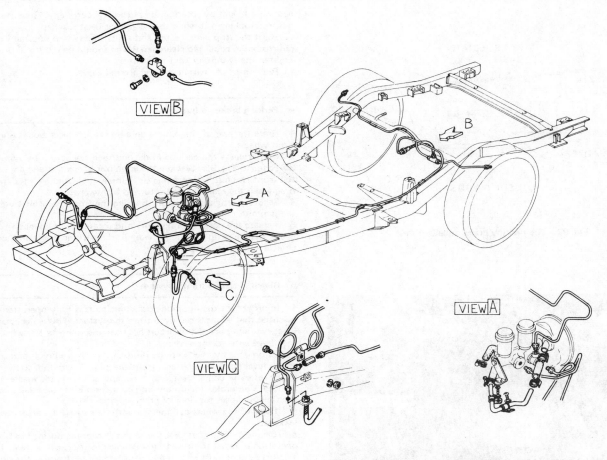

Fig. 9.1. General layout of the dual hydraulic circuit

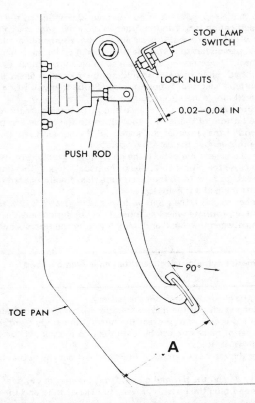

Fig. 9.2. Sectional view of the brake pedal adjustment and check points

3.5 The clutch and brake pedal (Series 1) layout, showing the pushrod clevises, return springs and switches

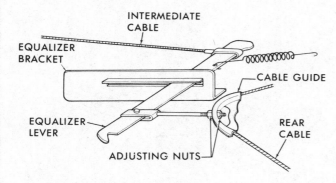

Fig. 9.3. The parking brake equalizer unit

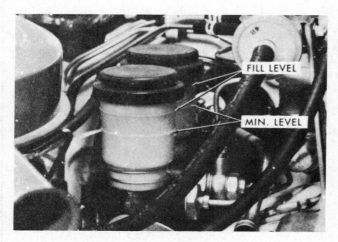

Fig. 9.4. The hydraulic fluid full and minimum level in the dual master cylinder reservoirs

Fig. 9.5. Bleeding a rear wheel circuit

necessary height by rotating the pushrod accordingly. Retighten the pushrod locking nut when the height has been reached.
10 Refit the stop lamp switch and adjust the switch housing (not the pin) to brake pedal tab clearance which should be 0.02 to 0.04 in. Tighten the switch locknut to secure.
11 Reconnect the wiring and the ground cable.

4 Parking brake - adjustment

1 Raise the rear of the vehicle on a hoist and place chocks under the front wheels.
2 Fully release the parking brake and then re-apply it by two notches.
3 Refer to Fig. 9.3. Slacken the equalizer check nut and adjust the front jam nut by loosening or tightening to obtain a moderate drag when the rear wheels are turned frontwards.
4 Securely retighten the nuts, holding the front nut whilst tightening the jam nut.
5 Release the parking brake fully and rotate the rear wheels to check that they do not drag. Re-apply the parking brake and lower the vehicle and remove the hoist.

5 Bleeding the hydraulic system

1 In order that the hydraulic braking circuit is fully operational, it is essential that all air is removed from the system. This is necessary whenever any part of the circuit has been disconnected or when air may have entered the system.
2 Commence by checking the fluid level in the reservoir and top up to the level mark if necessary. Refit the cap. Ensure that the vent hole in the reservoir cap is clear, and clean any dirt from the master cylinder.
3 Check all brake line unions and connections for seepage, and check the flexible hoses for signs of cracking or perishing.
4 If any of the wheel cylinders condition is suspect, check for signs of leakage.
5 It is most important that the engine is running during the bleeding operation in order to prevent any damage to the pushrod seal. The exhaust system must be suitably ventilated if performing the operation inside.
6 Ensure that a supply of clean non-aerated brake fluid of the correct specification is to hand in order to replenish the reservoir during the bleeding process. It is essential to have someone available to help, as one person has to pump the brake pedal while the other attends to each wheel. The reservoir level has also to be continuously watched and replenished. Fluid bled out should not be re-used. A clean glass jar and a suitable length of rubber or plastic tube that will fit tightly over the bleed nipple are also required.
7 Make sure the bleed nipple is clean and put a small quantity of fluid in the bottom of the jar. Fit the tube onto the nipple and place the other end in the jar under the surface of the liquid. Keep it under the surface throughout the bleeding operation.
8 Loosen the bleed nipple, push the pedal down slowly thru its full travel then close the nipple and release the pedal. Repeat this operation until air ceases to flow from the bleed tube then finally tighten the nipple whilst the pedal is held depressed.
9 Remember to check the reservoir level then repeat this procedure for the next appropriate wheel cylinder. On completion, ensure that the fluid is between the 'Max' and 'Min' levels on the reservoir wall.

6 Brake master cylinder (single and tandem) - removal and installation

1 Detach the ground cable from the battery.
2 Clean the exterior of the master cylinder (photo) and connecting lines and as a precaution to protect the surrounding body paint, place a sheet of paper or piece of cloth beneath the master cylinder to soak up any spillage of fluid.
3 Remove the reservoir caps and filters and extract the hydraulic fluid.
4 Carefully detach the hydraulic lines from the master cylinder connectors and plug the lines to prevent spillage of fluid and the ingress of dirt.
5 Remove the master cylinder retaining bolts and nuts and lockwashers.

Chapter 9/Braking system

Fig. 9.6. The hydraulic lines to the dual master cylinder connections

6.2 Clean the master cylinder and pipe connections prior to removal

Remove the master cylinder unit from the car taking care not to spill fluid onto the paintwork.

6 Installation is the reverse of the removal procedure. After topping up the reservoirs, bleed the master cylinder and brake system as described in Section 5.

7 Brake master cylinder (tandem) - dismantling and reassembly

1 Dismantle the components of the master cylinder on a clean work surface and lay the parts out in the order of dismantling.
2 Remove the reservoir caps and filters. Drain any remaining fluid from the reservoirs.
3 Locate the master cylinder in a vise which has soft jaws fitted but do not overtighten the jaws. Slacken the reservoir clamp screws and remove the reservoirs from the cylinder body.
4 Unscrew the connector bolt and remove the connector and gaskets from the rear outlet of the cylinder (front system).
5 Unscrew the end plug and remove with the gasket, check valve, return spring and spring seat.
6 Unscrew the connector and remove with the gasket, check valve, return spring and spring seat from the front outlet of the master cylinder (rear system).
7 Press the primary piston fully into the cylinder bore and then unscrew the stopper bolt and remove with the gasket from the right-hand side of the cylinder.
8 With a pair of snap-ring pliers extract the primary piston snap-ring from the cylinder and then withdraw the primary and secondary piston units from the cylinder. Make a special note of which way round they are installed.
9 Remove the seals and again note which way they are located.
10 Clean all parts thoroughly in clean brake fluid. Do not use gasoline, kerosene or any other type of solvent. After drying the items with a lint free cloth, inspect the seals for signs of distortion, swelling, splitting, or hardening although it is normal practice and also recommended that new seals are always fitted after dismantling.
11 Inspect the bore and piston for signs of deep scoring which, if evident, means a new cylinder should be fitted. Ensure that the cylinder bore parts are clean.
12 As the respective parts are fitted to the cylinder bore ensure that they are thoroughly lubricated with clean hydraulic fluid of the specified type.
13 Install the new seals to the pistons making sure that they are fitted the correct way round as noted during dismantling.
14 Take care not to scratch the piston cups when fitting the secondary and primary piston units. If damaged they must be renewed.
15 Tighten the piston stopper bolt and the connector bolt to the specified torque.

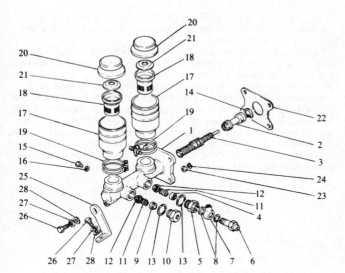

Fig. 9.7. The dual master cylinder component parts

1 Master cylinder
2 Primary piston
3 Secondary
4 Check valve
5 End plug
6 Connector bolt
7 Connector
8 Gasket
9 Check valve
10 Connector
11 Return spring
12 Spring seat
13 Gasket
14 Snap ring
15 Stopper bolt
16 Gasket
17 Reservoir
18 Filter
19 Clamp
20 Filler cap
21 Plate
22 Gasket
23 Nut
24 Spring washer
25 Bracket
26 Bolt
27 Spring washer
28 Washer

8 Brake master cylinder (single) - dismantling and reassembly

1 The basic procedures given in the previous Section are applicable to single master cylinders. Ensure that the new secondary and primary seals are located in their piston grooves with the flared faces inwards. Install the connecting bolt and fluid reservoir to the cylinder and tighten to the specified torque.

9 Brake hoses - inspection, removal and installation

1 Inspect the condition of the flexible hydraulic hoses. If they are swollen, damaged or chafed they must be renewed.
2 Wipe the top of the brake master cylinder reservoir and unscrew the cap. Place a piece of polythene sheet over the top of the reservoir and install the cap. This is to stop hydraulic fluid syphoning out during subsequent operations. Note that tandem master cylinders have two reservoirs but it is not necessary to cover both cylinders unless both front and rear hoses are being removed.

Front brake hose

3 First jack-up the vehicle to take the weight off the suspension. To remove the hose, wipe the unions and bracket free of dust, and undo the union nut from the metal pipe end.
4 Withdraw the metal clip securing the hose to the bracket and detach the hose from the bracket. Unscrew the hose from the wheel cylinder.
5 Installation is the reverse of the removal procedure but ensure that the hose is connected at the wheel cylinder end first, with the wheels in the 'straight-ahead' position. On completion, the front brakes must be bled of air, as described in the previous Section.

Rear brake hose

6 To remove a rear flexible hose, wipe the unions, bracket and three way adaptor free of dust, and undo the union nut from the metal pipe end.
7 Withdraw the metal clip securing the hose to the bracket and detach the hose from the bracket. Unscrew the hose from the three-way adaptor.
8 Installation is the reverse of the removal procedure. Ensure that the hose is connected at the three-way adaptor end first, and on completion bleed the rear brakes.

10 Front brake shoes (Series 1 to 4) - inspection, removal and installation

1 Apply the parking brake and raise the front end of the vehicle. Remove the roadwheel.
2 Unscrew the brake drum retaining screws (photo) and withdraw the brake drum. If it proves difficult to remove the drum, disconnect the rubber hole plugs from the flange plate and insert a screwdriver thru the flange hole and locate in the hole in the brake shoe, refer to Fig. 9.8. Lift the end of the brake shoe return spring, freeing it from the serration and move the brake shoe towards the wheel cylinder.
3 Before removing the drum, mark the relative position to the hub to ensure correct reinstallation. With the drum removed do not depress the brake pedal.
4 Inspect the linings for excessive wear or shoe damage, and contamination from brake fluid or hub grease. Brake linings which are worn down to within 0.059 in of the shoes should be renewed. Shoes which are impregnated with oil or grease should be renewed as it is not very satisfactory to use proprietary solvents which tend to clean the surface but then allow the deposits which have soaked into the linings to come to the surface and form a glazed film.
5 At this stage, also inspect the condition of the drum. If the drum diameter exceeds that specified, or is heavily scored, a replacement item must be installed. If it is necessary to remove the brake shoes proceed as follows.
6 With a pair of pliers, detach the wheel cylinder piston springs (photo) from the shoes and pistons.
7 The hold down spring retainers are now pressed down and turned thru 90°, again using a pair of pliers, so that the springs and retainers can be removed (Fig. 9.9).
8 The upper and lower brake shoes can now be removed and if they are to be refitted, they should be marked so that they are reinstalled in their correct locations.
9 If the self-adjusting retainer, spring, washer pin and adjuster lever are to be removed from the brake shoe, an arbor press will be needed to depress the spring retainers. Hold the retainer and rotate the shoe thru 90° to separate.
10 To prevent the pistons coming out of the wheel cylinders during cleaning and inspection, rubber bands can be used to hold them in place.
11 Any grease or dirt retaining on the backing plate can now be removed using gasoline or other proprietary solvents. Dry off afterwards with a lint free cloth.
12 Inspect the flange plate and securing bolts to ensure that they are tight. Check the return springs for distortion, and weakness. Renew any suspect or damaged parts.
13 Reassemble the self-adjusting retainer and adjusting lever to the brake shoe in the reverse manner of removal. Ensure that the pin end of the spring is located in the retainer groove. Fit the washer with its lining side towards the lever The right- and left-hand adjuster levers are not interchangeable so be sure to fit them to their respective sides.
14 The brake shoe contact surfaces to flange plate must be lubricated with Delco Brake Lube 5450032 or an equivalent and also the wheel cylinder contact faces, see Fig. 9.10 before assembly.

10.2 The brake drum retaining screw

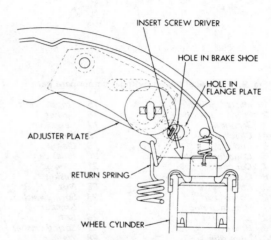

Fig. 9.8. The brake shoe and wheel cylinder assembly

Fig. 9.9. Press and turn the spring retainers to remove

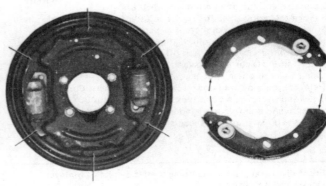

Fig. 9.10. Arrows point to the lubrication points

10.6 Remove the piston spring and brake shoe return spring

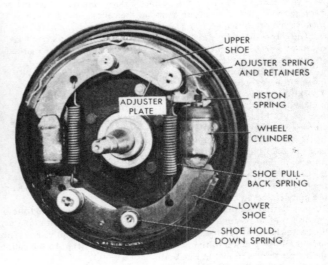

Fig. 9.11. General view of the brake assembly on refitting

Fig. 9.12. Withdraw the caliper stops

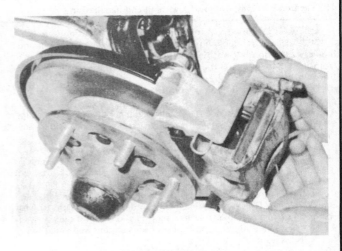

Fig. 9.13. Remove the caliper

15 Hook the right- and left-hand return springs to their respective brake shoes - they are not interchangeable. The left-hand side springs are colored light blue, the right-hand side black.
16 Engage the grooved section of the adjuster lever to the guide pin and then relocate the brake shoes into position. The shoes must be correctly fitted to the guide pin. If the brake shoe end is not located in the groove the shoes have lifted off of the ridged section. Check that the return spring is correctly located to the adjuster lever.
17 Using a pair of pliers, refit the piston springs.
18 Relocate the shoe hold down springs and retainers. Compress the spring with a pair of pliers and twist the retainer thru 90°, ensuring that the pin end is correctly seated in the retainer groove.
19 Refit the brake drum and readjust the brakes as described in Section 2.

11 Front disc brake, shoes and lining (Series 5 and 6) - removal, inspection and installation

1 Raise the vehicle on a hoist at the front and chock the rear wheels.
2 Remove the roadwheels.
3 With a pair of pliers extract the clip pins from the caliper stops and then withdraw the stops (Fig. 9.12).
4 Withdraw the caliper from the support as in Fig. 9.13 and then remove the stop plates from the caliper. Taking care not to distort or damage the flexible brake hose, suspend the caliper unit from the main frame or upper link using a piece of nylon cord or similar.
5 The respective shoe and lining assemblies with shims can now be detached (Fig. 9.14). If they are to be refitted, they must be marked so that they are refitted in their original location.
6 Clean off the respective components with a soft dry brush and inspect for wear and distortion.
7 The anti rattle springs, stop pins, caliper, support and adaptors must all be in good condition, if not, they must be renewed as necessary. Check the disc for run-out and if badly scored or cracked this must also be renewed.
8 The brake shoes and linings must be renewed when the lining is worn to 0.039 in (approx) thickness over the shoe table. The shoes and linings are renewed as sets.
9 Check the inner caliper for signs of leakage. The caliper seal must be in good condition and free of cracks, swelling or distortion. Never use compressed air to clean the inner caliper.
10 Installation is a reversal of the removal instructions but the following points must be adhered to. When fitting the shims and stop plates, lubricate them with Delco Brake Lube 5450032 or an equivalent. This should also be applied to the new stop plates when fitted to the caliper sliding portion as shown in Fig. 9.15.
When fitting the linings to the support the wear indicator must be located to the lower side of the support.

12 Disc brake caliper - removal, overhaul and installation

1 Follow instruction 1, 2 and 3 in the previous Section.
2 Wipe clean the flexible hose connections between the caliper and brake line. Undo the front flexible hose connection and disconnect it from the brake line. Plug the end of the line to prevent spillage of fluid and the ingress of dirt.
3 Disconnect the caliper unit from the support and remove the stop plates from the caliper.
4 Unscrew and remove the flexible hose connection from the caliper.
5 With a small screwdriver, prise the dust seal from the caliper - see Fig. 9.16.
6 Remove the piston from the caliper by applying compressed air into the flexible hose connection. Do not catch the piston by hand, but place a suitable block of wood into the caliper to check it on exit.
7 Detach and discard the piston square ring seal.
8 Clean all components in clean brake fluid and blow dry with compressed air. Check the cylinder bore and pistons for signs of wear, corrosion or surface defects such as scoring and renew if necessary.
9 The dust seal and piston seal must always be renewed when the caliper is dismantled.
10 To reassemble the caliper, lubricate the caliper bore and the piston square ring seal with Delco Silicone Lube 5459912 or an equivalent before inserting the piston seal into the bore.

Fig. 9.14. Remove the shoe and lining

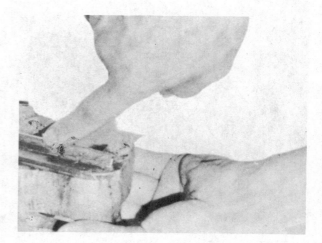

Fig. 9.15. Lubricate the stop plates

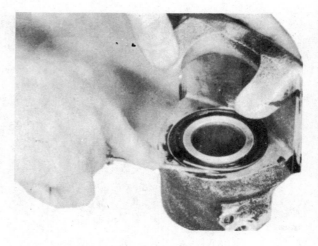

Fig. 9.16. Remove the dust seal

11 Carefully insert the piston into the caliper unit (Fig. 9.17) using only finger pressure.
12 Apply Delco Silicone Lube 5459912 or an equivalent to the piston and then fit the dust seal to the piston and caliper. Insert the seal ring into the dust seal, and then refit the flexible hose to the caliper together with new gaskets.
13 Smear the stop plates and sliding surfaces of the caliper with Delco Brake Lube 5450032 or an equivalent and insert the stop plates into the caliper. Now refit the caliper, stops and new stop pins.
14 Reconnect the flexible brake hose to the brake line, with the identification stripe in a straight line. The hose should be clear of any interference during the full movement of the steering and suspension components.
15 Refit the roadwheel and then bleed the brake system, topping up the master cylinder to replace any loss, as described in Section 5.

13 Brake disc/hub - removal, inspection and installation

1 Remove and suspend the caliper unit as described in Section 11, paragraphs 1 to 4.
2 Prise the hub grease cap free using a screwdriver and withdraw the cotter pin, spindle nut retainer and nut. The hub and disc rotor assembly can now be lifted clear. Take care not to drop the wheel bearings.
3 Clean off the component parts including the dust shield and adaptor. Inspect the disc and hub for signs of wear, cracks or score marks. The hub and disc are replaced as a unit if they are in need of renewal. The dust shield and adaptor must also be renewed if they are damaged. To remove them, unscrew the attachment bolts and detach them from the steering knuckle (Fig. 9.18).
4 Installation of the dust shield and adaptor, and the disc/hub and caliper assemblies is a direct reversal of the removal procedure, but note the following:

a) Tighten all bolts to the specified torque settings.
b) When the caliper has been refitted and the brake line reconnected, top up the master cylinder and bleed the brakes as described in Section 5.

14 Rear brake shoes - inspection, removal and installation

1 Raise the rear of the vehicle on a hoist and chock the front wheels. Remove the rear wheels.
2 Slacken off the parking brake equaliser check nuts to release the brake cable tension.
3 Unscrew the brake drum retaining screws and remove the drum/s. If the drums are held by the shoes, detach the rubber hole plugs in the flange plate and retract the brake shoes by inserting a screwdriver and turning the self-adjuster. Mark the drum location to hub, to ensure correct refitting position. Do not press the brake pedal whilst the drums are removed.

Fig. 9.17. Insert the piston using only finger pressure

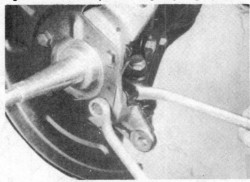
Fig. 9.18. Method of removing the dust shield and adaptor

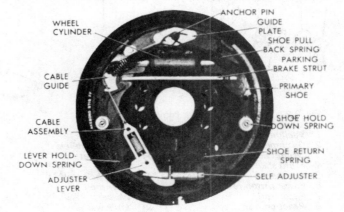
Fig. 9.19. The rear brake layout

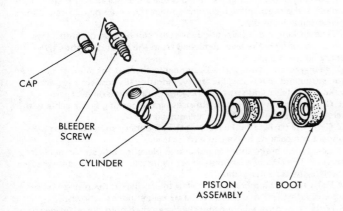
Fig. 9.20. The front wheel cylinder components

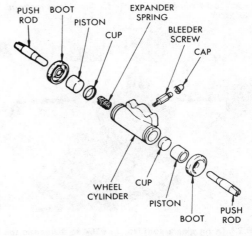
Fig. 9.21. The rear wheel cylinder components

Chapter 9/Braking system

14.5 The pull back springs on anchor pin

14.7 Detach the spring from the adjuster lever

14.16 Locate the cable over the anchor pin and round the guide

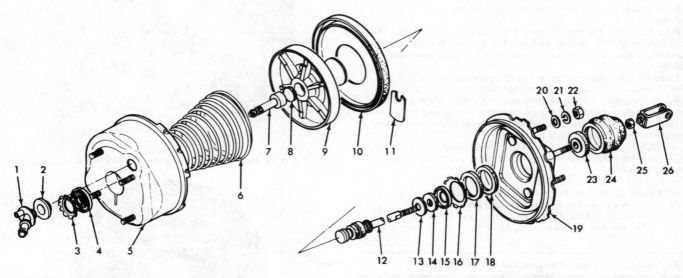

Fig. 9.22. The power cylinder component parts

1 Vacuum check valve
2 Grommet
3 Seal retainer
4 Front shell seal
5 Front shell
6 Power piston return spring
7 Piston rod
8 Reaction disc
9 Power piston
10 Diaphragm
11 Push rod retainer plate
12 Push rod assembly
13 Silencer
14 Silencer
15 Silencer retainer
16 Bearing retainer
17 Bearing
18 Rear shell seal
19 Rear shell
20 Washer
21 Lock washer
22 Nut
23 Air filter
24 Push rod boot
25 Clevis locknut
26 Clevis

Fig. 9.23. Using special tool No. J - 9504 to dismantle the power cylinder

4 The brake shoes and associate components can now be wiped clean and inspected as detailed in Section 10, paragraphs 4 and 5. If the shoes are to be removed, proceed as follows.
5 Unhook the brake pull-back springs from the anchor pin (photo) and detach the springs.
6 With a pair of pliers, remove the brake shoe hold down springs. Press in the spring retainer and twist it thru 90° to align the slot with the flanged end of the pin.
7 The self-adjuster cable is removed by detaching the spring from the adjuster lever (photo) and the cable end from the anchor pin.
8 Withdraw the adjuster lever and the hold down wire from the brake shoe pivot. Detach the brake shoes from the wheel cylinder pushrods and then separate the respective brake shoes, the adjuster, return spring and parking brake strut units. If the brake shoes are to be refitted, note and mark their respective locations so that they may be reinstalled into their original locations.
9 The parking brake lever and rear cable can now be disconnected and the parking brake lever detached from the secondary shoe by removing the dip and washer.
10 Having removed the self-adjuster cable check it for damage or elongation and renew it if necessary. Additionally inspect the parking brake cable, lever and strut and if worn or defective renew them.
11 Commence installation by lubricating the parking brake cable with Delco Brake Lube 5450032 or an equivalent.
12 Refit the parking brake lever to the secondary shoe and then connect the parking brake cable to its lever.
13 Locate the brake shoes together and place the adjuster screw into position. The adjuster assembly must be installed with the star wheel closer to the secondary shoe.
14 Refit the parking brake strut with the spring on the primary shoe end and refit the brake shoes to the wheel cylinder pushrods.
15 With a pair of pliers, relocate the brake shoe hold down springs. Press the springs and retainers and twist thru 90° to locate.

Chapter 9/Braking system

16 Fit the guide plate to the anchor pin, and reassemble the self-adjuster lever and lever hold down wire to the pivot pin in the secondary pin. Locate the adjuster cable over the anchor pin (photo) and run the cable round the shoe shield (guide). Attach the spring to the opposing end of the adjuster lever.
17 Using pliers, refit the pull-back springs: Lubricate the brake shoe contact surfaces with Delco Brake Lube 5450032 or an equivalent by prising the shoes from the flange plate.
18 Check that the actuator lever operates correctly by hand operating the self-adjustment.
19 Adjust the brakes as specified in Sections 2 and 4 prior to using the vehicle on the road.

15 Brake wheel cylinders - removal, servicing and installation

1 Raise the vehicle with a hoist and remove the roadwheel and brake drum. Mark the drum location to the hub for refitting.
2 Detach the hydraulic brake line connection from the wheel cylinder. Plug the end of the line to prevent spillage of fluid and the ingress of dirt.
3 Disconnect the brake pull-back springs using a pair of pliers.
4 Remove the wheel cylinder to flange plate retaining screws.
5 Disconnect the brake cylinder pushrod from the shoes and withdraw the cylinder.
6 Detach the boot/s from the ends of the wheel cylinder, and withdraw the piston/s and cup/s.
7 Using clean brake fluid, clean the wheel cylinder components. Do not use gasoline, kerosene or any other type of cleaning solvent, as they are detrimental to rubber.
8 Inspect the components carefully and if the cylinder bore/or piston scored or corroded, they will have to be renewed. The rubber cups and boots must always be renewed irrespective of their condition.
9 Assemble by first lubricating the cylinder bore with clean brake fluid. As each component is assembled they should be lubricated in a similar fashion.

Front wheel cylinders

10 On the front wheel cylinders insert the piston assembly into the cylinder taking great care not to distort or damage the boot.

Rear wheel cylinders

11 On the rear cylinders install the spring expander into the cylinder bore. Fit the new cups with their flat face towards the outer cylinder ends. They must be wiped clean before assembly with a lint free cloth. Insert the new pistons into the cylinder with the flat face to the cylinder center.
12 Fit the rubber boots onto the cylinder.
13 Locate the wheel cylinder to the brake flange plate and secure it with screws tightened to the recommended torque.
14 Refit the pushrods and the pull-back springs. Unplug the hydraulic hose and reconnect it to the cylinder. Tighten to the specified torque wrench figure.
15 Refit the brake pull-back springs, the drum and wheel.
16 Bleed the brake system as described in Section 5. Lower the vehicle and remove the hoist.

16 Power cylinder - removal and installation

1 Remove the master cylinder as described in Section 6 but leave attached to the power cylinder.
2 Slacken the vacuum line clamp screw from the check valve, and then unscrew the clip to sender skirt retaining screws. Move the vacuum line out of the way.
3 Detach the brake pedal return spring, and withdraw the cotter pin and washer from the pushrod clevis to brake pedal pin. Extract the pin.
4 Unscrew the power cylinder to dash panel nuts and remove with lockwashers and washers. The master cylinder and power cylinder can now be carefully removed as a unit from the vehicle.
5 Drain off any remaining fluid from the master cylinder reservoirs.
6 The master cylinder to power cylinder retaining nuts and washers can now be removed and the two components detached.
7 Installation is the reverse procedure of removal. On refitting, top up the master cylinder fluid level and bleed the system. Check and adjust the brake pedal travel as described in Section 3.

17 Power cylinder - dismantling, inspection and reassembly

1 Prior to dismantling the power unit, refer to Fig. 9.23 and 9.25, which show the special tool, Spanner Wrench J-9504, which is required to dismantle the unit. If this is available, proceed as follows.
2 Clean the outside of the power unit thoroughly before dismantling. It is of utmost importance that cleanliness be a prime factor during the overhaul.
3 Mark the relative positions of the front and rear shell units for correct alignment on reassembly.
4 Locate the master cylinder flange in a vise but do not overtighten. The power cylinder must be facing upwards.
5 Slacken the pushrod clevis locknut and remove the clevis and locknut.
6 Detach the pushrod boot.
7 Locate the special spanner wrench J-9504 over the rear shell studs and press downwards whilst unscrewing (the rear shell is under spring pressure). Remove the rear shell and also the piston rod, power piston and return spring and retainer, refer to Fig. 9.22.
8 Unscrew and remove the master cylinder to power cylinder retaining nuts/washer. Discard the gasket on separating the two units.
9 Prise the air silencer retainer from the power piston and detach the air silencer and idler.
10 Carefully remove the rubber diaphragm from the piston.
11 Turn the piston so that the pushrod retainer slot position is down, then press the pushrod and let the retainer fall out of the piston. Withdraw the pushrod assembly and the reaction disc. Do not dismantle the pushrod unit. If on inspection it is found to be defective, renew it as a unit.
12 Inspect the rear seal and if defective pry it out of the retainer using a small screwdriver, and then withdraw the spacer and seal assembly.

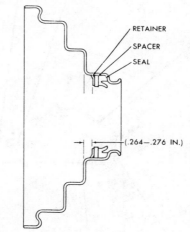

Fig. 9.24. Set the retainer to the depth shown

Fig. 9.25. Use special gauge J - 24568 to check the piston adjustment

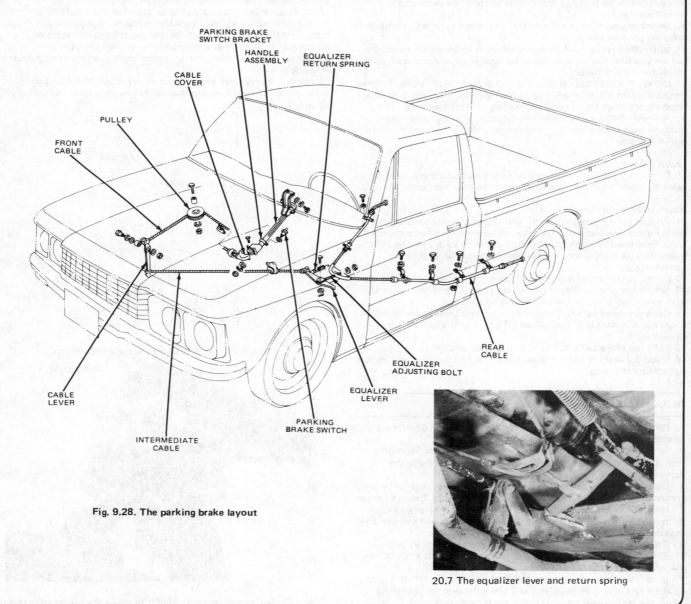

Fig. 9.26. If a height gauge is not readily available, it can be made from dimensions shown

Fig. 9.27. The clutch and brake pedal layout

Fig. 9.28. The parking brake layout

20.7 The equalizer lever and return spring

Chapter 9/Braking system

13 Inspect the front seal and, if defective, similarly pry it out with a small screwdriver and renew it.
14 If defective, the vacuum check valve can be removed by rotating it backwards and forwards. Remove the grommet.
15 All metal, plastic and rubber parts of the power cylinder must be cleaned with a solution of denatured alcohol. Do not use any mineral based cleaning solvents, ie, gasoline, kerosene etc.
16 Blow out the passages and valve holes and air dry. Any rust found on the inside of the shells can be removed using a fine emery or crocus cloth. Clean out all rust and dirt particles.
17 Renew all suspect components - do not attempt to repair them.
18 Reassemble in the reverse order of dismantling but note the following:
19 The front shell seal, if removed, must be smeared with a liberal amount of Delco Silicone number 5459912 lubricant or an equivalent. The lip of the seal faces forwards.
20 Similarly, lubricate the rear shell seal. The seal is fitted with the lip facing rearwards. Set the retainer to the depth specified in Fig. 9.24.
21 A thin coat of Delco Silicone lubricant must also be applied to the piston seal and inner surface lip of the piston housing. Take care when fitting the pushrod piston not to damage the seal. Should it be damaged on assembly, the pushrod assembly must be renewed.
22 On installing the diaphragm, turn it one half of a turn to ensure that it is correctly located. Then, temporarily retain the master cylinder to the front shell with nuts and washers. Locate the flange part of the master cylinder in the vise with the front shell side upwards. Lubricate the piston rod face, the reaction disc and outer rim of the diaphragm with Delco Silicone Lubricant (No. 5459912) and refit the reaction disc in the power piston unit.
23 The rear shell, power piston, piston rod, return spring retainer and spring, and finally the rear shell can now be assembled in order. Locate the front and rear shells using the special spanner wrench used for removal, and align the relative marks. Be certain that the rear shell is locked to the front shell by all of the tabs before releasing the pressure on the special tool.
24 Relocate pushrod boot, and loosely install the pushrod clevis locknut and clevis. Remove the master cylinder from the power cylinder and locate the power cylinder in the vise so that the piston rod is up. Do not overtighten the vise.
25 Locate the special gauge J-24568 over the piston rod with its 'legs' resting on the master cylinder mounting face - see Fig. 9.25. The piston rod must touch the cutaway portion of the gauge. Adjust the piston if required by holding the rod at the serrated portion with a pair of pliers and rotate the thread at one end accordingly. The piston rod must be fully bottomed in the cylinder before adjusting. A height gauge may be made from the dimensions shown in Fig. 9.26.

18 Brake/clutch pedal - removal and installation

1 A common bolt retains both the clutch and brake pedals in the pedal support bracket. Pedal removal is as follows.
2 Detach the ground cable from the battery.
3 Detach the clutch pedal return spring from the instrument panel, and the brake pedal return spring from the pedal.
4 Withdraw the cotter pins from the clevis pin, and remove the brake and clutch pedal clevis pins and washers.
5 Undo the pivot bolt nut and remove with the lockwasher. Withdraw the pivot bolt, and remove the pedals, bushes and plastic thrust washers.
6 Installation is the reverse procedure of removal but lubricate the bushes with Delco Brake Lube 5450032 or an equivalent before refitting the pedals. Adjust the brake light switch and clutch switch on completion.

19 Parking brake - removal and installation

1 On all Series models, detach the ground cable from the battery.
2 Remove the air cleaner from the carburetor. On Series 1 models only, detach the accelerator control rod from the dash panel.
3 Drain the cooling system and disconnect the heater hoses from the core outlet tubes in the dash panel.
4 Detach the parking brake front cable from the control lever on the right-hand side engine compartment, and then unscrew the right-hand pulley center bolt and remove it.
5 Unscrew and remove the cable cover nuts at the dash panel to remove the cover. Disconnect the windshield wiper switch from the instrument panel, and on Series 1 models the hazard warning light wiring connections from the switch. Remove the switch.
6 Unscrew the lever unit to instrument panel securing screws and detach the parking brake light switch wires from the switch. Remove the lever and front cable.
7 Installation is a direct reversal of removal. Refit the cooling system on completion and check for hose leaks at their connections. Also check the hazard warning light, brake light and parking brake warning lights on reconnecting the ground cable.

20 Parking brake cables - removal and installation

Front parking brake cable
1 Follow the instructions given in paragraphs 1 to 6 in Section 19 and disconnect the lever unit from the instrument panel.
2 Pull the assembly to the rear and lay it on the floor. Slacken the parking brake light switch bracket screw and then turn the switch and bracket thru 90°. Free the ratchet and press the handle fully in.
3 Withdraw the cotter pin and remove with washer pivot pin and pulley, then withdraw the cable.
4 To install the cable, attach it to the handle lower end and pull the handle to the rear several notches.
5 Turn the parking brake light switch and bracket back into position and tighten the retaining screw. Re-insert the pivot pin with washers and fit a new cotter pin to retain. The remaining assembly procedures are the reversal of removal.

Intermediate parking cable
6 Securely raise the vehicle on a hoist and chock the front wheels.
7 Detach the equalizer lever spring from the lever and then slacken the cable guide nut to remove the cable assembly (photo).
8 Refitting is a reversal of the above procedure; but on completion readjust the parking brake as given in Section 4.

Rear parking brake cable
9 Raise the vehicle securely at the rear on a hoist and chock the front wheels. Remove the right- and left-hand rear wheels.
10 Disconnect the rear cable retaining clamps on the right- and left-hand side. Detach the equalizer lever return spring from the lever and disconnect the equalizer adjusting bolt nut. Remove the rear cable from the adjustment bolt.
11 Remove the brake drums and shoes as described in Section 14, and detach the cable from the brake lever.
12 Using a suitable box end wrench, remove the rear cable spring as shown in Fig. 9.29.
13 Remove the cable assembly by withdrawing the cable ends from the right- and left-hand flange plates.
14 Installation is the reversal of the above procedure but on completion check and adjust the parking brake as described in Section 4.

Fig. 9.29. Removing the rear cable spring using a box end wrench

21 Fault diagnosis

Symptom	Reason/s
Pedal travels almost to floor before brakes operate	Brake fluid level too low. Wheel cylinder leaking. Master cylinder leaking (bubbles in master cylinder fluid). Brake flexible hose leaking. Brake line fractured. Brake system unions loose. Shoes linings excessively worn.
Brake pedal feels 'springy'	New linings not yet bedded-in. Brake drums badly worn or cracked. Master cylinder securing nuts loose.
Brake pedal feels 'spongy' and 'soggy'	Wheel cylinder leaking. Master cylinder leaking (bubbles in master cylinder reservoir). Brake pipe line or flexible hose leaking. Unions in brake system loose. Blocked reservoir cap vent hole.
Excessive effort required to brake vehicle	Shoe linings badly worn. New shoes recently fitted - not yet bedded-in. Harder linings fitted than standard resulting in increase in pedal pressure. Linings and brake drums contaminated with oil, grease or hydraulic fluid. Servo unit inoperative or faulty. Scored drums.
Brakes uneven and pulling to one side	Linings and drums contaminated with oil, grease or hydraulic fluid. Tire pressure unequal. Radial ply tires fitted at one end of the vehicle only. Brake shoes fitted incorrectly. Different type of linings fitted at each wheel. Anchorages for front or rear suspension loose. Brake drums badly worn, cracked or distorted. Incorrect front wheel alignment. Incorrectly adjusted front wheel bearings.
Brakes tend to bind, drag or lock-on	Air in hydraulic system. Wheel cylinders seized. Handbrake cables too tight. Weak shoe return springs. Incorrectly set foor pedal or pushrod. Master cylinder seized. Brakes over-adjusted.

Chapter 10 Electrical system

For information relating to Series 8 thru 12 models, including 4WD, refer to Chapter 13.

Contents

Air conditioning components ... 42	Heater controls - removal and refitting ... 37
Alternator - dismantling, inspection, testing and reassembly ... 9	Heater unit - removal and refitting ... 35
Alternator - general description, maintenance and precautions ... 6	Heater water valve - removal and refitting ... 38
Alternator - removal and installation ... 7	Horn ... 31
Alternator - testing ... 8	Ignition switch ... 26
Back-up (reverse) lamps ... 22	Instrument panel - dismantling and reassembly ... 25
Battery - charging ... 5	Instrument panel - removal and installation ... 24
Battery - electrolyte replenishment - ... 4	Licence plate light ... 20
Battery - maintenance ... 3	Radio equipment - removal and installation ... 34
Battery - removal and installation ... 2	Rear combination lamps ... 19
Blower motor and heater core unit - removal and replacement ... 36	Regulator - description and testing ... 10
Blower resistor - removal and replacement ... 43	Regulator - removal, adjustment and installation ... 11
Dash outlet grille - removal and refitting ... 41	Side marker lights - front and rear ... 21
Direction signal and dimmer switch - removal and installation ... 32	Starter motor - dismantling, servicing and reassembly ... 14
Dome (interior) lamp ... 23	Starter motor - general description ... 12
Fault diagnosis - electrical system ... 45	Starter motor - removal and installation ... 13
Front combination lamps ... 18	Thermostatic switch - removal and replacement ... 44
Fuses ... 15	Turn signal and dimmer switch - testing ... 27
General description ... 1	Ventilation air inlet valve - removal and refitting ... 39
Hazard warning switch - removal and installation ... 33	Windshield washer ... 30
Headlamps - adjustment (beam alignment) ... 17	Windshield wiper motor - dismantling and inspection ... 29
Headlamps - removal and installation ... 16	Windshield wiper motor - removal and installation ... 28
Heater control cable - adjustment ... 40	

Specifications

System type ... 12 volt, negative ground (earth)

Battery
Model ...	N50
Capacity ...	12V - 50AH

Alternator
Make ...	Hitachi
Drive ...	V-belt
Model - Series 1 to 3 models ...	LT 130-83
Voltage output ...	12V - 30A
Series 4 models ...	LT 135-26
Voltage output ...	12V - 35A
Series 5 and 6 models ...	LT 135-30
Voltage output ...	12V - 35A
Weight ...	8.6 lb (3.9 kg)
Rated speed ...	5000 rpm
Operating speed ...	1000 - 13500 rpm
Pulley diameter ...	2.756 in (70 mm)
Pulley ratio ...	1.96
Brushes	
Standard length ...	0.57 in (14.5 mm)
Minimum length ...	0.276 in (7.0 mm)
Brush spring tension ...	0.66 lb (2.9 n)

Shaft diameter
 Front ... 0.6 in (15 mm)
 Rear ... 0.47 in (12 mm)
Rotor coil resistance @ 20°C (68°F) ... 4.3 ohm
Stator single coil resistance @ 20°C (68°F) ... 0.13 ohm

Regulator
Model
 Series 1 to 3 models ... TL 1Z 66
 Series 4 to 6 models ... TL 1Z 87
Fold relay actuates voltage ... 5 volts
Regulated voltage
 Series 1 to 3 ... 13.5 to 14.5 volts
 Series 4 to 6 ... 13.8 to 14.8 volts
Core gap ... 0.024 to 0.039 in (0.6 to 1.0 mm)
Point gap ... 0.012 to 0.016 in (0.3 to 0.4 mm)
Field charge relay
 Voltage coil resistance ... 31.9 ohms
 Core gap ... 0.032 to 0.039 in (0.8 to 1.0 mm)
 Point gap ... 0.016 to 0.024 in (0.4 to 0.6 mm)

Starter motor
Make ... Hitachi
Model
 Series 1 to 3 models ... S114 - 118
 Series 4 model ... S114 - 136
 Series 5 and 6 models ... S114 - 202
Rating ... 30 sec
Output ... 12V - 1.0 HW
Rotation direction ... Clockwise from pinion side
Clutch type ... Overrunning clutch
Engagement method ... Magnetic shifting
Pinion activating voltage ... 8V or less
No load performance
 Terminal voltage ... 12V
 Current ... 60A or less
 Speed ... 600 rpm or more
Loaded performance
 Voltage (potential drop characteristics) ... 12V/OA - 5V/300A
 Current ... 200A or less
 Torque ... 3.6 ft lbs (4.9 Na)
 Speed ... 1,100 rpm or more
Constrained
 Voltage ... 12V/OA - 5V/300A
 Current ... 330A or less
 Torque ... 5.8 ft lbs (7.8 Na) or more

Fuses
Fuse box location ... Inner wing panel of engine compartment
Fuse rating ... Marked on fuse box cover

Windshield wiper motor
Operating voltage ... 10 - 16V
Minimum operating voltage ... 8V
Wiper operating angle
 Driver side ... 90°
 Passenger side ... 110°

Windshield washer
Rating ... 12V - 2.5A
Operating voltage ... 10 - 15V
Delivery pressure ... 13 psi (90 kpa)
Nozzle hole diameter ... 0.03 in (0.8 mm)
Tank capacity ... 0.4 gall (1.5 liters)

Horn
Rated voltage ... 12V
Rated current ... 3A (max)
Sound energy ... 105 ± 5 dB
Standard frequency ... 370 ± 20 H2

Bulbs
Headlamp Type 1 ... 12V - 37.5W
 Type 2 ... 12V - 37.5/50W

Chapter 10/Electrical system

Front combination lamp	12V - 23/8W
Turn signal lamp	12V - 23/8W
Clearance lamp	12V - 23/8W
Interior (dome) lamp	12V - 5W
Stop/tail lamp	12V - 23/8W
Rear turn signal lamps	12V - 23W
Back up (reverse) lamp	12V - 23W
Licence plate lamp	12V - 7.5W
Front side marker lamps	12V - 8W
Rear side marker lamps	12V - 8W
Indicator - headlamp high beam lamp	12V - 3.4W
Oil pressure indicator warning lamp	12V - 3.4W
Seat belt warning lamp	12V - 3.4W
Catalytic converter warning lamp	12V - 3.4W
Turn signal indicator lamp	12V - 3.4W
Instrument panel lamps	12V - 3.4W
Wiper illumination switch lamp	12V - 3.4W
Heater illumination switch lamp	12V - 1.3W
Ash tray lamp	12V - 3.4W
Generator indicator lamp	12V - 3.4W
Brake system warning lamp	12V - 3.4W

Heater blower motor

Series 1 models

Type	Outside and recirculating air
Voltage	12V
Blower speed	
High	3,500 rpm
Low	2,500 rpm
Blower motor loaded current	Below 7 amps (high blower)

Series 2 to 6

Type	Full time outside air
Voltage	12V
Blower speed	
High	3,800 rpm
Medium	3,400 rpm
Low	1,800 rpm
Blower motor loaded current	Below 7 amps (high blower)

1 General description

The electrical system is of 12 volt negative ground (earth) type. The major components comprise the battery, a belt-driven alternator and a pre-engaged starter motor.

The battery supplies a steady current to the ignition system (see Chapter 4), and for the operation of the electrical accessories.

The alternator maintains the charge in the battery, and the voltage regulator adjusts the charging rate according to the system demands. Silicon diodes within the alternator rectify the alternating current produced into direct current. A cut-out prevents the battery discharging when the engine is switched off or when it is running at idle speed.

2 Battery - removal and installation

1 The battery is in a special carrier fitted on the right-hand wing valance of the engine compartment. It should be removed once every three months for cleaning. Disconnect the leads from the battery terminals by slackening the clamp retaining nuts and bolts, or by unscrewing the retaining screws if terminal caps are fitted instead of clamps.

2 Unscrew the clamp bar retaining nuts, then remove the clamp. Carefully lift the battery from its carrier. Hold the battery vertical to ensure that none of the electrolyte is spilled.

3 Installation is a direct reversal of this procedure. **Note:** Install the negative lead before the positive lead and smear the terminals with petroleum jelly to prevent corrosion. **Never** use an ordinary grease as applied to other parts of the car.

3 Battery - maintenance

1 Keep the top of the battery clean by wiping away dirt and moisture.

2 Remove the plugs or lid from the cells and check that the electrolyte level is just above the separator plates. If the level has fallen, add only distilled water until the electrolyte level is just above the separator plates.

3 As well as keeping the terminals clean and covered with petroleum jelly, the top of the battery, and especially the top of the cells, should be kept clean and dry. This helps prevent corrosion and ensures that the battery does not become partially discharged by leakage through dampness and dirt.

4 Once every three months, remove the battery and inspect the battery securing bolts, the battery clamp plate, tray and battery leads for corrosion (white fluffy deposits on the metal which is brittle to touch). If any corrosion is found, clean off the deposits with an ammonia or soda solution and paint over the clean metal with a fine base primer and/or underbody paint.

5 At the same time inspect the battery case for cracks. If a crack is found, clean and plug it with one of the proprietary compounds marketed for this purpose. If leakage through the crack has been excessive then it will be necessary to refill the appropriate cell with fresh electrolyte as detailed later. Cracks are frequently caused to the top of the battery case by pouring in distilled water in the middle of winter *after* instead of *before* a run. This gives the water no chance to mix with the electrolyte and so the former freezes and splits the battery case.

6 If topping-up the battery becomes excessive and the case has been inspected for cracks that could cause leakage, but none are found, the battery is being over-charged and the voltage regulator will have to be checked and reset.

7 With the battery on the bench at the three monthly interval check, measure its specific gravity with a hydrometer to determine the state of charge and condition of the electrolyte. There should be very little variation between the different cells and if a variation in excess of 0.025 is present it will be due to either:

   a) *Loss of electrolyte from the battery at some time caused by spillage or a leak, resulting in a drop in the specific gravity of*

electrolyte when the deficiency was replaced with distilled water instead of fresh electrolyte.
b) An internal short circuit caused by buckling of the plates or a similar malady pointing to the likelihood of total battery failure in the near future.

8 Fully charged, the battery will have a specific gravity reading of approximately 1.270 at an electrolyte temperature of 80°F. For every 10°F above 80°F, add four specific gravity points (0.004) and for every 10°F below 80°F subtract four specific gravity points (-0.004).

4 Battery - electrolyte replenishment

1 If the battery is in a fully charged state and one of the cells maintains a specific gravity reading which is 0.025 or lower than the others, and a check of each cell has been made with a voltage meter to check for short circuits (a four to seven second test should give a steady reading of between 1.2 and 1.8 volts), then it is likely that electrolyte has been lost from the cell with the low reading at some time.
2 Top-up the cell with a solution of 1 part sulphuric acid to 2.5 parts of distilled or de-ionized water. If the cell is already fully topped-up draw some electrolyte out of it with a hydrometer.
3 When mixing the sulphuric acid and water **never add water to sulphuric acid** - always pour the acid slowly onto the water in a glass container. **If water is added to sulphuric acid it will explode.**
4 Continue to top-up the cell with the freshly made electrolyte and so recharge the battery and check the hydrometer readings.

5 Battery - charging

Note: If the battery is to remain in the vehicle when being charged always disconnect the battery leads.
1 In winter time when a heavy demand is placed on the battery, such as when starting from cold, and virtually all of the electrical equipment is continually in use, it is a good idea to occasionally have the battery fully charged from an external source at a rate of approximately 4 amps.
2 Continue to charge the battery at this rate until no further rise in specific gravity is noted over a four hour period.
3 Alternatively, a trickle charger, charging at the rate of 1.5 amps can be safely used overnight.
4 Special rapid 'boost' charges which are claimed to restore the power of the battery in 1 to 2 hours are not recommended unless they are thermostatically controlled as they can cause serious damage to the battery plates through overheating.

6 Alternator - general description, maintenance and precautions

1 Briefly, the alternator comprises a rotor and stator. Current is generated in the coils of the stator as soon as the rotor revolves. This current is three-phase alternating, which is then rectified by positive and negative silicon diodes; the charging current required to maintain the battery charge is controlled by a regulator unit.
2 Maintenance consists of occasionally wiping away any oil or dirt which may have accumulated on the outside of the unit.
3 No lubrication is required as the bearings are grease-sealed for life.
4 Check the drivebelt tension periodically to ensure that its specified deflection is correctly maintained (see Chapter 2).
5 Take extreme care when making circuit connections to a vehicle fitted with an alternator and observe the following.

When making connections to the alternator from a battery, always match correct polarity.

Before using electric-arc welding equipment to repair any part of the vehicle, disconnect the connector from the alternator and disconnect the battery cables.

Never start the car with the battery charger connected.

Always disconnect both battery cables before using a mains charger.

If boosting from another battery, always connect in parallel using heavy cable.

7 Alternator - removal and installation

1 Detach the ground cable from the battery.
2 Remove the air pump as described in Chapter 3, but do not detach the refrigerant feed lines to and from the pump.
3 Detach the alternator wires from the connector and the cable from 'A' terminal.
4 Unscrew and remove the lower alternator mounting bolts and the adjustment bracket bolts (photo).
5 Disconnect the fan belt and remove the alternator.
6 Installation is the reversal of removal but retension the fan belt as given in Chapter 2, Section 7 and the air pump belt as given in Section 41.

8 Alternator - testing

1 Testing and repair of the alternator is a job best left to your GM dealer or local auto-electrician who has the specialized knowledge and equipment necessary to undertake this task. If you have electrical knowledge and test equipment available, proceed as follows.
2 With a resistance tester, measure the rotor coil resistance across terminals 'F' and 'E' as in Fig. 10.1. The normal resistance is 5 ohms.
3 Should the resistance be higher, then there is a poor contact between the brushes and commutator. If there is no continuity between these terminals the problem is an open rotor coil circuit, a sticking brush or a broken lead wire. A lower resistance may indicate a rotor coil layer short or the circuit is being grounded.
4 Test the rectifying diodes to indicate continuity, irrespective of the diode conditions, as follows.

a) Connect the test negative (-) lead to the alternator 'A' terminal and the positive (+) lead of the tester to the alternator 'N' terminal. If continuity exists, one or more of the three diodes in the positive side are shorting.
b) Connect the tester negative (-) lead to the alternator 'N' terminal and the positive (+) lead to the alternator 'E' terminal. If there is continuity one or more of the three diodes in the negative side are shorted.

9 Alternator - dismantling, inspection, testing and reassembly

Refer to Fig. 10.2
1 Unscrew and remove the pulley locknut and washer and withdraw the pulley and fan.
2 Melt the solder on the brush lead and extract the brush holder assembly and remove the three thru bolts. Taking care, separate the rotor assembly and front cover and ensure that the stator unit is not detached from the rear cover.
3 Remove the diode cover and lift the stator coil lead clip. Melt the

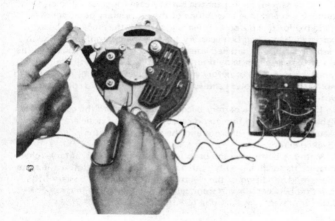

Fig. 10.1. Measuring the resistance across terminals F and E

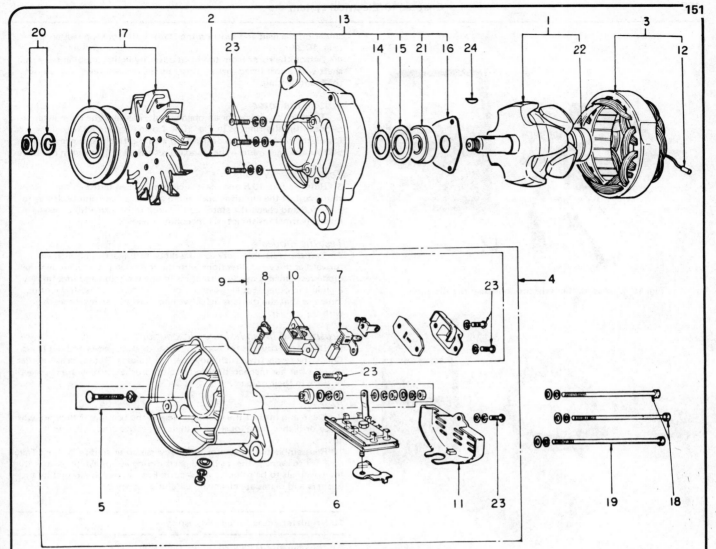

Fig. 10.2. The generator component parts

1 Rotor assembly	7 Brush assembly (F)	13 Front cover assembly	19 Through bolt
2 Spacer	8 Brush assembly (E)	14 Seal	20 Nut and lockwasher assembly
3 Stator assembly	9 Brush holder assembly	15 Seal retainer	21 Front ball bearing
4 Rear cover assembly	10 Brush holder	16 Bearing retainer	22 Rear ball bearing
5 Terminal bolt assembly	11 Diode cover	17 Pulley assembly w/Fan	23 Screw
6 Diode assembly	12 Terminal	18 Through bolt	24 Key

7.4 The generator showing the adjustment bracket bolt

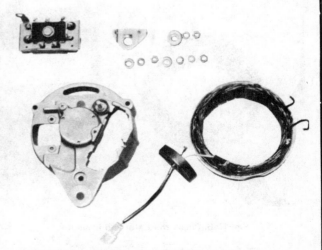

Fig. 10.3. Separate the stator unit from the rear cover

Fig. 10.4. Measure the resistance between the slip rings

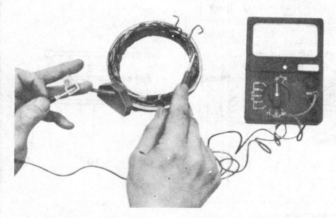

Fig. 10.5. Test the stator coil continuity

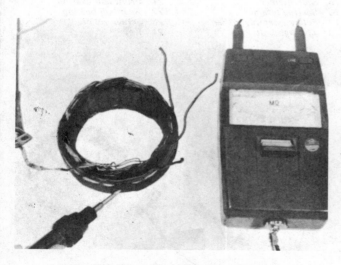

Fig. 10.6. Check the stator and insulation

Fig. 10.7. Test the resistance between the diode terminals

solder on the lead and separate the stator unit from the rear cover (Fig. 10.3).
4 Detach the rotor from the front cover by lightly tapping the rotor shaft with a soft faced mallet. Unscrew the cover screws and remove the front bearing.

Testing the rotor
5 To check the rotor for an open circuit, measure the resistance between the slip rings as in Fig. 10.4. Normal resistance is 5 ohms. Check the rotor core and positive (+) side slip ring resistance using a 500 volt megger meter. The standard resistance is 1 megohm or above.

Test the starter coils
6 Refer to Fig. 10.5 and make a continuity test between the stator coil. If the circuit is open there will be no continuity. Refer to Fig. 10.6 and check the stator coil to core resistance with a megger meter. Normal resistance is 1 megohm or above.

Test the diodes
7 Test the resistance between the diode terminals and holder in forward and reverse directions with the tester leads switched. Normal diode resistance is nearly zero ohms in one direction and indefinitely high in the other. Equal resistance or no resistance in both directions indicates that the diode is defective and must be renewed together with the holder.

Brushes and brush holders - inspection
8 Check the brushes for excessive wear or damaged or broken leads. Check the respective brush movement in holders. Renew brush holder unit if wear has reached the limit line on brushes or they are damaged or slack in their holders.

Bearings
9 Clean and inspect the bearings. Renew the bearings if they are worn, noisy or if they are a loose fit in their respective covers, renew the covers.
10 Reassemble the alternator in the reverse order of dismantling. Take care not to damage the insulated parts during assembly and ensure that the terminals to be soldered are secure. Ensure that the insulating tubes washers and plates are clean and carefully installed.

10 Regulator - description and testing

1 The regulator is located on the right side of the engine compartment, and incorporates a separate voltage regulator and cut-out (photo).
2 The voltage regulator controls the output from the alternator depending upon the state of the battery and the demands of the vehicle electrical equipment, and it ensures that the battery is not overcharged. The cut-out is virtually an automatic switch which

Chapter 10/Electrical system 153

completes the charging circuit as soon as the alternator starts to rotate and isolates it when the engine stops so that the battery cannot be discharged through the alternator. One visual indication of the correct functioning of the cut-out is the ignition warning lamp. When the lamp is out, the system is charging.
3 Before testing, check that the alternator drivebelt is not broken or slack, and that all electrical leads are secure. Where possible test the regulator when it is cold, not immediately after driving.
4 Start the engine and run at 2,500 rpm for a few minutes and then check the ammeter reading which should be 5 amps or less. If the reading is higher fit a fully charged battery in place of the existing one and recheck the reading.
5 Run the engine at idle speed and increase gradually to 2,500 rpm and take a voltmeter reading. If normal, the reading should be within 13.8 to 14.8 volts. If the reading deviates from these figures, the regulator is in need of adjustment or defective. If adjustment is made with the regulator in position on the vehicle, disconnect the connecting leads to prevent the battery circuit from being shorted.

11 Regulator - removal, adjustment and installation

1 If the regulator is still under guarantee by the manufacturer it should not be disturbed, but must be reset or exchanged by the manufacturer.
2 Detach the battery ground cable from the terminal.
3 Disconnect the regulator leads taking note of their positions.
4 Unscrew the mounting screws and remove the regulator.
5 Detach the regulator cover.
6 Clean the contact points with 500 or 600 grade sand paper.
7 Refer to Fig. 10.8 and check the core gap which should be 0.024 to 0.039 in (0.60 to 0.99 mm). To adjust, loosen the screws securing the contact set to yoke and move the set up or down as required.
8 Refer to Fig. 10.9 and check the point gap which should be between 0.012 and 0.016 in (0.30 to 0.40 mm). To adjust loosen the upper contact screw and move the contact as required.
9 Always adjust the core gap first followed by the point gap. Yoke gap adjustment is not necessary.
10 To increase the regulator voltage turn the adjustment screw inwards (Fig. 10.10). To decrease the voltage, screw it out. Lock the screw in position by tightening the locknut.
11 Refit the cover and relocate the regulator in position on the vehicle and recheck the voltage.

12 Starter motor - general description

1 The starter motor comprises a four-pole, four-brush direct current motor, with an ignition switch which is fitted with a safety lock. The starter motor circuit has a negative ground polarity.
2 The starter motor engagement mechanism is integral. The main switch is fitted into the magnetic switch which actuates the starter pinion. An overrun clutch is fitted to the pinion.

13 Starter motor - removal and installation

Series 1 to 4 models
1 Detach the ground cable from the battery.
2 Disconnect the wiring from the starter motor terminals.
3 Unscrew and remove the starter motor securing bolts/nuts and withdraw the starter motor.

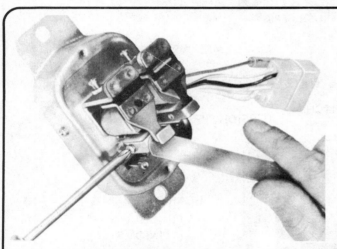

Fig. 10.8. Adjusting the regulator core gap

Fig. 10.9. Adjusting the regulator point gap

Fig. 10.10. Adjusting the regulator voltage

10.1 The regulator

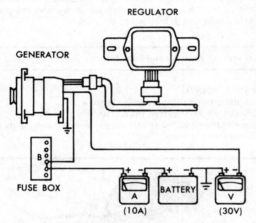

Fig. 10.11. The voltage regulator showing the core gap and points gap adjustment points

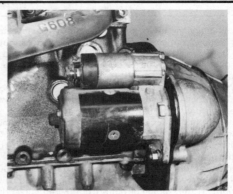

13.1 The starter motor

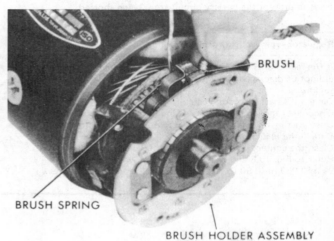

Fig. 10.12. The regulator testing circuit

Fig. 10.13. Lifting the brush spring to remove the brush

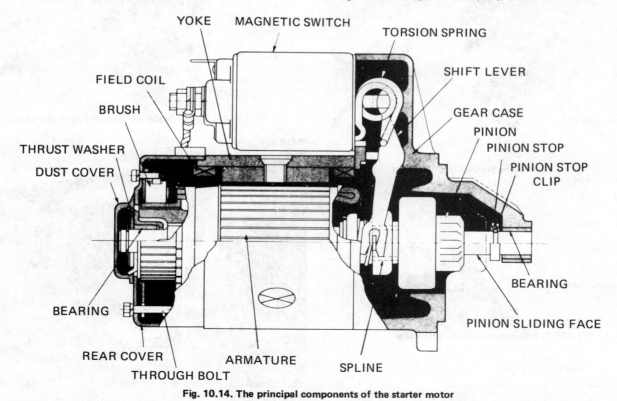

Fig. 10.14. The principal components of the starter motor

Series 5 and 6 models

4 Detach the ground cable from the battery.
5 Detach the EGR pipe from the EGR valve and exhaust manifold. Remove the EGR pipe.
6 Disconnect the wiring from the starter motor.
7 Unscrew and remove the starter motor retaining bolts, and withdraw the starter motor thru the clearance beneath the inlet manifold.
8 Installation of the starter motor is the reverse of removal.

14 Starter motor - dismantling, servicing and reassembly

1 Detach the lead magnetic switch terminal marked 'M'.
2 Unscrew the magnetic switch retaining bolts and remove the switch.
3 Withdraw the torsion spring from the magnetic switch.
4 Remove the dust cover and snap-ring washer.
5 Unscrew the thru bolts and the two rear cover screws and remove the rear cover.
6 Lift the brush spring and withdraw the brush and remove the brush holder assembly.
7 Withdraw the yoke from the gearcase and remove the dust cover. Withdraw the armature and shift lever.
8 Withdraw the pinion stop clip and detach the pinion stop. Remove the pinion unit.
9 Clean the respective parts ready for inspection.
10 Inspect the armature. Check the commutator face for burning or roughness. Reface the commutator with fine sand paper if required - do not use emery cloth as the fine carborundum particles will become embedded in the copper surfaces. If the commutator is very rough or out of round it may be lightly cleaned down on a lathe, but refer to the specifications for the outside diameter limits.
11 Undercut the commutator separators using an old hacksaw blade ground to suit, to a depth of 0.02 to 0.03 in (0.5 to 0.8 mm).
12 If an ohmmeter is available, test the field coil for continuity. To do this, connect one probe of the meter to the field coil positive terminal and the other to the positive brush holder. If no reading is indicated then the field coil circuit has a break in it.
13 Connect one probe of the meter to the field coil positive lead and the other one to the yoke. If there is a low resistance, then the field coil is earthed due to a breakdown in insulation. When this fault is discovered, the field coils should be renewed by an automotive electrician as it is very difficult to remove the field coil securing screws without special equipment. In any event, it will probably be more economical to exchange the complete starter motor for a reconditioned unit.
14 The armature may be tested for insulation breakdown again using the ohmmeter. To do this, place one probe on the armature shaft and the other on each of the commutator segments in turn. If there is a reading indicated at any time during the test then the armature must be renewed.
15 Check the brushes for wear. The standard brush lengths are:

Series 1 to 4 models 0.73 in (18.5 mm) - minimum limit 0.05 (12.70 mm)
Series 5 and 6 models 0.629 (16.0 mm) - minimum limit 0.47 (11.9 mm)

If the brushes are worn beyond this limit they must be renewed.
16 Inspect the drive gear components for wear or damage, particularly the pinion teeth, and renew as required.
17 Reassembly of the starter motor is a reversal of dismantling but note the following:

   a) Grease the bearings and sliding faces before assembly.
   b) When the shift lever is set to the guide on the pinion, do not rotate the armature as the lever may be detached from the guide on the pinion.
   c) When installing the yoke align the jointing portion groove with the corresponding projection.
   d) Ensure that the brushes are free in their holders and do not bind or stick.
   e) Fit the torsion spring with the adjustment plate on the negative switch.

Locate the shift lever end into the plunger fitting hole in the switch, and set the torsion spring end into the groove in the intermediate part of the shift lever.

15 Fuses

1 A comprehensive fuse box is fitted to the inner wing panel within the engine compartment, and the fuse cover is labelled to denote the particular fuse circuit and fuse valve (photo).
2 In the event of a fuse 'blowing', always establish the cause before installing a replacement. The most likely cause is faulty insulation in the circuit.
3 Always carry a spare fuse for each rating and do not substitute a piece of wire or nail for the fuse as a fire may result or at least, the electrical component ruined.

16 Headlamps - removal and installation

1 Detach the headlamp rim (photo).
2 Slacken the three sealed beam unit retaining screws (photo).
3 Turn the sealed beam unit by turning it counterclockwise and withdraw.
4 Detach the wire connector (photo) to remove the sealed beam unit completely.
5 Installation is a reversal of the removal procedure but ensure that the headlamp lens is correctly located with the 'TOP' mark upwards.

17 Headlamp - adjustment (beam alignment)

1 Before adjusting the headlights, ensure that the tire pressures are as specified, and that the vehicle is standing on level ground and unladen.
2 Refer to Fig. 10.16 and adjust the respective screws accordingly.

Screw A adjusts the vertical alignment.
Screw B adjusts the horizontal alignment.

3 The most accurate method of checking and setting the beam alignment is by using a special optical tester as used by most garages.
4 If this equipment is not available, the vehicle should be parked approximately 32 feet (10 m) from a vertical wall and standing level.
5 Measure the height from the ground to the center of the headlight lens and chalk a line at this height horizontally across the wall in line with the vehicle.
6 Measure the distance between the headlight centers and correspondingly mark these on the wall dissecting the horizontal line.
7 Switch the headlights on and adjust each light beam to the corresponding mark on the wall.

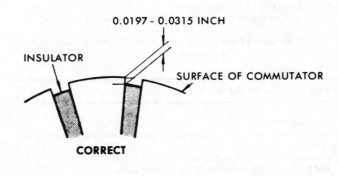

Fig. 10.15. The starter motor commutator insulator undercut

156 Chapter 10/Electrical system

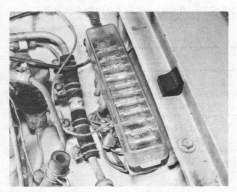
15.1 The fuse box and cover indicating the various circuits and ratings

16.1 Remove the headlamp rim

16.2 Loosen the sealed beam unit retaining screws

16.4 Detach the wire connector

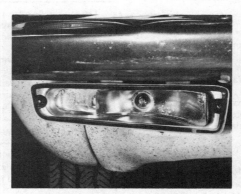
18.1 The front combination lamp with lens removed

19.1 The Series 1 model rear light lens removed

18 Front combination lamps

Bulb renewal
1 Remove the two screws retaining the lamp lens, and remove the lens (photo).
2 Press and turn the bulb to remove it from its socket.
3 Install the new bulb and ensure that it is correctly located.
4 Refit the lens and retaining screws.

Lamp removal and installation
5 Undo and remove the lamp housing to front bumper retaining nuts.
6 Detach the wiring from the connector and remove the lamp unit.
7 Refitting is the reversal of removal.

19 Rear combination lights

Bulb replacement
1 Remove the lens retaining screws and detach the lens (photo) and Fig. 10.17.
2 Press and turn the bulb to remove.
3 Reinstall the bulb and lens in the reverse order.

Lamp unit removal
4 *Series 1 models:* Unscrew the four retaining nuts securing the lamp housing to the lamp panel and remove the cover. Detach the wiring from the connector and partly pull the socket out. Installation is the reverse of removal.
5 *Series 2 to 6 models:* Remove the rear gate bumper rubber and unscrew the four nuts securing the light housing to the panel. Detach the wiring from the connector and withdraw the light unit. Installation is the reverse of removal.

20 Licence plate light

Bulb - removal and installation
1 On Series 1 models, prise the lens from the body as in photo. On Series 2 to 6, unscrew the lens retaining screws and remove the lens (Fig. 10.18) (photo).
2 Press and turn the bulb to remove. Renew the bulb and install in the reverse order.

Light unit - removal and installation
3 *Series 1 models:* Unscrew the lamp housing to body panel retaining screws and remove the lamp unit. Disconnect the wiring. Install in the reverse order.

21 Side marker lights - front and rear

Bulb renewal
1 Unscrew and remove the lens retaining screws. Remove the lens, press and twist the bulb to remove. Install in the reverse order.

Light unit - removal and installation
2 Unscrew the lens retaining screws and detach the wiring from the connector. Remove the lamp unit. Reverse the removal procedure to install.

22 Back up (reverse) lamps

Bulb renewal
1 Withdraw the grommet on the rear face of the lamp housing and press and twist the bulb to remove. Install in the reverse order.

Fig. 10.16. The headlight adjustment screws

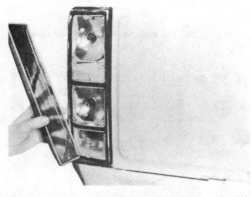

Fig. 10.17. The rear combination light, Series 2 to 6 lens removal

Fig. 10.18. Removing the licence plate lens - Series 2 to 6

20.1 The Series 1 model license plate light with lens removed from body and light unit

Lamp unit - removal and installation

2 Unscrew the lamp bracket retaining nuts and remove the bracket and lamp unit. Unscrew the lamp to bracket retaining nuts and remove the lamp unit. Install in the reverse order.

23 Dome (interior) lamp

1 To renew the bulb, unclip the lamp cover (Fig. 10.19) and withdraw the bulb from the connectors. Fit the new bulb, check that it works and install the cover.
2 To remove the lamp unit, unclip the cover and unscrew the self-tapping screws and withdraw the lamp unit. Detach the wiring. Install in the reverse order.

24 Instrument panel - removal and installation

1 From the rear of the instrument panel detach the speedo cable.
2 Unscrew the instrument panel retaining wing nuts and withdraw the panel sufficiently to disconnect the harness from the connector.
3 Remove the panel.
4 Installation of the panel is the reversal of removal but take care not to trap the wiring harness between the fascia panel and the instrument panel.

25 Instrument panel - dismantling and reassembly

1 With the instrument panel removed, as described in Section 24, the gauges or lights can be removed for renewal, but special care must be taken to handle the gauges with great care during removal and installation. Do not attempt to repair gauges as this must be entrusted to an instrument specialist or simply renewed. To dismantle the complete panel proceed as follows.
2 Remove the screws from the rear of the instrument panel and separate the meter case from the panel (Fig. 10.20).
3 Disconnect and remove the speedometer and lights that are not print wired.
4 Remove the combination fuel and temperature gauge.
5 Rotate the instrument light and indicator lights counterclockwise and remove them.
6 Remove the retaining screws and separate the print wiring board from the case (Fig. 10.21).
7 Reassembly of the instrument panel is the reversal of the removal sequence.

26 Ignition switch

1 When the ignition switch is in the OFF position, all of the electrical circuits except the lighting circuit are de-energised.
2 With the ignition switch in the second stop (clockwise) the starter motor is energized to crank the engine. On releasing the ignition key, the switch returns to the ON position.
3 In the ON position, lead wires 'B', 'L' and 'BY' are connected.
4 The ACC position is the first stop in the counterclockwise direction

Fig. 10.19. Unclip the lens and hinge up to remove the bulb

from the OFF position, When engaged, lead wire 'L' is connected to 'B'.
5 The ignition switch is a pre-sealed and set unit and therefore cannot be adjusted or repaired. However, continuity of the terminals can be tested using a voltmeter. Continuity must only exist between those indicated on the chart (Fig. 10.23).

27 Turn signal and dimmer switch - testing

1 The turn signal and dimmer switch are an integrated unit, and actuation is by a single lever.
2 Using a tester, the switch terminals may be tested for continuity to determine a possible switch fault. Continuity should only exist between those terminals shown in Fig. 10.25 and Fig. 10.26.
3 The turn signal and dimmer switch is a fully integrated unit and therefore if faulty must be renewed.

28 Windshield wiper motor - removal and installation

1 Detach the wiper blades and arms from the pivots by unscrewing the wiper arm retaining nut (photo).
2 Pull the arm from the pivot shaft to remove (photo).
3 Unscrew and remove the two bolts retaining the pivot.
4 Unscrew the wiper motor mounting bolts and detach the motor unit and linkage. Disconnect the wiring.
5 Installation is a reversal of removal but when fitting the wiper arms and blades, check their position on the pivot before tightening the retaining nut. Also ensure that the wiper motor linkage is not fouling any adjacent parts during the wiping action.

29 Windshield wiper motor - dismantling and inspection

1 Unscrew the cover securing bolts and turn the cover assembly to remove.
2 Withdraw the rotor unit turning in a clockwise direction (Fig. 10.28) taking care not to damage the brushes.
3 Clean the components before inspection. If the commutator segment grooves are blocked, remove the carbon from the grooves between the segments using a small knife blade or other such suitable tool. Lightly reface the commutator with fine glass paper.
4 Inspect the brushes and springs for wear, signs of damage, corrosion or distortion. Renew if required.
5 The self-parking device is faulty if, when the wiper motor switch is turned off, the motor continues to run, or if the motor stops immediately when the switch is de-activated.
6 Renew parts as required and reassemble the motor in the reverse sequence of dismantling.

Fig. 10.20. The instrument cluster and case

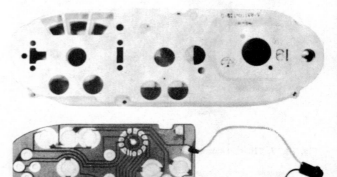

Fig. 10.21. The printed circuit board and case

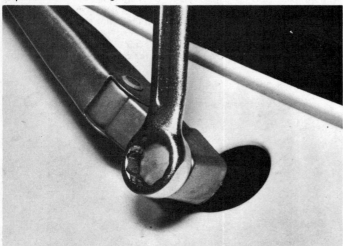

28.1 Unscrew the wiper arm retaining nut

Fig. 10.22. The ignition switch positions

Switch position \ Circuit / Cable color	Battery	Accessories	Ignition	Starter
	B	L	BY	BW
On	O	O	O	
Start	O			O
Acc.	O	O		

Fig. 10.23. The starter switch connections

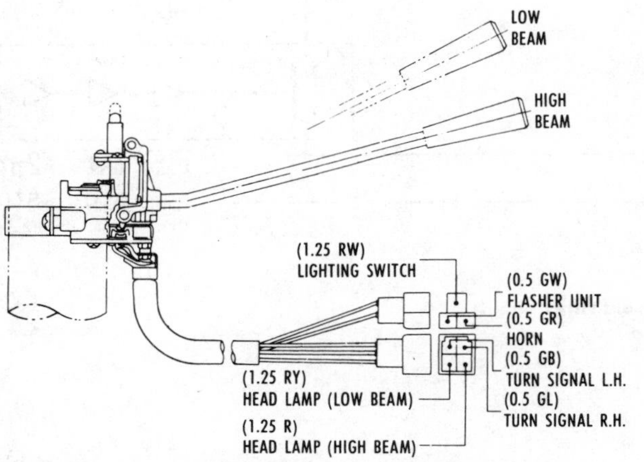

Fig. 10.24. The turn signal and dimmer switch operation diagram

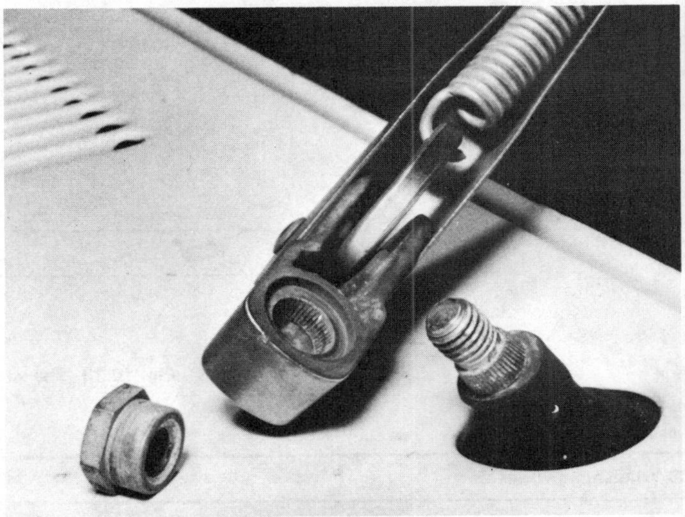

28.2 Remove the arm from the splined pivot

S Switch position \ Cable color	GW	GL	GB
RH	O	O	
LH	O		O

Fig. 10.25. Test for continuity in the turn signal switch terminals

Switch position \ Cable color	RY	RW	R
Low Beam		O	O
High Beam	O	O	

Fig. 10.26. Test for continuity in the dimmer switch terminals

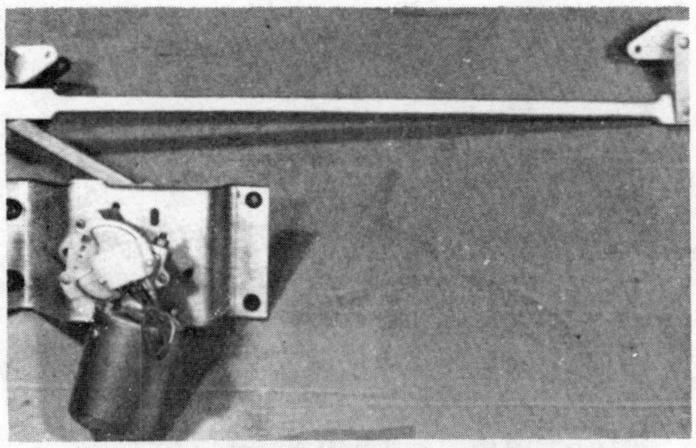

Fig. 10.27. The windshield wiper motor and linkage

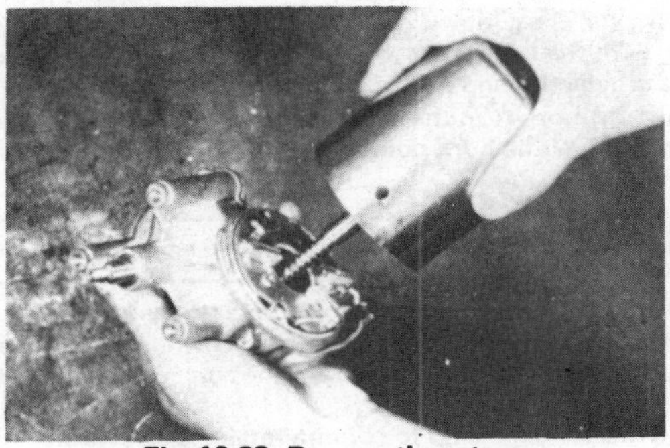

Fig. 10.28. Remove the rotor

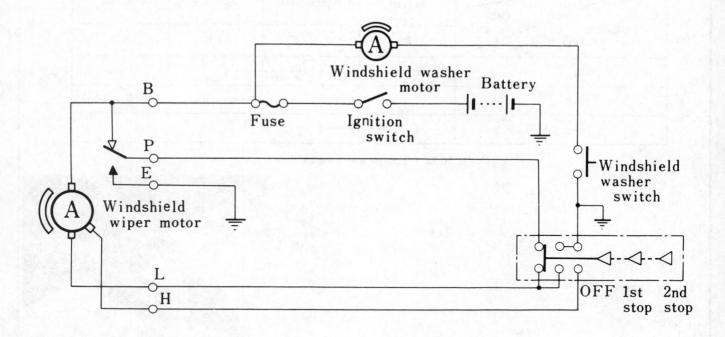

Fig. 10.29. The windshield wiper circuit

30 Windshield washer

1 The windshield washer fluid level must be checked regularly for the fluid level. To protect against freezing use GM Optikleen or an equivalent fluid. This also increases cleaning action.
2 The windshield washer is actuated by the windshield wiper switch and, if the washer unit should fail to operate, check the continuity between the switch terminal and washer motor. Check the fuse.
3 If the above items are in order, the washer motor is defective or the feed pipes/nozzles are blocked.
4 Detach the interconnecting pipes from the washer motor and check that they are clear, and, if this is the case, the motor is defective and must be removed for repair or renewal.

31 Horn

1 If the horn fails to operate when the button is depressed, check the fuse and wiring for good connections.
2 Check that the horn ground wire is connecting properly.
3 To test the horn button, remove it from the steering wheel and short the lower section of the button to the steering shaft. If the horn does not operate, the contact in the combination switch is not connecting with the lower part of the steering wheel. If the horn sounds, the button is defective and must be renewed.
4 To adjust the horn turn the adjustment screw (Fig. 10.31) as required to readjust the horn point gap.

32 Direction signal and dimmer switch - removal and installation

1 Detach the battery ground cable from its terminal.
2 Unscrew and remove the steering column cowling retaining screws.
3 Disconnect the wire connectors from the combination switch to harness.
4 Unscrew the switch clamp to column retaining screws and withdraw the switch.
5 Installation is a direct reversal of the removal procedure and on completion check the switch for correct functioning.

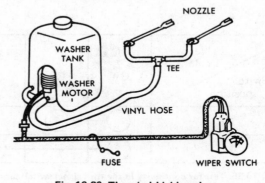

Fig. 10.30. The windshield washer

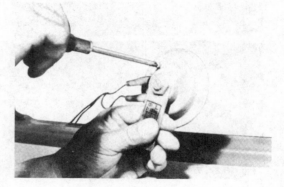

Fig. 10.31. Adjust the horn

Chapter 10/Electrical system

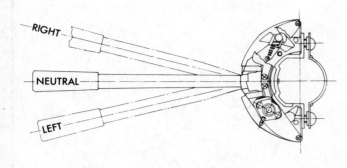

Fig. 10.32. The direction signal and indicator switch

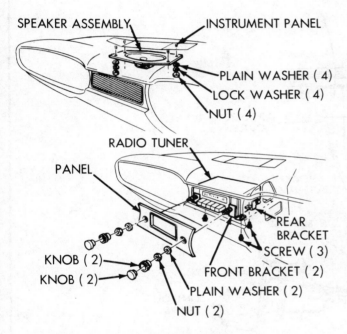

Fig. 10.33. The radio and speaker location

33 Hazard warning switch - removal and installation

1 Detach the ground cable from the battery terminal.
2 Unscrew and remove the steering column cowling screws.
3 Unscrew the hazard warning switch retaining screws, and detach the wires from the connector. Withdraw the switch.
4 Installation is the reversal of removal but check the switch for correct functioning on completion.

34 Radio equipment - removal and installation

Radio
1 Detach the ground cable from the battery.
2 Detach the ash tray and plate.
3 Unscrew and remove the volume control knob and tuner knob together with nuts, washers and face panel (Fig. 10.33).
4 Unscrew the rear and front mounting bracket screws.
5 Detach the electrical connections and the antenna lead from the radio.
6 Withdraw the radio and remove the front mounting brackets.
7 Installation is a direct reversal of removal.

Radio speaker
8 Remove the radio as above.
9 Unscrew the speaker retaining nuts, washers and lockwashers and withdraw the speaker and disconnect the wire connections. Remove the speaker unit.
10 Installation is the reverse of removal.

Antenna and lead
11 Dsiconnect the antenna lead from the radio.
12 Unscrew the antenna retaining cap and remove the antenna carefully pulling the cable thru the body panel grommets.
13 Installation is the reversal of removal but be sure to refit the grommets securely to the body panels.

Radio suppressors
14 If the radio operates but suffers from interference, check the suppressors shown in Fig. 10.36.
15 Disconnect the capacitor cable and remove the capacitor retaining screw. Remove and renew the capacitor.
16 Another possible cause for interference of the radio is an insecure antenna. Check the antenna and cable connections and ensure that they are secure.

35 Heater unit - removal and refitting

Series 1 models
1 Disconnect the ground cable from the battery, and drain the cooling system (see Chapter 2).

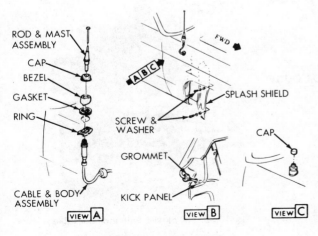

Fig. 10.34. The antenna location

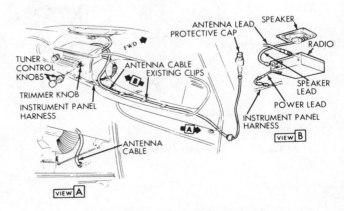

Fig. 10.35. The antenna lead connection

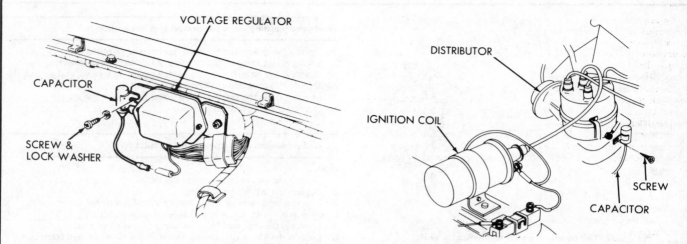

Fig. 10.36. The radio interference suppression components

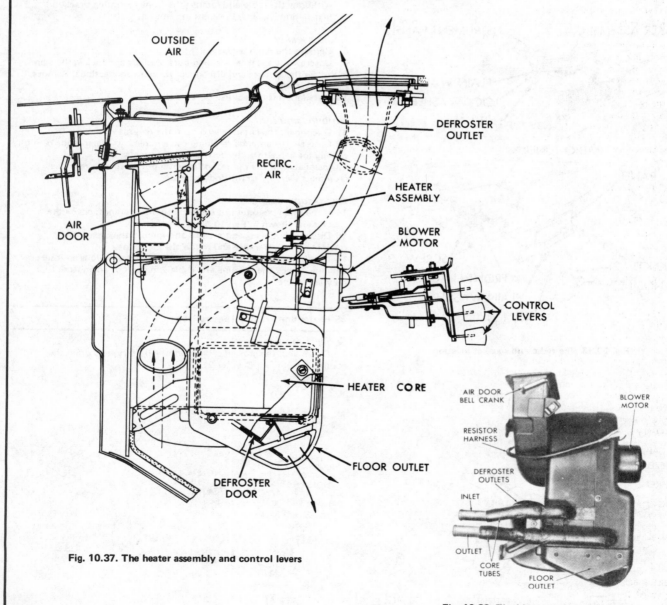

Fig. 10.37. The heater assembly and control levers

Fig. 10.38. The blower unit and core tubes

Chapter 10/Electrical system

2 Detach the heater hoses from the heater unit and blank off the connection ends to prevent spillage.
3 Unscrew the two instrument panel to heater control screws via the access holes in the lower flange of the control unit, and lower the control. The blower switch leads can now be disconnected and the heater case to dash panel screws removed.
4 Carefully lower the unit and detach the resistor leads. From the door bell crankarm disconnect the defroster door and air door Bowden cables. Detach the defroster hoses from the floor outlet and remove the unit from the vehicle.
5 Refitting is a reversal of the above procedure. When reconnecting the defroster door and air cables adjust and tighten the clamp screws so that the doors can fully close. Refill the system with coolant and check for leaks.

Series 2

6 Disconnect the ground cable from the battery.
7 Locate a drain pan under the heater core tubes and slacken the hose clamp screws to the tubes. Detach the tubes and blank off the hose and tube ends to prevent further spillage. Position the hoses in a vertical position.
8 Unscrew the five screws retaining the parcel shelf and remove it.
9 Slacken the air diverter and defroster door cable clamps from the heater case and detach the cables.
10 Detach the blower and resistor leads, and then unscrew the control unit to instrument panel retaining screws. Hinge the control to the left being careful not to distort the water valve Bowden cable. Lay the control unit on the floor.
11 Unscrew the heater unit to dash screws and withdraw the heater unit to the rear so that the core tubes clear the dash opening and then move the heater downwards and to the right and remove it.
12 Refitting is the reversal of removal but ensure that the cowl seal is correctly positioned and finally that the control levers can operate. Readjust the cables if required. Check the hose connections for signs of leakage.

Series 3 to 6 models

13 Follow the instructions for Series 2 models in paragraphs 6 and 7 then proceed as follows.
14 Remove the interior of the glove box.
15 Detach the Bowden cables from the heater unit, and remove the ash tray.
16 Disconnect and remove the right dash outlet hose unit.
17 Unscrew the control to instrument panel retaining screws and lower the control. Detach the electric connections and the water valve Bowden cables.
18 Disconnect the defroster hoses.
19 From the right frame lever, disconnect the parking brake cable. Detach the parking brake cable pulley cover in the engine compartment and remove the parking brake to dash and instrument panel attachments. Place out of the way.
20 Disconnect the windshield wiper switch.
21 Unscrew the heater unit to dash panel retaining screws and pull the heater rearwards until the core tubes are clear of the dash opening. Move the heater to the left and downwards and withdraw it.
22 Undo the core tube clamp screw and remove the clamp. The two heater case halves may now be separated and the core removed.
23 Installation is the reversal of removal but ensure that the core seals are repositioned and are intact. Check that the cowl and evaporator unit seals are correctly positioned. Adjust the operating cables as described in Section 39.

36 Blower motor and heater core unit - removal and installation

1 Remove the heater unit as described in Section 23.
2 Undo the six screws retaining the floor outlet unit to the heater case and carefully prise the floor outlet from the case. Unscrew the front/rear case screws to separate the casing.
3 With a pair of pliers, bend the blower motor ground cable tab in half and push it thru the rear case.
4 Remove the blower motor from the rear case by unscrewing the retaining screws. The heater core unit can now be removed from the front case half if necessary.

5 Refitting of the blower motor is a direct reversal of removal, but insert the blower motor ground cable thru the rear casing hole, and then refit the motor to the casing. If the ground cable tab breaks or is too distorted to reconnect, a new one must be soldered to the cable.

37 Heater controls - removal and refitting

1 Disconnect the ground cable from the battery.
2 Remove the ash tray (some models) and remove the side outlet hoses from the control unit.
3 Remove the screws retaining the control to the instrument panel thru the access hole in the control lower flange.
4 Detach the electrical harness from the blower switch and disconnect the Bowden cables from the control levers. Remove the control unit.
5 Refit in the reverse order and adjust the control cables.

38 Heater water valve - removal and installation

1 Drain the cooling system (see Chapter 2), slacken the heater hose to valve clips and remove the hoses from the valve. Position the hose vertically to prevent spillage of coolant.
2 Slacken the water valve Bowden cable clamps and remove the cable from the valve.
3 The valve retaining screw can now be removed and the valve lifted clear.
4 Refit in the reverse order, refill with coolant and check for leaks.

39 Ventilation air inlet valve - removal and refitting

1 Disconnect the ground cable from the battery.
2 Slacken the hose clip screw and detach the hose from the valve.
3 From the inlet, detach the floor outlet operating rod, and then remove the valve extension clamp (lower) from the right-hand side and withdraw the extension.
4 Unscrew the valve to dash panel retaining screws to remove the valve unit.
5 Refitting is a reversal of the above procedure.

40 Heater control cable - adjustment

Vent door

1 If the control unit is already in position loosen the retaining screws to the instrument panel.
2 Position the select lever to the vent position.
3 Retain the vent door in vent position and align the vent cable and loop with the hook on the lever at the side of the heater by moving the panel fore and aft.
4 Tighten the nuts retaining the control to instrument panel and insert the cable onto the vent door lever.

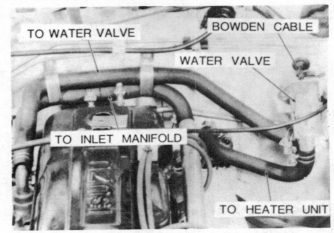

Fig. 10.39. The water valve and heater hoses

Air inlet door cables

5 Position the selector lever on A/C, and connect the A/C cable to the link lever hook on the evaporator unit lower portion.
6 Adjust the cable to give a clearance of 0.3 in (7.6 mm) between the link lever hook end and the link lever bracket as in Fig. 10.42.
7 Tighten the cable clamp screw and close the outside heater air door.
8 Fit the A/C recirculation door cable to the heater outside air door shaft and tighten the cable clamp screw.

Defrost door cable

9 Position the selector lever to DEF.
10 Position the defroster door of the heater unit on the defroster door shaft and tighten the cable clamp.

Water valve cable

11 Position the temperature lever to cold.
12 Position the water valve in the closed position and connect the cable to the lever.
13 The cable is fitted in the clamp to give a clearance of 0.3 in (7.6 mm) between the clamp and the end of the cable sheath. Tighten the clamp and water valve screw.

Thermostatic switch cable

14 Position the temperature cable at cold.
15 Locate the switch in the cold position and insert the switch hook. Tighten the cable clamp.

41 Dash outlet grille - removal and installation

1 Undo the three screws retaining the outlet grille to the instrument panel and withdraw the grille.
2 Slacken the hose clip at the air inlet and remove the outlet unit to instrument panel screws. The outlet and hose are removed as a unit and then separated.
3 Refit in the reverse order, aligning the outlet unit so that the levers will not bind on the grille bars when the grille is refitted.

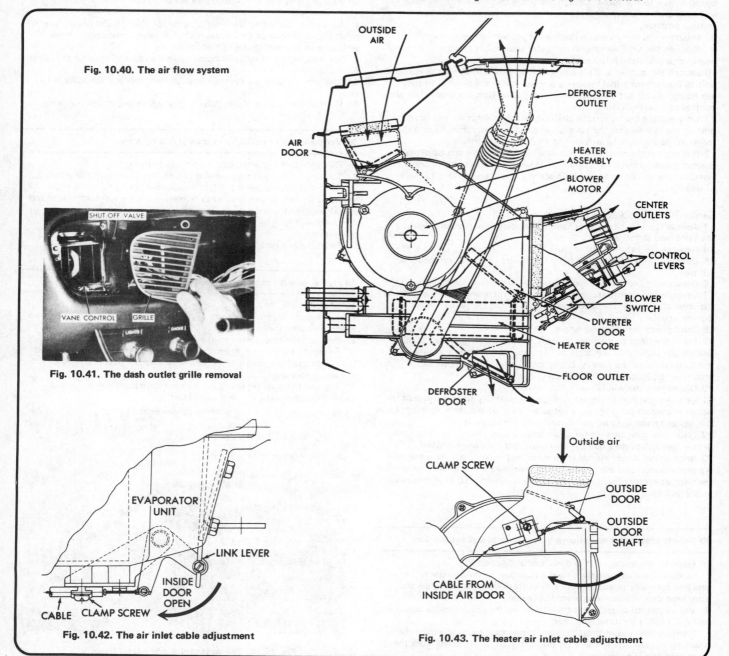

Fig. 10.40. The air flow system

Fig. 10.41. The dash outlet grille removal

Fig. 10.42. The air inlet cable adjustment

Fig. 10.43. The heater air inlet cable adjustment

Chapter 10/Electrical system

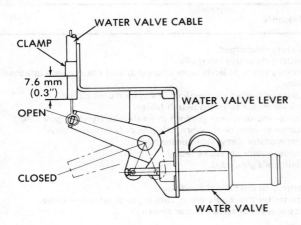

Fig. 10.44. The water valve cable adjustment

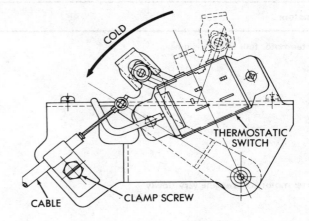

Fig. 10.45. The thermostatic switch cable adjustment

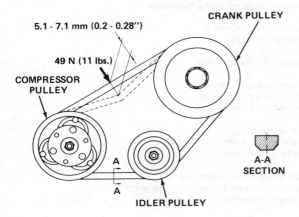

Fig. 10.46. Check the air pump drive belt adjustment

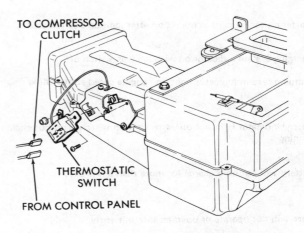

Fig. 10.47. Thermostatic switch

42 Air conditioning components

1 Should any part of the air conditioning system fail to operate correctly it is not recommended that the 'home mechanic' repair or renew any of the defective components. Specialised tools, equipment and most important, knowledge are required to service and repair items within the system.
2 Do not disconnect any of the air conditioning supply lines during overhaul or repair procedures, see special note in Chapter 1.

Maintenance

3 Ensure that the engine exhaust is suitably ventilated.
4 Check the compressor drivebelt for correct tension and general condition. Renew the belt and retension as shown in Fig. 10.46. by slackening the center bolt and tightening the adjustment screw until 11 lb of force is required to deflect the belt 0.2 to 0.28 in. Retighten the pulley center bolt to 10 lb f ft (1.3 kg f m).
5 Set the controls for system operation and run the engine at 2000 rpm for ten minutes. Check the clutch pulley bolt and ensure that the compressor and clutch pulley are operating at the same speed. A speed variation indicates clutch slippage and the compressor unit will have to be serviced by your GM dealer.
6 Prior to switching off the engine, check the sight glass to ensure that the refrigerant charge is sufficient. The glass is normally clear although during mild weather traces of bubbles may be present. Foam in the charge is indicative of a low charge. No charge is indicated by there being no liquid visible and no temperature difference between the compressor inlet and outlet lines. If recharging is necessary, have this done by your local GM dealer.
7 Check the refrigerant line connections for secureness.
8 Check the compressor unit for signs of excessive oil leakage. A small leakage from the front seal is permissible but if a serious leak is apparent, get your GM dealer to check it out.

43 Blower resistor - removal and installation

1 Detach the ground cable from the battery.
2 Disconnect the right-hand defroster duct.
3 Detach the wires from the resistor and unscrew the mounting screws. Remove the resistor.
4 Installation is the reversal of removal

44 Thermostatic switch - removal and installation

1 Detach the ground cable from the battery.
2 Remove the interior of the glove box.
3 Detach the thermostatic switch wires and loosen the switch cable clamp and remove the cable from the switch.
4 Unscrew and remove the switch to bracket screws. Remove the bracket and then the switch. The bracket has to be removed as the switch capillary tube is coiled inside the evaporator.
5 Installation is the reversal of removal.

45 Fault diagnosis - electrical system

Symptom	Reason/s
Starter motor fails to turn engine	Battery discharged. Battery defective internally. Battery terminal leads loose or ground lead not securely attached to body. Loose or broken connections in starter motor circuit. Starter motor solenoid switch faulty. Starter motor pinion jammed in mesh with flywheel gear ring. Starter brushes badly worn, sticking or brush wire loose. Commutator dirty, worn or burnt. Starter motor armature faulty. Field coils grounded.
Starter motor turns engine very slowly	Battery in discharged condition. Starter brushes badly worn, sticking or brush wires loose. Loose wires in starter motor circuit.
Starter motor operates without turning engine	Pinion or flywheel gear teeth broken or worn.
Starter motor noisy or engagement excessively rough	Pinion or flywheel teeth broken or worn. Starter motor retaining bolts loose.
Starter motor remains in operation after ignition key released	Faulty ignition switch. Faulty solenoid.
Charging system indicator on with ignition switch off	Faulty alternator diode.
Charging system indicator light on - engine speed above idling	Loose or broken drivebelt. Shorted negative diode. No output from alternator.
Charge indicator light not on when ignition switched on but engine not running	Burnt out bulb. Field circuit open. Lamp circuit open.
Battery will not hold charge for more than a few days	Battery defective internally. Electrolyte level too weak or too low. Battery plates heavily sulphated.
Horn will not operate or operates intermittently	Loose connections. Defective switch. Defective relay. Defective horn.
Horns blow continually	Faulty relay. Relay wiring grounded. Horn button stuck (grounded).
Lights do not come on	If engine not running, battery discharged. Light bulb filament burnt out or bulbs broken. Wire connections loose, disconnected or broken. Light switch shorting or otherwise faulty.
Lights come on but fade out	If engine not running battery discharged. Light bulb filament burnt out, or bulbs or sealed beam units broken. Wire connections loose, disconnected or broken. Light switch shorting or otherwise faulty.
Lights give very poor illumination	Lamp glasses dirty. Lamps badly out of adjustment.
Lights work erratically - flashing on and off, especially over bumps	Battery terminals or ground connection loose. Lights not grounding properly. Contacts in light switch faulty.
Wiper motor fails to work	Blown fuse. Wire connections loose, disconnected, or broken. Brushes badly worn. Armature worn or faulty. Field coils faulty.
Wiper motor works very slowly and takes excessive current	Commutator dirty, greasy or burnt.

Chapter 10/Electrical system

	Armature bearings dirty or unaligned. Armature badly worn or faulty.
Wiper motor works slowly and takes little current	Brushes badly worn. Commutator dirty, greasy or burnt. Armature badly worn or faulty.
Wiper motor works but wiper blades remain static	Wiper motor gearbox parts badly worn or teeth stripped.

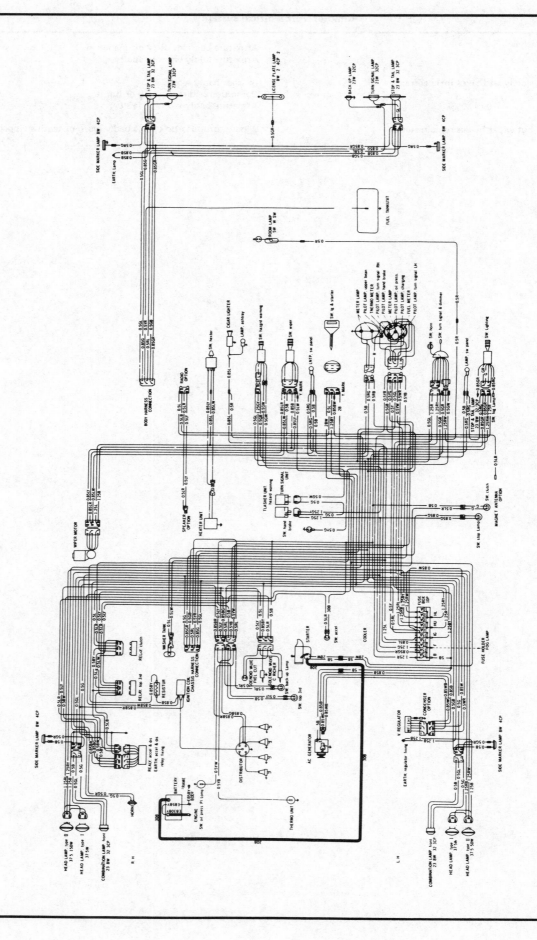

Fig. 10.48. Wiring diagram - Series 1 models

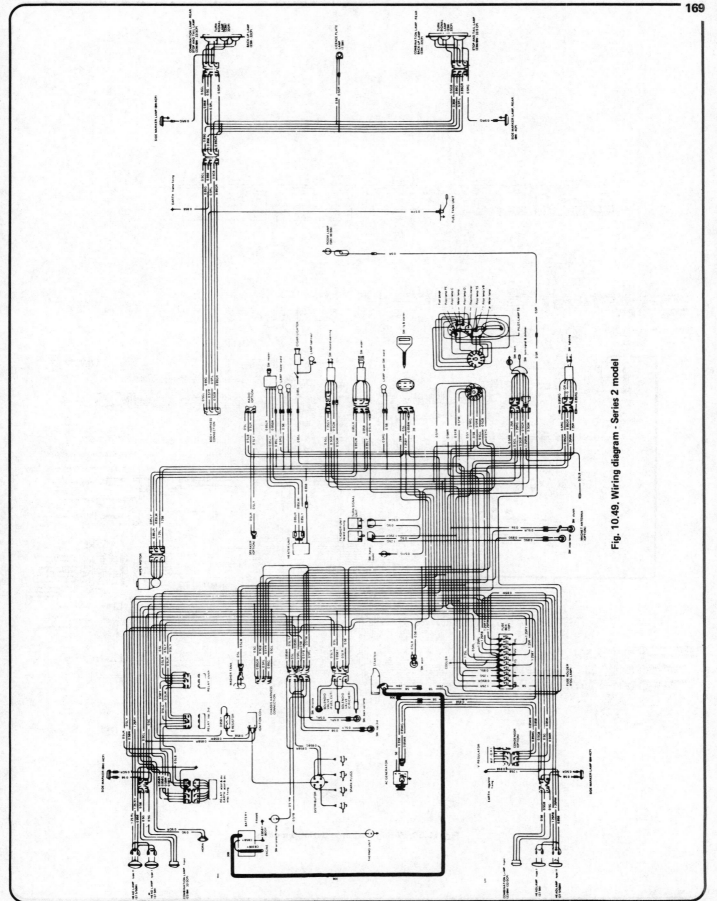

Fig. 10.49. Wiring diagram - Series 2 model

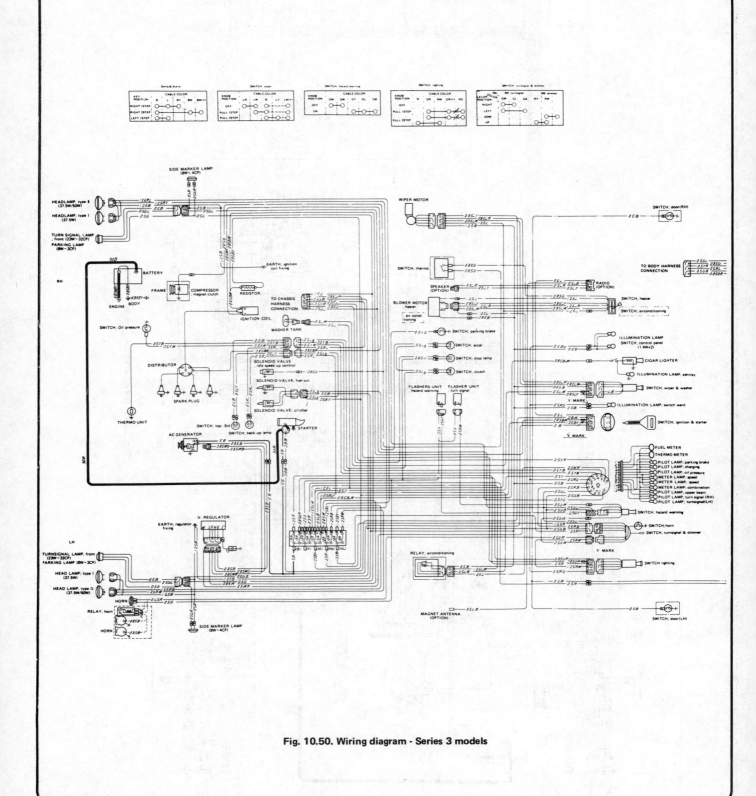

Fig. 10.50. Wiring diagram - Series 3 models

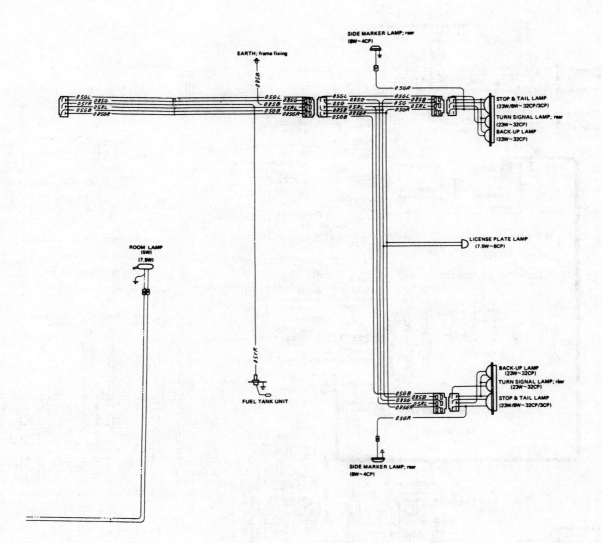

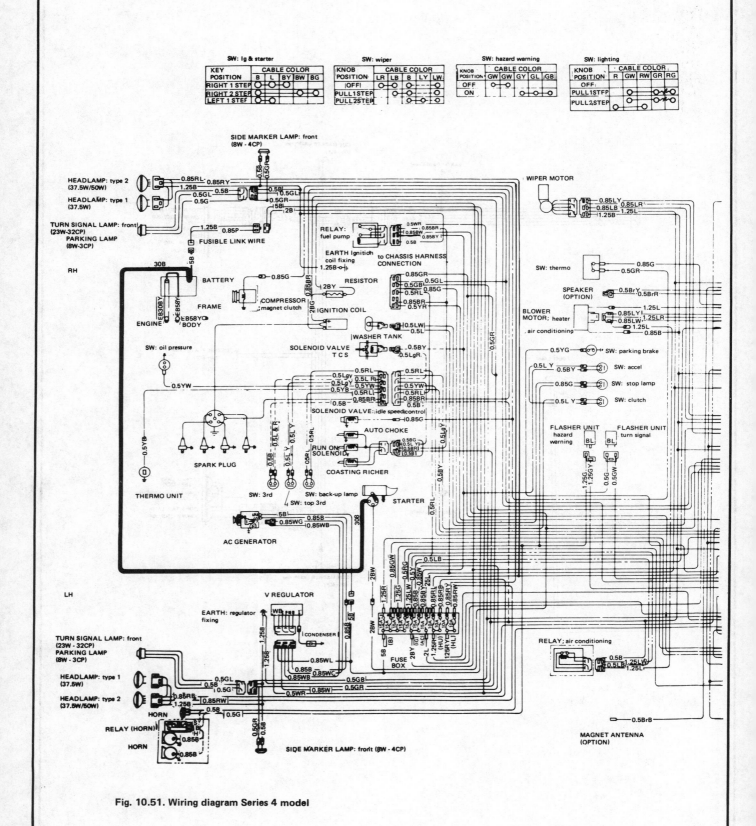

Fig. 10.51. Wiring diagram Series 4 model

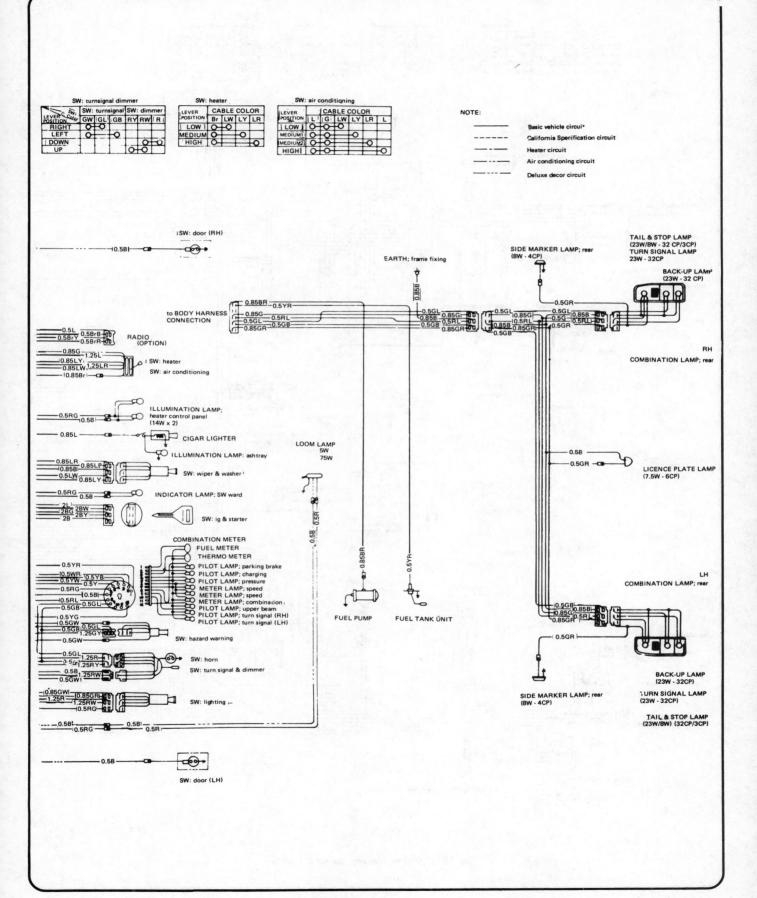

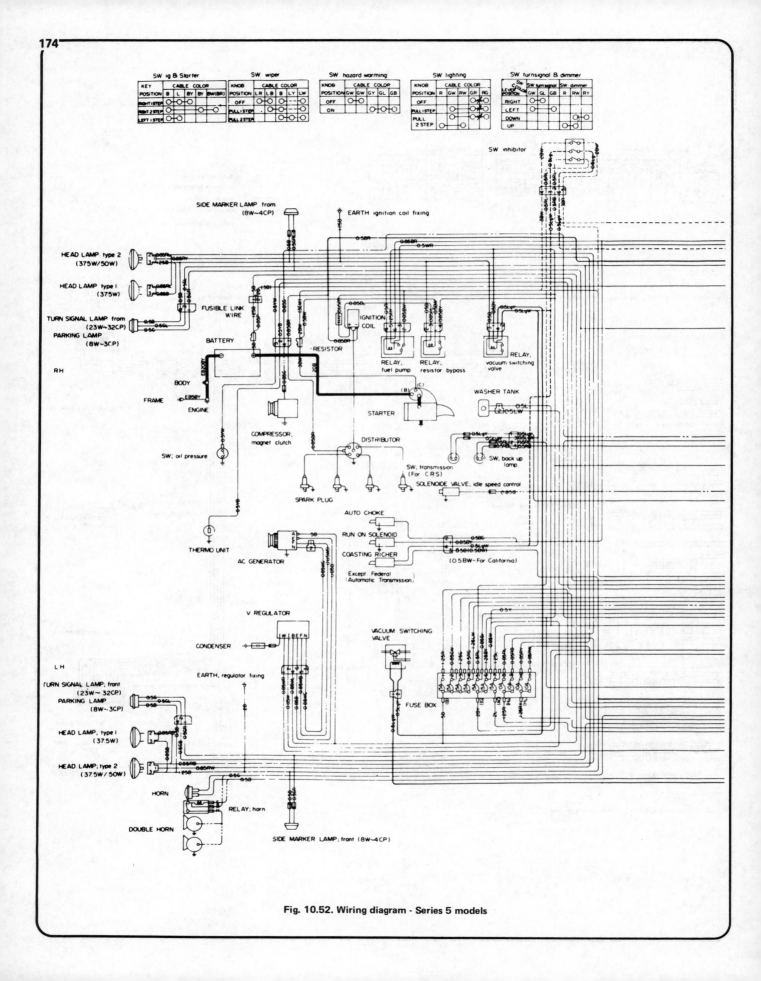

Fig. 10.52. Wiring diagram - Series 5 models

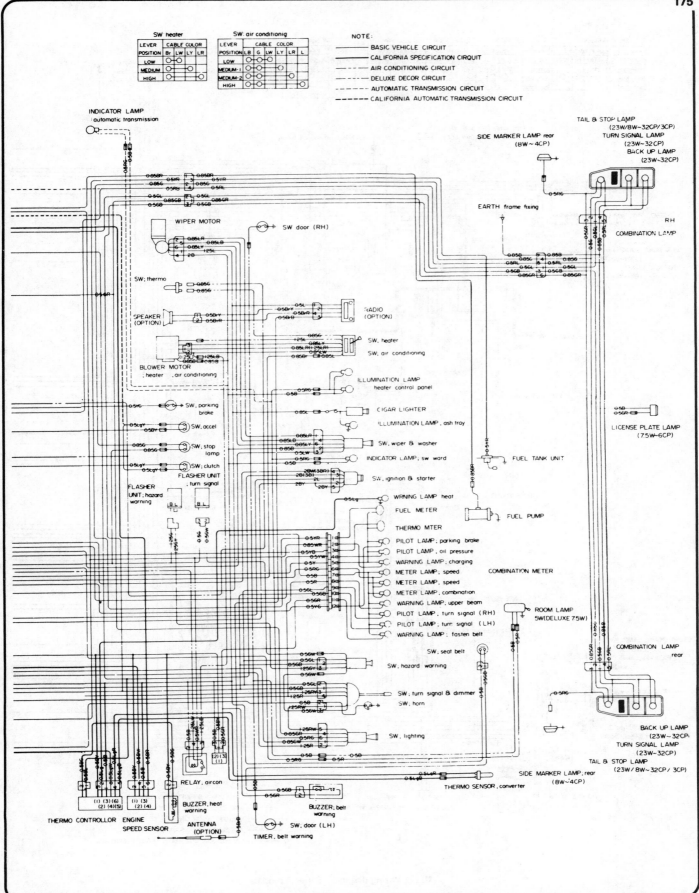

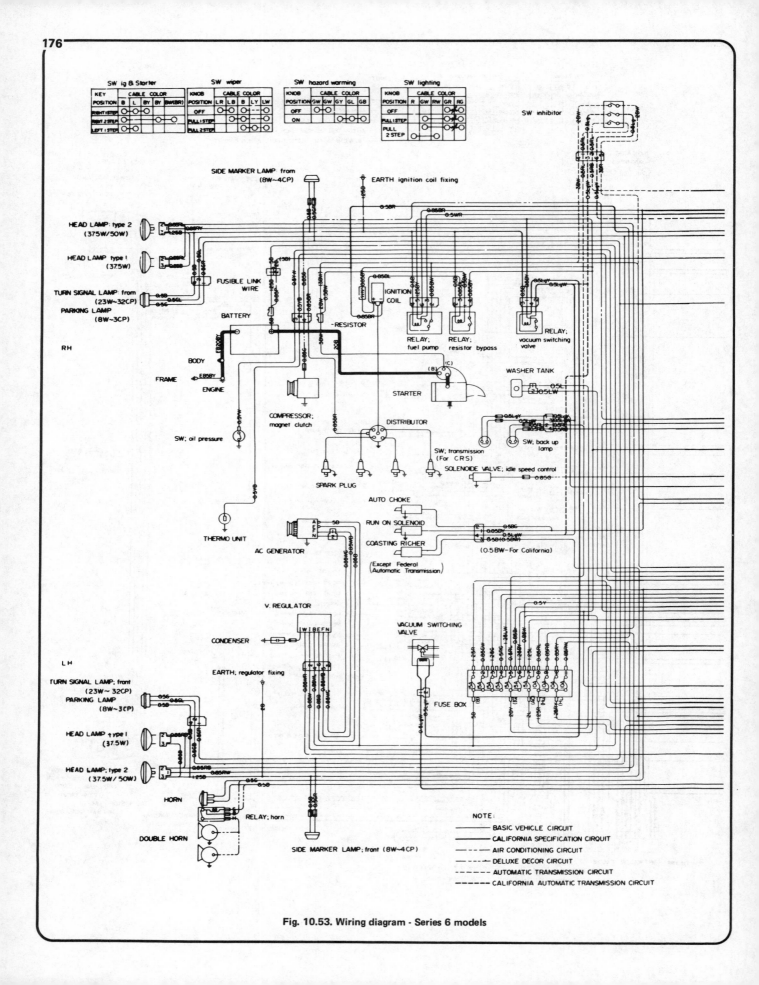

Fig. 10.53. Wiring diagram - Series 6 models

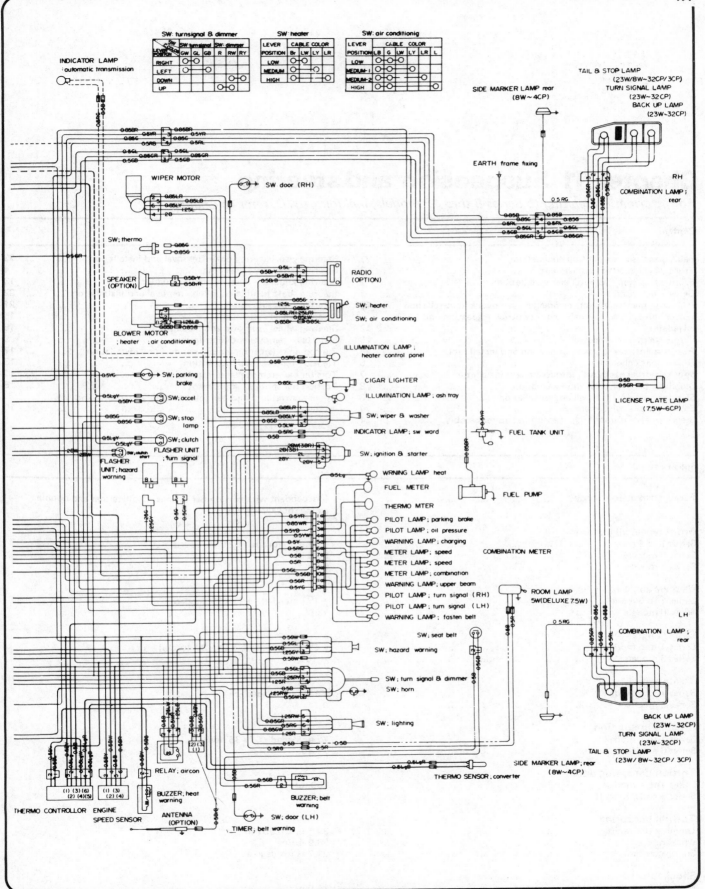

Chapter 11 Suspension and steering

For information relating to Series 8 thru 12 models, including 4WD, refer to Chapter 13.

Contents

Fault diagnosis - suspension and steering ... 27	Steering gear (steering box) - removal and installation ... 23
Front wheel bearing - adjustment ... 3	Steering geometry - checking ... 4
Front wheel hub - removal and installation ... 2	Steering knuckle - removal and installation ... 13
General description ... 1	Steering shaft flexible coupling - removal and installation ... 22
Idler arm, pivot shaft and/or bracket - removal and installation ... 19	Steering shaft - renewal ... 21
Lower control arm and ball joint - removal, inspection and installation ... 12	Steering wheel - alignment ... 15
	Steering wheel - removal and installation ... 16
Pitman shaft seal - renewal ... 25	Strut bar - removal and installation ... 9
Rear suspension - removal, inspection and installation ... 26	Tie rod links (steering arms) - removal and installation ... 17
Ride height - checking ... 5	Tie rod - removal and installation ... 18
Shockabsorber - removal, inspection and installation ... 7	Torsion bar - removal and installation ... 11
Stabilizer bar - removal and installation ... 8	Torsion bar springs - adjustment ... 6
Steering column - removal and installation ... 20	Upper control arm and ball joint assembly - removal and installation ... 10
Steering gear - adjustment ... 14	
Steering gear - dismantling, inspection and reassembly ... 24	

Specifications

Front suspension type ... Independent with torsion bar springs, stabilizer bar and double acting shockabsorbers

Front wheel alignment
Series 1 to 4 models ... 52.7 in unloaded - 52.8 in loaded
Series 5 models ... 53.8 in
Series 6 models ... 54 in

Camber angle
Series 1 to 5 models ... + 1° unloaded - + 1° - 25' loaded
Series 6 models ... + 30'

Caster angle
Series 1 to 5 models ... + 0° 20' unloaded, + 1° - 20' loaded
Series 6 models ... − 10'

Toe-in
Series 1 to 5 models ... + 1/8 in unloaded - + 1/8 in loaded
Series 6 models ... 0

Kingpin inclination
Series 1 to 5 models ... 7° unloaded - 6° - 35' loaded
Series 6 models ... 7° 30'

Torsion bar spring data
Series 1 to 4 models ... 800 lbs
Series 5 and 6 models ... 830 lbs

Torsion bar spring
Length x diameter ... 38.1 in x 0.851 in
Wheel rate/side ... 96.9 lbs/in
Spring constant ... 21.7 ft lbs/degree

Shockabsorbers
Maximum length ... 14.0 in

Minimum length	8.86 in
Stroke	5.14 in

Damping force
Rebound side	232 lb/11.8 in/sec
Compression side	79.4 lb/11.8 in/sec

Rear suspension type	Semi-elliptic leaf springs and hydraulic shockabsorbers

Steering
Type	Ball-screw
Gear ratios	
Series 1 to 5	22.4 overall 17.8
Series 6	23.5 to 27.5
Steering wheel diameter	15.7 in
Steering wheel play	0.394 in (outside diameter)
Ball diameter	¼ in
Number of balls x number of ball circuits	
Series 1 to 5	29 x 2
Series 6	26 x 2
Steering angles	
Outer wheel	30° - 15'
Inner wheel	39°
Oil capacity	0.635 US pints, 10 fl oz
Oil type	1051052 GM Soec. GM 4673M - SAE 90 or equivalent
Worm bearing pre-load	
Series 1 to 5	3.0 to 5.6 in lbs
Series 6	2.6 to 5.2 in lbs
Sector lash adjustment	
Series 1 to 5	3.5 to 8.5 in lbs
Series 6	4.3 to 8.7 in lbs
Sector lash locknut	
Series 1	25 ft lbs
Series 2 to 6	20 ft lbs
Total steering gear pre-load	
Series 1	3.5 to 8.5 in lbs
Series 2 to 6	3.5 to 10.5 in lbs

Wheels
Type	Pressed steel
Size	
Series 1 to 4	4J - 14
Series 5 and 6	5J - 14

Tires
Size	
Series 1 to 4	6.00 - 14 - 6 ply
Series 5 and 6	E78 - 14 - 4 ply
Tire inflation pressures	
Front - Series 1 to 4	21 lbs sq in
Rear (loaded)	42 lbs sq in
(unloaded)	25 lbs sq in
Front - Series 5 and 6	20 lbs sq in
Rear (loaded)	32 lbs sq in
(unloaded)	20 lbs sq in

Torque wrench settings	lb f ft	kg f m
Steering and front suspension		
Tie rod ball joint nuts		
Series 1	40	5.5
Series 2	50	6.9
Series 3 to 6	43.5	5.9
Idler arm mounting bolts		
Series 1 models		
Large	70	9.6
Small	30	4.1
Series 2 models		
Large	55	7.6
Small	20	2.7
Series 3 to 6 models		
Large	50.5	6.9
Small	20	2.7

Chapter 11/Suspension and steering

Idler arm/intermediate rod nut
Series 1	40	5.5
Series 2 to 4	50	6.9
Series 5 and 6	43.5	6.0
Idler arm to idler arm pivot shaft	87	12.0

Pitman arm/intermediate rod nut
Series 1	45	6.2
Series 2	50	6.9
Series 3 to 6	43.5	5.9

Intermediate rod locknuts
Series 1	50	6.9
Series 2	90	12.4
Series 3 to 6	87	12.0

Pitman arm to sector shaft nut
Series 1 and 2	165	22.8
Series 3 to 6	160	22.1

Steering gear mounting bolts
Series 1 Large	70	9.6
Series 2 to 6 Large	55	7.6
Series 1 Small	30	4.1
Series 2 to 6 Small	20	2.7
Steering wheel nut	25	3.4
Coupling bolts	18	2.4
Shaft coupling damper bolts	20	2.7
Lower coupling/wormshaft clamp bolt	20	2.7
Column to cowl panel screws	10	1.3
Combination switch screws	Snug	

Ball joint stud nut
Lower and upper studs Series 1	80	11.0
Series 2 to 6	75	10.3
Ball joint to lower arm	45	6.2

Control arm pivot to frame
Upper - Series 1	55	7.6
Series 2 to 6	50	6.9
Lower - Series 1	135	18.6
Series 2 to 6	130	17.9

Upper control arm pivot shaft bushings
Series 1	250	34.5
Series 2	220	30.4

Shockabsorber
Upper - Series 1	100	13.8
Series 2 to 6	18	2.4
Lower - Series 1	35	4.8
Series 2 to 6	45	6.2

Stabilizer bar
Nuts	7	0.9
Locknuts	18	2.4

Lower control arm to crossmember
Series 1	135	18.6
Series 2 to 6	130	17.9

Backing plate to knuckle
Large bolts - Series 1	55	7.6
Series 2 to 4	50	6.9
Small bolts - Series 1 to 4	35	4.8

Adapter and dust shield to knuckle bolts - Series 5 and 6
Large	55	7.6
Small	35	4.8
Wheel stud nuts	65	8.9

Wheel bearing adjustment	... ...	22	3.0
Strut bar to lower control arm	... ...	45	6.2

Strut bar to frame

Nut - Series 1	... ...	175 in lbs	24.1 kg cm
Series 2 to 6	... ...	15	2.0
Locknut - Series 1	... ...	55	7.6
Series 2 to 6	... ...	50	6.9

Rear suspension

Axle to 'U' bolts	... ...	40	5.5
Shackle pins - Series 1 to 4	... ...	130	17.9
Series 5 and 6	... ...	95	13.1

1 General description

The fully independent front suspension is of the short and long arm type incorporating torsion bar springs.

The control arms are located to the vehicle with bolts and bushings on their inner pivot points and at the outer points to the steering knuckle. The steering knuckle is an integral part of the front wheel spindles. Each control arm has a ball joint to enable the knuckle to pivot for steering.

Each front wheel hub is located on the knuckle spindle by two roller bearings, and the brake drum (Series 1 to 4) or disc rotor (Series 5 and 6) is located to the hub.

Height control is governed by the third crossmember to which the forged ends of the torsion bar links are bolted at the rear and the control arms in front.

Strut bars regulate the fore and aft movement of the front suspension and these are located between the chassis frame and lower control arms. A torsion bar type stabilizer is fitted by shackle rods to the lower control arm.

The steering system consists of the steering gear unit, the steering column and steering linkage.

The steering gear is of the ball-screw type and is connected to the steering column at one end and the steering linkage at the lower end. On the lower end of the steering gear is a worm shaft gear which engages with a ball nut thru a number of recirculating balls. The turning motion of the steering shaft is transferred onto the sector shaft thru the balls. Provision is made for backlash adjustment between the sector gear and rack by means of an adjustment screw which bears upon the tapered sector gear. The steering wheel rotates 3.3 turns from lock to lock.

The energy absorbing steering column is designed to collapse during a frontal collision under pre-determined loads.

Steering linkage consists of a splined Pitman arm which is connected to an intermediate rod and tie rod, which are adjustable. The intermediate rod is also interconnected with the idler arm, which is located to the other tie rod, and an idler arm pivot shaft fastened to the frame via a bracket.

The non-adjustable tie rods are connected to the brake back plates by tie rod links. Lubrication facilities are located on all ball joints and also the idler arm pivot shaft.

The rear suspension consists of semi-elliptic leaf springs and double-acting telescopic shockabsorbers.

2 Front wheel hub - removal and installation

Series 1 to 4

1 Apply the handbrake firmly and raise the roadwheel clear of the ground.
2 Remove the hub cap, wheel nuts and wheel.
3 Prise the hub grease cap from the hub using a suitable screwdriver.
4 Withdraw the cotter pin (photo) and unscrew the spindle nut retainer and nut (photo).
5 The outer bearing can now be extracted from the hub (photo).
6 If the wheel hub is to be removed, then the brake drum must be removed first (see Chapter 9) or alternatively, it can be withdrawn complete with the hub and then separated.
7 Remove the roller bearing cups from the hub using a suitable drift or puller, and prise out the grease seal using a screwdriver.
8 Clean all of the old grease from the hub and bearing, and also the hub grease cap. Inspect the hub and bearing for signs of excessive wear, scratches, pits or corrosion or any other signs of damage. If necessary they should be renewed as required.
9 Clean the spindle and adjacent components and inspect for signs of wear and damage. Renew as required.
10 Commence assembly by packing the hub with the recommended wheel bearing grease. Lubricate the wheel bearing cups and cones with the same type of grease.
11 Refit the bearing cups to the hub and the new grease seal.
12 Refit the hub to the spindle and install the outer bearing cone, retaining it with the nut and washer.
13 Adjust the hub as described in the following Section.
14 Refit the brake drum and roadwheel.
15 Pad the grease cap with the recommended wheel bearing grease and tap into position. Refit the hub cap and lower the vehicle.

Series 5 and 6

16 Follow the instructions given in paragraphs 1 and 2.
17 Undo the disc brake support unit retaining bolts to the adaptor and withdraw the support and caliper unit.
18 Prise the hub grease cap from the hub with a screwdriver and withdraw the cotter pin. Remove the spindle nut retainer and nut. The rotor and hub units can now be removed.
19 Undo the bolts to separate the rotor from the hub.
20 Clean and inspect the hub/bearing unit as described in paragraphs 7 to 10.
21 Assemble in the reverse sequence and adjust the hub as given in the following Section.

3 Front wheel bearing - adjustment

1 Apply the handbrake and raise the front of the vehicle so that the front wheel is clear of the ground.
2 Remove the hub cap and carefully tap or prise out the hub grease cap.
3 Wipe any excess grease from the end of the spindle and then withdraw the cotter pin and castellated nut retainer.
4 Spin the roadwheel and tighten the hub nut to a torque of 22 lb f ft.
5 Rotate the hub two or three complete turns and loosen the nut sufficiently to allow the hub to be turned with the fingers, and check that there is no free play in the hub. Ensure when rotating the hub that the brake linings are not in contact with the drum.
6 Refit the castellated nut retainer and insert a new cotter pin, then bend over the ends to retain it in position.
7 Check that the wheel rotates freely and then install the grease cap and hub cap. **Note:** If the wheel does not rotate freely, then the bearings are probably due for renewal - see previous Section.

4 Steering geometry - checking

1 In order to obtain satisfactory tire wear and good steering stability, it is essential that the steering geometry is correctly set. This is not really a do-it-yourself job and it is always preferable to have the job undertaken by your vehicle main dealer who will have the correct

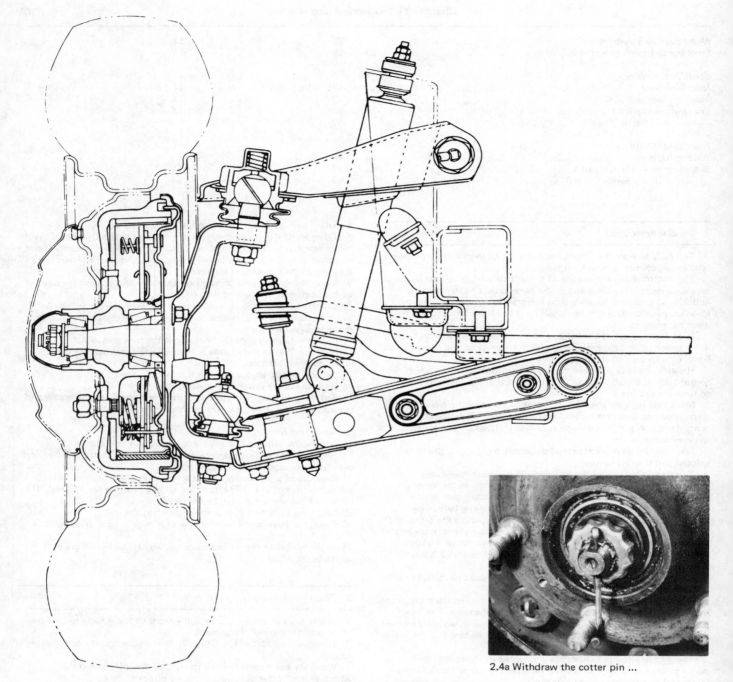

Fig. 11.1. The front suspension and wheel hub assembly

2.4a Withdraw the cotter pin ...

2.4b ... remove the retainer, nut and washer

2.5 Extract the bearing

7.5 The upper shock absorber location nut and bushes

alignment gauges. However, if you feel competent to do the job yourself, the basic procedure is given below.
2 Before commencing any checks, the vehicle must be standing on a level floor, the tires must be inflated to the correct pressures (cold), excessive mud must be removed from the vehicle underframe and the fuel tank, engine oil level and coolant level must be correct.
3 In addition to the above, it must be known that the following components are in good condition and correctly tightened and/or adjusted:

    a) Wheel bearings.
    b) Steering shaft coupling and steering box.
    c) Tie rod and steering connections.
    d) Front torsion bars.
    e) Front torsion bar heights and shockabsorbers.
    f) Control arms and stabilizer bar.

Before adjustment is made to the caster angle and camber angle, lift the front bumper and release it to allow the vehicle to return to its normal height.

Camber angle

4 The camber angle is the angle by which the wheel tilts outwards when the weight is on the suspension.
5 To obtain the specified camber angle, change the shims at the front and rear of the upper control arm pivot shaft and frame (Fig. 11.4). Add an equal amount of shims at the front and rear of the pivot shaft to decrease the positive camber.

Caster angle

6 Fit shims front to rear and rear to front. The transfer of one shim to the front bolt from the rear increases the positive caster.
7 Both caster and camber angles can be adjusted in one operation and on completion the toe-in must be checked as follows.

Toe-in

8 Raise the front end of the vehicle so that the wheels are clear of the ground, and in the straight ahead position.
9 Turn each wheel by hand and scribe a chalk line in the center of the tire tread around its circumference.
10 Measure between the chalk marks at the front and rear of the tire at equal heights from the ground. If the toe-in is correct, the distance between the chalk marks at the rear should be greater than the distance between them at the front by the amount given in the Specification.
11 If adjustment is required, loosen the tie-rod clamp bolts and rotate the tie rod as necessary. Each tie rod is threaded right-hand at one end and left-hand at the other end in order to retain the correct balance of the linkage; the two tie rods must be the same length after any adjustment.

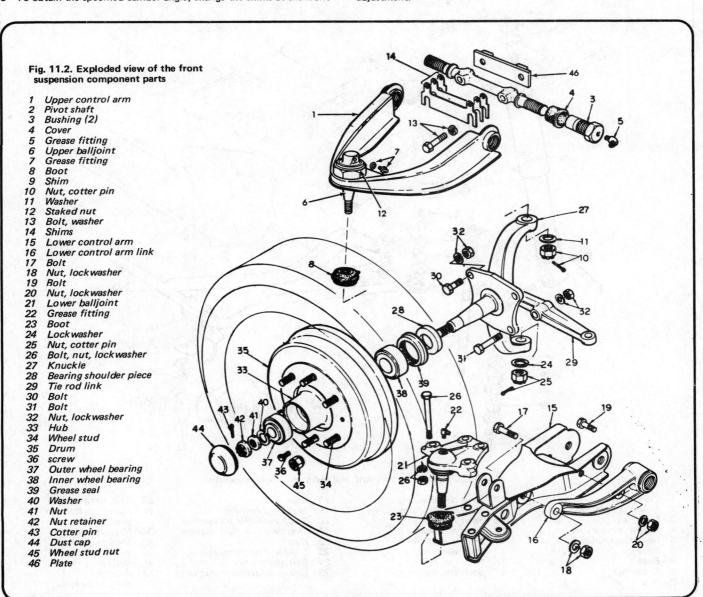

Fig. 11.2. Exploded view of the front suspension component parts

1 Upper control arm
2 Pivot shaft
3 Bushing (2)
4 Cover
5 Grease fitting
6 Upper balljoint
7 Grease fitting
8 Boot
9 Shim
10 Nut, cotter pin
11 Washer
12 Staked nut
13 Bolt, washer
14 Shims
15 Lower control arm
16 Lower control arm link
17 Bolt
18 Nut, lockwasher
19 Bolt
20 Nut, lockwasher
21 Lower balljoint
22 Grease fitting
23 Boot
24 Lockwasher
25 Nut, cotter pin
26 Bolt, nut, lockwasher
27 Knuckle
28 Bearing shoulder piece
29 Tie rod link
30 Bolt
31 Bolt
32 Nut, lockwasher
33 Hub
34 Wheel stud
35 Drum
36 screw
37 Outer wheel bearing
38 Inner wheel bearing
39 Grease seal
40 Washer
41 Nut
42 Nut retainer
43 Cotter pin
44 Dust cap
45 Wheel stud nut
46 Plate

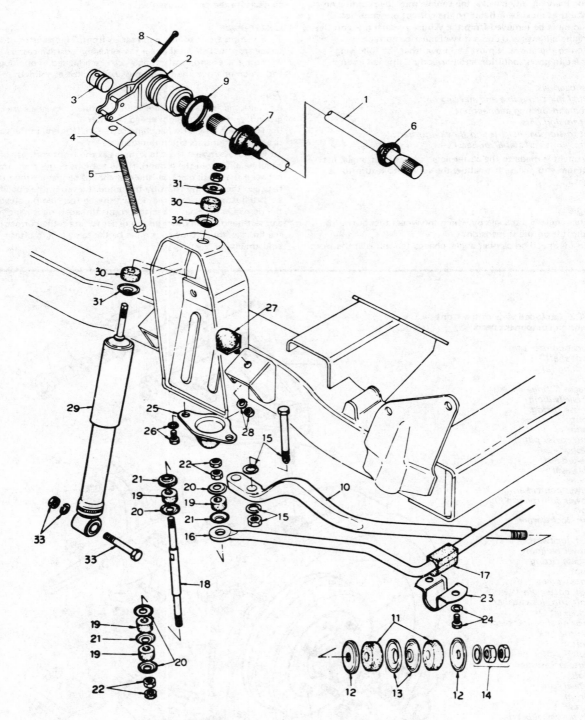

Fig. 11.3. The torsion bar, strut rod and stabilizer components

1 Torsion bar	10 Strut rod assembly	18 Link stud	26 Bolt, washer
2 Height control arm	11 Strut rod bushings	19 Link stud bushings	27 Upper control arm bumpers (2)
3 Pivot nut	12 Strut rod washer	20 Stabilizer link stud washers	
4 Height control seat	13 Strut rod washer	21 Stabilizer link stud washers	28 Nut, washer
5 Height control bolt	14 Nuts and washer	22 Nuts	29 Shock absorber
6 Boot	15 Bolt, washer, nut	23 Stabilizer bar bracket	30 Bushing
7 Boot	16 Stabilizer bar	24 Bolt and washer	31 Retainer
8 Cotter pin	17 Stabilizer bushings	25 Lower control arm bumper	32 Retainer
9 Seal			33 Bolt, lockwasher, nut

Chapter 11/Suspension and steering 185

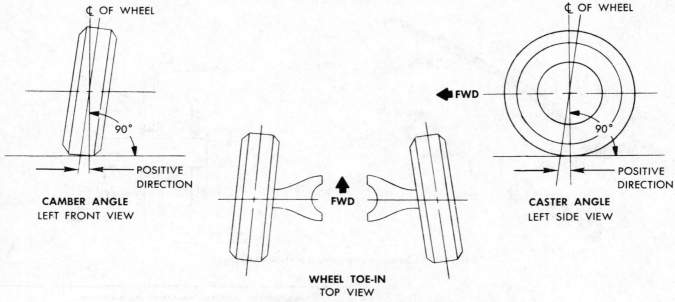

Fig. 11.4. The caster and camber angles and toe in

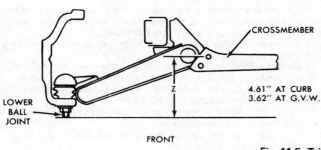

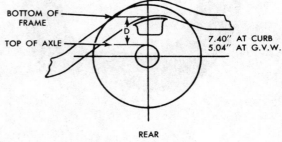

Fig. 11.5. Trim height points

5 Ride height - checking

1 Park the vehicle on a smooth level surface. Bounce the front end of the vehicle up and down several times then lift at the front and allow to settle to normal height.
2 Refer to Fig. 11.5 and measure the indicated distances Z or D.
3 The distances should be as indicated in the Figures with the vehicle at kerb weight ie. no passengers, full tank of gas and spare tire and jack fitted.
4 The measurements should not differ beyond ½ in for both sides of the vehicle. If adjustment is needed proceed as follows.

6 Torsion bar spring - adjustment

1 The vehicle height is adjusted by means of an adjustment bolt on the height control arms.
2 Follow the instructions given in the preceding Section and ensure that the tire pressures are as specified.
3 Slacken the front end of the strut bar nuts and calculate the buffer clearance between the rubber bumper and lower control arm.
4 Adjust the clearance bolt on the height control arms to set the buffer clearance at 7/8 in (22.2 mm).
5 To increase the height, turn the bolt inwards.
6 Ensure that the trim heights are as specified and tighten the strut bar nuts to the specified torques.

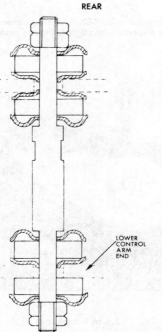

Fig. 11.6. The stabilizer bar link assembly

Chapter 11/Suspension and steering

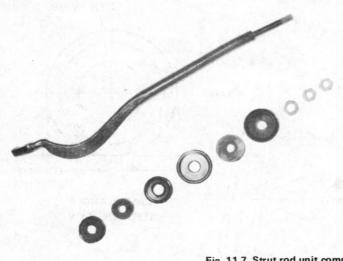
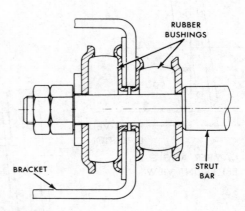

Fig. 11.7. Strut rod unit component parts and location

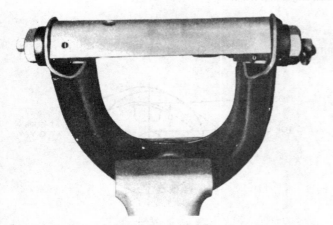

Fig. 11.8. Special tool no. J.24258 located between the control arm forks acts as a spacer when tightening the pivot shaft bushing

Fig. 11.9. The height control unit

7 Shockabsorber - removal, inspection and installation

1 Apply the parking brake and jack up the vehicle.
2 To prevent the shockabsorber upper stem from turning when undoing the retaining nut, retain it with an open-ended wrench. Undo and remove the upper stem retaining nut, retainer and rubber grommet (photo).
3 Undo and remove the lower shockabsorber pivot to lower control arm retaining bolt and remove the shockabsorber unit.
4 Clean the shockabsorber outer surface and inspect for signs of oil leakage and/or weak damping tension. Renew the unit if necessary. It is not possible to repair faulty shockabsorbers.
5 If the lower or upper bushes are worn, damaged or defective, they must be renewed.
6 Fully extend the shockabsorber and with the lower retainer and rubber grommet in position over the upper stem, install the shock-absorber upwards thru the control arm and ensure that the upper stem passes thru the frame bracket mounting hole.
7 Fit the upper rubber grommet, retainer and attachment nut over the upper stem of the shockabsorber.
8 Retain the upper stem with an open ended wrench to prevent it turning and tighten the retaining nut.
9 Fit the retainers that attach the lower pivot to the lower control arm, and tighten.
10 Lower the vehicle and remove the jack.

8 Stabilizer bar - removal and installation

1 Apply the parking brake and jack up the front of the vehicle.
2 Remove the stabilizer bar bracket retaining bolts and detach the brackets and bushings.
3 Unscrew and remove the stabilizer bar link bolts, spacers and rubber grommets from the lower control arms.
4 Clean the respective components and renew as required.
5 Note before reassembly that the washers with the center ring must be fitted with the stabilizer bar and lower control arm brackets interposed between the washers.
6 Slide the new frame bushes into position and fit the stabilizer brackets over them. Reconnect them to the frame but do not fully tighten yet.
7 Reconnect the stabilizer ends to the link bolts of the lower control arms.
8 Tighten the bracket bolts and link bolt nuts to the specified torque.

9 Strut bar - removal and installation

1 Apply the parking brake and jack up the vehicle.
2 Unscrew and remove the double nuts, washer and rubber bushing

Chapter 11/Suspension and steering

from the front of the strut bar and remove the frame side bracket.
3 Undo the strut bar to lower control arm bolts and remove with the strut bar.
4 Clean and inspect the strut bar and bushings for defects or wear. Renew as required.
5 Fit the washers and bushes to the strut rod and insert thru the frame bracket.
6 Fit the second set of washers and bushings to the strut rod and one nut. Do not tighten the nut at this stage.
7 Fit the strut rod to the lower control arm and tighten the bolts to the specified torque.
8 Lower the vehicle and tighten the first nut, to the specified torque to the frame bracket. Fit and tighten the second nut to the specified torque.

10 Upper control arm and ball joint assembly - removal and installation

1 Apply the parking brake and jack up the front of the vehicle. Place chassis stands or blocks under the lower control arm. Remove the roadwheel.
2 Withdraw the cotter pin from the nut retaining the upper control arm and upper ball joint unit. Detach the upper control arm from the steering knuckle, but support the knuckle unit by tying it with wire or cord and hang it out of the way. Do not allow it to hang from the brake hose!
3 Remove the upper pivot shaft bolts and remove the control arm from the bracket, noting the position and number of camber and caster angle shims.
4 Undo the bush nuts from the pivot shaft, loosening in an alternate manner, and remove the pivot from the control arm.
5 Clean and inspect the components for excessive wear, and for signs of deterioration or damage. Renew any defective parts.
6 Fit the boots to the pivot shaft, and lubricate the inner bushings with molybdenum disulphide grease.
7 Screw the bushings into the pivot shaft, in an alternate manner and avoid getting grease onto the outer face of the bushings.
8 Fit a suitable length bar or, if available, special tool number J.24258 between the control arm forks as a spacer (Fig. 11.8) and then tighten the pivot shaft bushing.
9 Ensure that the bushes are centered correctly and that the control arm rotates with resistance but does not bind on the pivot shaft when tightened to the specified torque.
10 Refit the grease fittings and lubricate.
11 Fit the ball joint stud thru the knuckle and locate the castellated nut which must be tightened to the specified torque and then to the

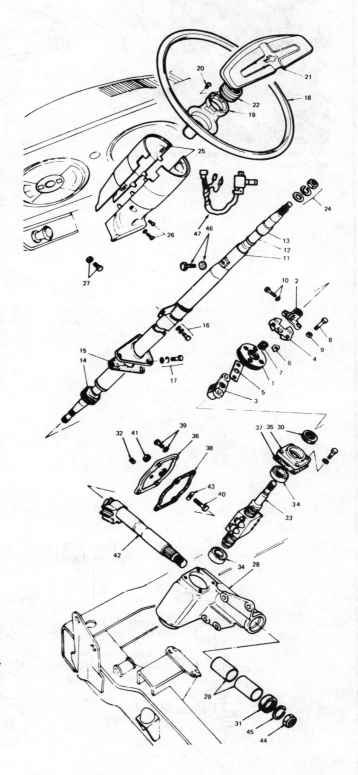

Fig. 11.10. The steering column and gearbox component parts

1 Coupling
2 Flange, upper coupling
3 Flange, lower coupling
4 Cross strap
5 Cross strap
6 Thrust washer
7 Spring
8 Thru bolt
9 Lock nut
10 Pinch bolt, lockwasher
11 Mast jacket
12 Shaft assembly
13 Bushing
14 Grommet
15 Gasket
16 Bolt and washers
17 Screw and washer
18 Wheel assembly
19 Horn shroud seat
20 Screw
21 Horn shroud
22 Spring
23 Nut
24 Shaft nut and washers
25 Column cowling
26 Cowling screws and washers
27 Bolt, washer
28 Steering gear housing
29 Sector shaft bushings
30 Wormshaft seal
31 Sector (Pitman) shaft seal
32 Filler plug
33 Worm and ball nut assembly
34 Wormshaft bearing
35 End cover
36 Top cover
37 Worm preload shims
38 Gasket
39 Bolt, lockwasher
40 Sector adjuster screw
41 Locknut
42 Sector shaft
43 Adjusting shim
44 Pitman shaft nut
45 Lockwasher
46 Bolt, nut (stopper)
47 Hazard warning switch assembly

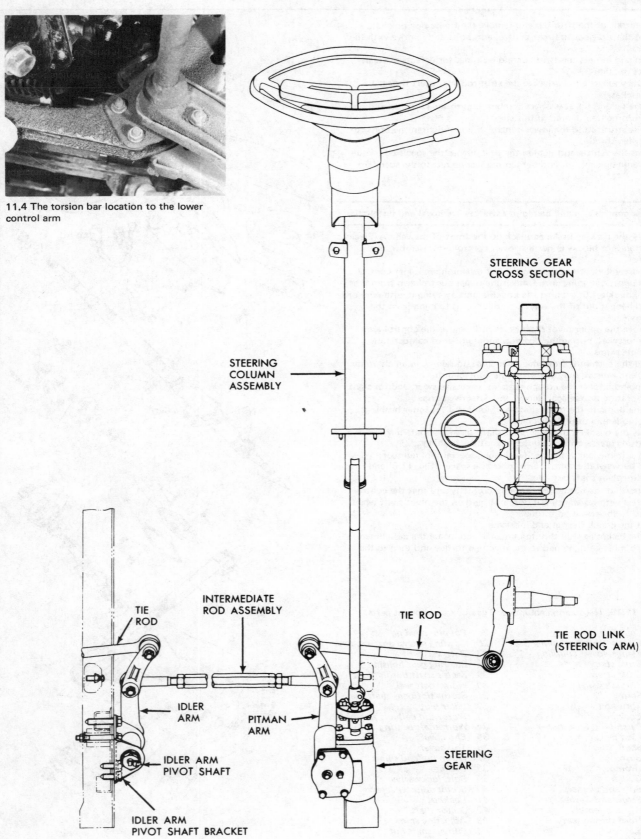

11.4 The torsion bar location to the lower control arm

Fig. 11.11. The steering column, gear and connecting rod layout

nearest hole alignment for the cotter pin. Install the new cotter pin and bend over the ends to retain it.
12 Locate the upper control arm into position in the chassis frame and insert the camber and caster angle shims to their respective locations as removed.
13 Tighten the pivot shaft to the specified torque. For good shaft to frame clamping force and retention, tighten the thin shim pack bolt first.

11 Torsion bar - removal and installation

1 Apply the handbrake and jack up the front of the vehicle. Support the front of the vehicle with chassis stands or blocks.
2 Unscrew and remove the height control arm adjustment bolt. Mark its relative location and remove the height control arm from the torsion bar and third crossmember.
3 Mark the torsion bar's relative location and withdraw it from the lower control arm (photo).
4 Clean the torsion bars and inspect for distortion or damage. Renew, if required.
5 Smear some grease to the torsion bars serrated ends, and then, locating the rubber bumpers in contact with the lower control arm (with the front of the vehicle jacked up under the lower control arm), insert the front of the torsion bar into the arm.
6 Fit the height control arm into position with, and contacting, the adjustment bolt. Grease the part of the height control arm which fits within the chassis frame.
7 Insert a new cotter pin and bend over to retain.
8 The adjustment bolt is now tightened to its original location which was marked before removal.
9 Check and adjust the vehicle ride height as in Section 5.

12 Lower control arm and ball joint - removal, inspection and installation

Refer to Fig. 11.2
1 Apply the parking brake and jack up the vehicle at the front. Make secure by placing chassis stands under the frame.
2 Remove the wheel.
3 Detach the strut bar and withdraw the stabilizer from the rod - see Section 9.
4 Remove the torsion bar - see Section 11.
5 Unscrew and remove the shockabsorber to lower control arm retaining nut.
6 Undo and remove the two lower balljoint bolts and detach from the control arm.
7 Remove the lower control arm.
8 Remove the ball joint to steering knuckle retaining nut and cotter pin. Separate the knuckle and ball stud.
9 Unscrew and remove the lower balljoint to strut rod bolts and remove the ball joint.
10 Clean the respective components and inspect for signs of wear and damage. The ball joints must have no visible play when checked, the maximum play allowable being 0.06 in (1.5 mm). Renew any defective parts.
11 Fit the lower ball joint to the lower control arm and tighten the retaining bolts to the specified torque.
12 Locate the ball joint stud into the steering knuckle, fit the castellated nut and tighten to the specified torque, plus a sufficient amount to align the cotter pin holes. Insert the pin and bend it over to retain.
13 Locate the lower control arm to the chassis and drive the bolt carefully into position. Use a soft drift to drive the bolt home in order not to damage the serrated bolt head. Fit the nut and tighten to the specified torque.
14 Refit the torsion bar (see Section 11).
15 Remove the chassis stands/blocks and lower the jack. 'Bounce' the front of the vehicle up and down and then lift at the front, and allow the vehicle to settle at normal height. The strut bars can now be refitted but do not tighten the strut bar to frame nuts until the vehicle ride height has been checked (Section 5).
16 Ensure that the lower control arm side nuts are tightened to the specified torque.

13 Steering knuckle - removal and installation

1 Apply the parking brake and jack up the front of the vehicle to raise the wheels clear of the ground, and support with chassis stands or blocks.
2 Remove the roadwheel.
3 Detach the flexible brake hose from the wheel cylinder - see Chapter 9.
4 Remove the wheel hub - see Section 2 - and brake drum/rotor.
5 Unscrew and remove the tie rod end retaining bolts and detach the rod.
6 Remove the remaining backing plate bolts and detach the backing plate.
7 Withdraw the cotter pins and remove the upper and lower ball joint retaining nuts.
8 Detach the knuckle from the ball joints and remove the knuckle.
9 Clean the respective components and inspect for wear or damage and renew if defective.
10 Reassembly is a direct reversal of removal but note the following.
   a) *On reassembly of the wheel hub, adjust the wheel bearing as described in Section 3.*
   b) *On reconnecting the brake hose, the brakes will require bleeding as described in Chapter 9.*
   c) *Tighten the respective bolts and nuts to the specified torque and always fit new cotter pins on reassembly.*

14 Steering gear - adjustment

1 Adjustment of the steering gear should only be made when items such as the wheel alignment and suspension geometry are known to be correct. The shockabsorbers must be in good condition and the tire pressures as specified. The wheels must be correctly balanced.
2 Detach the ground cable from the battery.
3 Apply the parking brake and jack up the front of the vehicle.
4 Unscrew the retaining nut and remove with washer from the Pitman (idler) arm. Mark the Pitman arm in relation to the shaft and detach it. If special tool no. J-5504-1 is available the arm should be detached by the method shown in Fig. 11.12.
5 Detach the horn shroud and spring.
6 Slowly turn the steering wheel in one direction until it stops and then turn it back halfway. Do not force the steering wheel against the stops when the linkage is detached as the ball guides may be seriously damaged.
7 Using a torque wrench with a maximum torque reading of 50 in lbs, measure the bearing drag when rotating the steering wheel nut thru 90°. Note the torque reading.

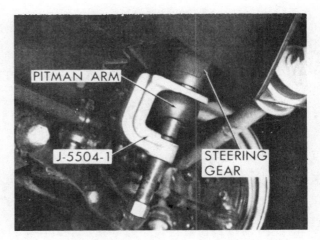

Fig. 11.12. Using special tool J.5504 to remove the Pitman arm

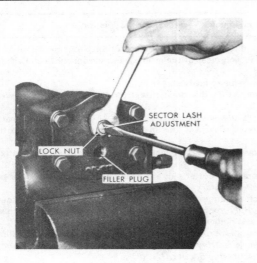

Fig. 11.13. Adjust the sector to ballnut backlash

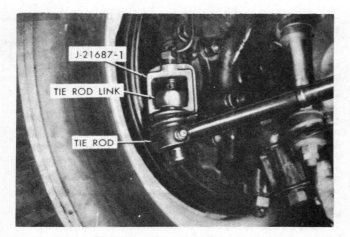

Fig. 11.14. Separating the ball stud from the link using special tool J.21687-1

Fig. 11.15. The idler arm fitting dimension

8 To adjust the over center pre-load, turn the steering wheel from lock to lock and then turn it back to exactly halfway to centralize it.
9 Slacken the locknut and turn the sector (lash) adjuster screw clockwise to eliminate the ball nut to Pitman shaft sector teeth lash. Tighten the locknut.
10 Check the steering wheel rotational torque and note the highest reading in the over center position. Compare this reading with that given in the Specifications and if necessary, readjust the lash adjustment screw to acquire the correct torque reading.
11 Refit the Pitman arm to the shaft with the relative marks in alignment. Fit the nut and tighten to the specified torque.
12 Refit the horn spring and shroud. Remove the jack and reconnect the battery ground cable.

15 Steering wheel - alignment

1 Set the front roadwheels in the straight ahead position and check that the spot mark on the wormshaft which designates the steering high point, is on the top side of the shaft in the 12 o'clock position.
2 If it is not, loosen the left- and right-hand intermediate rod locknuts, and detach the rod ends. Refit turning both ends equally in the same direction to bring the gear back to the high point position.
3 Check and adjust the toe-in as given in Section 4.
4 Now check the steering wheel alignment and if the spokes are not as specified, remove the wheel and center it - see following Section.

16 Steering wheel - removal and installation

1 Detach the ground cable from the battery.
2 Push the horn shroud and turn it counterclockwise. Withdraw the shroud and spring. Disconnect the horn contact ring and wire.
3 Unscrew and remove the steering wheel retaining nut, washer and lockwasher.
4 Mark the relative position of the steering wheel and steering shaft.
5 Remove the four steering column cowling retaining screws and detach the cowling.
6 Using a suitable steering wheel puller, remove the steering wheel.
7 Installation of the steering wheel is the reversal of the above procedure but ensure that the wheel to shaft alignment marks are correct and check the horn operation when the battery is reconnected. Torque the steering wheel nut to that given in the Specifications.

17 Tie rod links (steering arms) - removal and installation

1 Apply the parking brake and jack up the front of the vehicle.
2 Remove the roadwheel.
3 Remove the rotor and hub (Series 5 and 6 models) or brake drum and hub (Series 1 to 4 models) - See Section 2.
4 Unscrew and remove the tie rod link to brake backing plate (Series 1 to 4) or steering knuckle (Series 5 and 6).
5 Withdraw the tie rod ball stud cotter pin and undo the nut.
6 Separate the ball stud from the link - use special tool no. J-21687-01 if available - and detach the tie rod link (Fig. 11.14).
7 Installation is the reverse of removal but note the following.

   a) *Ensure that the tie rod threads are clean before assembly.*
   b) *Lubricate the tie rod and ball joint before assembling.*
   c) *Fit new cotter pins and tighten the respective nuts to the specified torques.*
   d) *Readjust the wheel bearings and bleed the brakes.*

18 Tie rod - removal and installation

1 Apply the parking brake and jack up the front of the vehicle.
2 Withdraw the cotter pins from the ball stud retaining nuts. Undo the nuts and separate the ball studs.
3 Install in the reverse order and note the following.

   a) *Ensure that the ball stud threads are clean before refitting.*
   b) *Lubricate the ball studs.*
   c) *Tie rods are fitted with the lubrication fittings to the front.*
   d) *Tighten the ball stud nuts to the torque specified.*

19 Idler arm, pivot shaft and/or bracket - removal and installation

1 Apply the parking brake and jack up the vehicle.
2 Unscrew the idler arm to pivot shaft retaining nut and remove with washer.
3 Detach the idler arm from the pivot shaft - if available use special tool no. J-5825.
4 Withdraw the cotter pin and unscrew the nut that retains the intermediate rod and the tie rod to the idler arm. Detach the idler arm.
5 To remove the idler arm pivot shaft and/or bracket, follow the above instructions nos. 1 to 3 and then proceed as follows.
6 Detach the pivot shaft by unscrewing it from the bracket. The bracket is removed by unscrewing the three bolts, nuts and washers from the bracket to frame.
7 Installation is a direct reversal of the removal procedure, but tighten all nuts/bolts to the specified torque. When fitting the idler arm to the bracket and shaft unit, align the lower ends and tighten the nut to the specified torque, and check that the fitting dimensions are as shown in Fig. 11.15.

20 Steering column - removal and installation

1 Detach the ground cable from the battery terminal.
2 Press the horn shroud inwards, twist it counterclockwise and remove the shroud and spring.
3 Detach the steering column cowling and hazard warning switch.
4 Remove the clamp screws and withdraw the direction signal and dimmer switch.
5 Unscrew the steering wheel retaining nut, and withdraw the steering wheel from the shaft using a suitable puller or, if available, special tool no. J-24292. Do not attempt to hammer the wheel from the shaft.
6 Jack up the front of the vehicle and raise the wheels clear of the ground.
7 Slacken the steering shaft flexible coupling clamp bolts, working from the engine compartment, and spread the clamps to allow for the removal of the shaft.
8 Detach the heater hose from its clip on the column. Bend the clip flush to the column.
9 Remove the mast jacket bracket to cowl retaining screws, and the two steering column to instrument panel bolts.
10 Separate the flexible coupling from the shaft and then carefully withdraw the steering column towards the cab.
11 To install, fit the column thru the cowl and locate in position. Insert the steering shaft end into the flexible coupling clamp and install the coupling to gear sector shaft.
12 Fit and tighten the pinch bolts to the specified torque.
13 Fit the column to instrument panel bolts/nuts, but do not tighten yet.
14 Lower the vehicle and remove the jack.
15 Fit the steering column to cowl screws and tighten.
16 Tighten the steering column to instrument panel bolts. It should be noted that the instrument panel must not be removed or installed with the vehicle jacked up.
17 Re-locate the heater hose and retain with the clamp. Refit the direction indicator and dimmer switch to the column.
18 Fit the steering column cowling and locate the hazard warning switch wiring thru the cowling and tighten the setscrews.
19 Refit the steering wheel, washers and nut which must be tightened to the specified torque. Refit the horn shroud and spring, and reconnect the battery ground cable.
20 Check that the horn, combination light switch and hazard warning switch are in operation.
21 Check the interference between the steering column cowling and steering wheel, and also between the column cowling and combination switch.
22 Check the steering wheel free play which should be approximately 0.4 in (10.1 mm) measured at the outside diameter of the steering wheel.

21 Steering shaft - renewal

1 Refer to the previous Section and remove the steering column.
2 Remove the retaining wire from the rubber cup seal and separate the shaft from the mast jacket.
3 Check the ground wire connection on the new shaft and reassemble in the reverse order, and install the column.

22 Steering shaft flexible coupling - removal and installation

1 Apply the parking brake and jack up the front of the vehicle so that the wheels are just clear of the ground.
2 Disconnect and remove the coupling thru bolts (photo). Only two bolts can be removed.
3 Loosen the two pinch bolts to give the necessary clearance and remove the coupling unit.
4 If the coupling clamps are to be removed the steering column must be loosened, so therefore remove the column to cowl mounting screws (photo).
5 Lower the jack and remove the column to instrument panel mounting bolts. Slide the column back 1 inch and support it at the steering wheel.
6 Withdraw the column flexible coupling clamps.
7 Reassembly is a reversal of the removal sequence but note the following.

20.2 The flexible coupling

Fig. 11.16. Press and twist the horn shroud to remove

a) *Tighten the pinch bolts and other securing bolts/nuts to the specified torques.*
b) *When reinstalling the column to instrument panel bolts, ensure that the vehicle is not raised on the jack.*

23 Steering gear (steering box) - removal and installation

1 Apply the parking brake and raise the front of the vehicle.
2 Remove the nut and washer securing the Pitman arm to the shaft. Mark the relative positions of the arm and shaft and remove the arm.
3 Detach the engine stone shield.
4 Disconnect the lower clamp to flexible coupling bolts and then the steering gear to frame bolts (photo). Withdraw the steering gear.
5 Installation is a direct reversal of removal but note the following.
   a) *Ensure that the Pitman shaft and arm are refitted in the marked positions.*
   b) *Tighten all securing bolts/nuts to the specified torques.*

24 Steering gear - dismantling, inspection and reassembly

1 Remove the steering gear as detailed in Section 22.
2 Clean off the outside of the steering gearbox and drain the lubricant thru the oil filler hole, by removing the plug.
3 Remove the coupling from the worm shaft by withdrawing the pinch bolts.
4 Position the steering shaft in the straight ahead position and remove the top cover bolts, and the adjustment screw locknut.
5 Turn the adjustment screw clockwise and remove the top cover, but hold the sector shaft in the straight ahead position during removal.
6 Remove the adjuster screw and withdraw the sector shaft. Do not remove the shaft by driving out with a hammer or similar harsh methods.
7 Remove the end cover bolts and detach the cover and shims (Fig. 11.20).
8 Extract the worm and ball nut from the gearbox and remove the lower bearing. Retain the worm and ball nut in the horizontal position and do not hold it vertically or the ball nut will fall onto the worm gear end and damage the ball tubes.
9 Clean the respective components for inspection, but do not disturb the clamp plate on the ball tube which is sealed with white paint.
10 Check the following items for wear, distortion, pitting or damage.
   a) *Steering shaft*
   b) *Ball nut teeth*
   c) *Bearing*
   d) *Gear casing*

11 If, on inspection, the worm shaft components are found to be faulty, then the worm and ball nut unit must be renewed, as they are a matched pair. Oil seals must always be renewed on assembly and bearings, seals, gearbox and covers can be individually renewed as required.
12 To check the ball nut, position the worm and nut vertically and check that the ball nut lowers in a smooth twisting motion. If the motion is not smooth or the ball nut is noisy when lowering, the complete steering shaft unit must be renewed.
13 Check that the sector shaft teeth and serrated portion are not worn or damaged. Measure the reduction in the outside diameter of the shaft. The standard diameter is 1.181 in and the wear limit is 0.001 in. Renew the sector shaft if necessary.
14 Commence reassembly by installing the lower bearing into position in the gearbox.
15 Insert the worm shaft unit and ensure that the lower end of the worm shaft is fitted correctly into the lower bearing.
16 Locate the upper bearing into position on the worm shaft and fit the adjustment shims (Fig. 11.23) between the gearbox and end cover. Smear the end cover and gearbox joint faces with a liquid gasket solution when fitting. Shim adjustment is as follows.
17 Position the gear case in a vise as shown in Fig. 11.24, but do not overtighten the jaws. Locate a pull scale to the flange to coupling bolt and measure the starting torque required. The normal reading should be 3.0 to 5.6 in lbs (2.6 to 5.2 in lbs Series 6) when the coupling starts to turn. Shims are added or subtracted to obtain this reading.

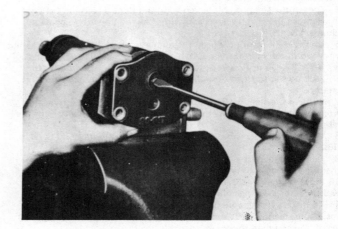

Fig. 11.17. Turn the adjustment screw clockwise

Fig. 11.18. Top cover removed showing gear

22.4 The steering gearbox, showing the column coupling, mounting bolts and adjustment screw. The oil filter is located above the adjustment screw

Fig. 11.19. Withdraw the sector shaft

Fig. 11.20. Remove the end cover

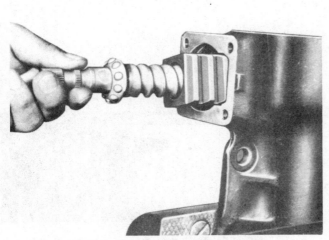

Fig. 11.21. Remove the worm and ball nut

Fig. 11.22. Insert the worm and ball nut

Fig. 11.23. Fit the end cover with shims

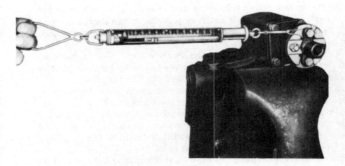

Fig. 11.24. Using a pull scale to calculate the starting torque for shim adjustment to the end cover

Chapter 11/Suspension and steering

Fig. 11.25. Install the sector shaft

Fig. 11.26. Fit the sector lash adjuster

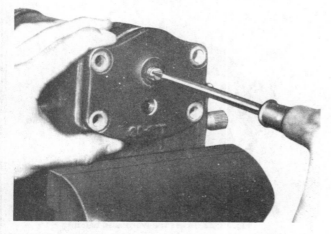

Fig. 11.27. Locate the top cover while turning the adjustment screw counterclockwise

Fig. 11.28. Check the total gear preload

Shims are available in thicknesses of 0.002, 0.003, 0.004 and 0.008 inch.

18 Locate the ball nut to the center part of the worm and install the sector shaft. Engage the ball nut, rock with the sector shaft taper gear (Fig. 11.25). The center teeth of the rack and shaft must be engaged in alignment.

19 Fit the adjustment screw into the sector shaft 'T' slot so that there is no clearance between the adjuster screw head and sector shaft. The adjuster screw must slide freely in the slot (Fig. 11.26). If the adjustment screw to sector shaft clearance exceeds 0.001 inch, adjustment can be made by shims available in thickness of 0.059 in.

20 Locate the top cover and gasket smeared with sealant into position while turning the adjustment screw counterclockwise as in Fig. 11.27.

21 Ensure that the ball nut is correctly located with the sector shaft, then fit and tighten the cover retaining bolts.

22 Locate the relative parts in the straight ahead position and pull the flange to coupling bolt with a pull scale as shown in Fig. 11.28. The starting torque required must be 3.5 (4.3 Series 6) to 10.5 in lbs, to rotate the coupling. Adjustment to this torque can be made by turning the adjuster screw accordingly.

23 Fit the adjustment screw locknut and tighten whilst retaining the screw in the set position (Fig. 11.13).

24 Reconnect the sector shaft with the Pitman arm and align the relative marks. Tighten the retaining nut to the specified torque of 160 lb f ft (22.1 kg fm).

25 Align the worm shaft keyway with the key hole in the coupling flange and fit the pinch bolt. Tighten the pinch bolt to 20 lb f ft (2.7 kg fm).

26 Refill the gearbox to the specified level with clean oil of the correct type given in the Specifications.

25 Pitman shaft seal - renewal

1 Apply the parking brake and raise the vehicle.
2 Unscrew the Pitman arm to shaft nut and remove with washer.
3 Mark the arm to shaft relative positions and disconnect the arm.
4 Wipe the seal and surrounding area clean and prise it out using a screwdriver. Clean out the seal recess.
5 If the oil in the steering gear is contaminated, the steering gear should be removed as described in Section 22 and dismantled for inspection as in Section 23 and repaired as required.
6 Smear the new seal with clean steering gear lubricant and locate in the Pitman shaft bore. Tap it carefully into position using a suitable size pipe or socket.
7 Refit the Pitman arm aligning the relative mark to the shaft. Locate the washer and nut and tighten to the specified torque.
8 Lower the vehicle and then check the steering gearbox lubricant level and top up as required, but do not overfill.

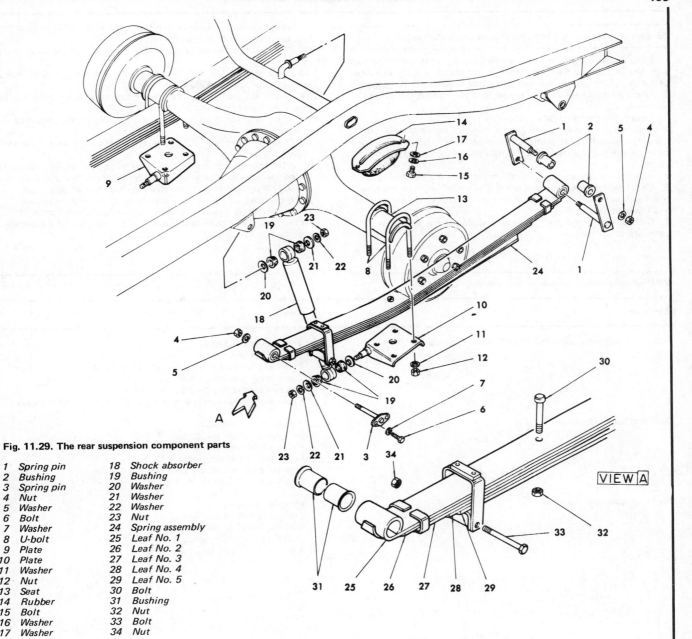

Fig. 11.29. The rear suspension component parts

1	Spring pin	18	Shock absorber
2	Bushing	19	Bushing
3	Spring pin	20	Washer
4	Nut	21	Washer
5	Washer	22	Washer
6	Bolt	23	Nut
7	Washer	24	Spring assembly
8	U-bolt	25	Leaf No. 1
9	Plate	26	Leaf No. 2
10	Plate	27	Leaf No. 3
11	Washer	28	Leaf No. 4
12	Nut	29	Leaf No. 5
13	Seat	30	Bolt
14	Rubber	31	Bushing
15	Bolt	32	Nut
16	Washer	33	Bolt
17	Washer	34	Nut

Fig. 11.30. Remove the 'U' bolt nuts

Fig. 11.31. Undo the spring shackle pin bolts

26 Rear suspension - removal, inspection and installation

1 Place blocks under the front wheels to prevent vehicle movement and then jack up the rear axle. Place chassis stands or suitable blocks under the frame near to the rear spring brackets.
2 Undo the shockabsorber retaining nuts and remove the shockabsorbers.
3 Detach the parking brake cable clips.
4 Unscrew the retaining nuts from the leaf spring to axle 'U' bolts. Clean the threaded part of the 'U' bolts and lubricate with penetrating oil to ease removal of the nuts. Remove the 'U' bolts and location plates.
5 Jack up the rear axle sufficiently to separate the leaf spring units.
6 Undo and remove the spring shackle pin retaining nuts from the front and rear. Use a soft drift and hammer out the rear shackle pin and lower the leaf spring.
7 Drift out the front shackle pin and withdraw the leaf spring assembly to the rear.
8 Undo the retaining nut and drift out the shackle pin from the rear bracket and remove the shackle.
9 Clean off the respective components and inspect for wear, and defects.
10 Check the shackles for distortion and the shackle pins for wear, and renew them if necessary.
11 Inspect the leaf springs for signs of cracks, wear or distortion and check the bushes. Renew if required.
12 If the 'U' bolts are distorted, worn or damaged they must be renewed.
13 Check that the jounce bumper is in good condition and firmly seated.
14 If the shockabsorbers show signs of oil leakage which indicates that the piston rod seal is wearing then the shockabsorber is in need of renewal. It is essential that the shockabsorbers have good compression, tension and make no abnormal noises when operated.
15 Installation is a reversal of removal but be sure to tighten all retaining nuts and bolts to the specified torques. Loosely assemble all parts before tightening.

27 Fault diagnosis - suspension and steering

Symptom	Reason/s
Steering feels vague, car wanders and floats at speed	Tire pressures uneven. Shockabsorbers worn. Steering gear balljoints badly worn. Suspension geometry incorrect. Steering mechanism free play excessive. Front suspension and rear axle pick-up points out of alignment.
Stiff and heavy steering	Tire pressures too low. No grease in steering joints. Front wheel toe-in incorrect. Suspension geometry incorrect. Steering gear incorrectly adjusted too tightly. Steering column badly misaligned.
Wheel wobble and vibration	Wheel nuts loose. Front wheels and tires out of balance. Steering knuckle badly worn. Hub bearings badly worn. Steering gear free play excessive.

Chapter 12 Bodywork and fittings

For information relating to Series 8 thru 12 models, including 4WD, refer to Chapter 13.

Contents

Door glass - removal and installation ... 15	Front fender - removal and installation ... 21
Door lock - removal and installation ... 12	General description ... 1
Door lock cylinder - removal and installation ... 13	Maintenance - bodywork and underframe ... 2
Doors - rattles and their rectification ... 7	Maintenance - hinges and locks ... 6
Doors - removal and installation ... 8	Maintenance - upholstery and carpets ... 3
Door striker - adjustment, removal and installation ... 14	Major body damage - repair ... 5
Door trim - removal and installation ... 9	Minor bodywork damage - repair ... 4
Door ventilator - removal and installation ... 10	Rear view mirrors - removal and installation ... 18
Door ventilator seal and inner frame - removal and installation ... 11	Seat - removal and installation ... 19
Engine hood/latch/control cable - removal and installation ... 20	Window regulator - removal and installation ... 16
Front and rear bumpers - removal and installation ... 22	Windshield/rear window - removal and installation ... 17

Specifications

Overall length	173.4 ins
Overall width	63.0 ins
Overall height	60.8 ins
Minimum road clearance	8.1 ins
Inside bed length	73.0 ins
Inside bed width	57.5 ins
Inside height of bed	15.6 ins
Width between wheel housings	39.4 ins
Tailgate opening width	53.7 ins
Bed load height	27.2 ins
Bed capacity (approx.)	38 cu ft
Shipping capacity	380 cu ft
Vehicle weight (gross)	3950 lbs
Curb weight	2450 lbs
Payload - less passengers	1100 lbs
Gross axle weight rating	
Front	1600 lbs
Rear	2650 lbs

Chapter 12/Bodywork and fittings

1 General description

The integral all steel cab and pick up body are mounted on a steel box section frame having six crossmembers. The engine compartment lid, the cab doors and rear tailgate are fitted with removable hinges which incorporate provision for adjustment.

The release catch for the engine compartment hood is located beneath the dash panel. The cab doors have over-riding door locks so that when locked, the latch mechanism becomes inoperative. This is a safety device to prevent accidental opening of the door by movement of the inside handle. The doors can be locked from the outside using either the ignition key or by pressing the lock button and pulling the outside handle whilst shutting the door. The doors are locked from inside the cab by pushing down the lock button and unlocked by lifting it.

The seat unit is adjustable by pushing the adjustment lever sideways, which is located on the left side at the front of the seat. To the rear of the hinged backrest there is a small stowage area.

A heater and ventilator system are fitted as standard equipment and are fully adjustable from the dash mounted control panel.

The spare tire is located under the 'cargo box' and when required can be lowered from its stowed position by means of a chain and winch system operated by a crank handle inserted at the rear of the body.

2 Maintenance - bodywork and underframe

1 The general condition of a vehicle's bodywork is the one thing that significantly affects its value. Maintenance is easy but needs to be regular and particular. Neglect, particularly after minor damage, can lead quickly to further deterioration and costly repair bills. It is important also to keep watch on those parts of the vehicle not immediately visible, for instance the underside, inside all the wheel arches and the lower part of the engine compartment.
2 The basic maintenance routine for the bodywork is washing - preferably with a lot of water, from a hose. This will remove all the loose solids which may have stuck to the vehicle. It is important to flush these off in such a way as to prevent grit from scratching the finish.

The wheel arches and underbody need washing in the same way to remove any accumulated mud which will retain moisture and tend to encourage rust. Paradoxically enough, the best time to clean the underbody and wheel arches is in wet weather when the mud is thoroughly wet and soft. In very wet weather the underbody is usually cleaned of large accumulations automatically and this is a good time for inspection.

3 Periodically it is a good idea to have the whole of the underside of the vehicle steam cleaned, engine compartment included, so that a thorough inspection can be carried out to see what minor repairs and renovations are necessary. Steam cleaning is available at many garages and is necessary for removal of accumulation of oily grime which sometimes is allowed to cake thick in certain areas near the engine, gearbox and back axle. If steam facilities are not available, grease solvents are available which can be brush applied. The dirt can then be simply hosed off.
4 After washing the paintwork, wipe off with a chamois leather to give an unspotted clear finish. A coat of clear protective wax polish will give added protection against chemical pollutants in the air. If the paintwork sheen has dulled or oxidised, use a cleaner/polisher combination to restore the brilliance of the shine. This requires a little effort, but is usually caused because regular washing has been neglected. Always check that the door and ventilator opening drain holes and pipes are completely clear so that water can drain out. Bright work should be treated the same way as paintwork. Windscreen and windows can be kept clear of the smeary film which often appears, if a little ammonia is added to the water. If they are scratched, a good rub with a proprietary metal polish will often clear them. Never use any form of wax or other body or chromium polish on glass.

3 Maintenance - upholstery and carpets

1 Mats and carpets should be brushed or vacuum cleaned regularly to keep them free of grit. If they are badly stained removed them from the vehicle for scrubbing or sponging and make quite sure they are dry before refitting. Seats and interior trim panels can be kept clean by a wipe over with a damp cloth. If they do become stained (which can be more apparent on light colored upholstery) use a little liquid detergent and a soft nail brush to scour the grime out of the grain of the material. Do not forget to keep the head lining clean in the same way as the upholstery. When using liquid cleaners inside the car do not over-wet the surfaces being cleaned. Excessive damp could get into the seams and padded interior causing stains, offensive odours or even rot. If the inside of the vehicle gets wet accidentally it is worthwhile taking some trouble to dry it out properly particularly where carpets are involved. **Do not** leave oil or electric heaters inside the cab for this purpose.

4 Minor bodywork damage - repair

The photographic sequence on pages 206 and 207 illustrates the operations detailed in the following sub-Sections.

Repair of minor scratches in the vehicle's bodywork

If the scratch is very superficial and does not penetrate the metal of the bodywork, repair is very simple. Lightly rub the area of the scratch with a paintwork renovator, or a very fine cutting paste, to remove loose paint from the scratch and to clear the surrounding bodywork of wax polish. Rinse the area with clean water.

Apply touch-up paint to the scratch using a thin paintbrush, continue to apply thin layers of paint until the surface of the paint in the scratch is level with the surrounding paintwork. Allow the new paint at least two weeks to harden; then, blend it into the surrounding paintwork by rubbing the paintwork, in the scratch area with a paintwork renovator, or a very fine cutting paste. Finally apply wax polish.

Where the scratch has penetrated right through to the metal of the bodywork, causing the metal to rust, a different repair technique is required. Remove any loose rust from the bottom of the scratch with a penknife, then apply rust inhibiting paint to prevent the formation of rust in the future. Using a rubber or nylon applicator fill the scratch with bodystopper paste. If required, this paste can be mixed with cellulose thinners to provide a very thin paste which is ideal for filling narrow scratches. Before the stopper paste in the scratch hardens, wrap a piece of smooth cotton rag around the top of a finger. Dip the finger in cellulose thinners and then quickly sweep it across the surface of the stopper paste in the scratch; this will ensure that the surface of the stopper paste is slightly hollowed. The scratch can now be painted over as described earlier in this Section.

Repair of dents in the vehicle's bodywork

When deep denting of the vehicle's bodywork has taken place, the first task is to pull the dent out, until the affected bodywork almost attains its original shape. There is little point in trying to restore the original shape completely, as the metal in the damaged area will have stretched on impact and cannot be reshaped fully to its original contour. It is better to bring the level of the dent up to a point which is about 1/8 inch (3 mm) below the level of the surrounding bodywork. In cases where the dent is very shallow anyway, it is not worth trying to pull it out at all.

If the underside of the dent is accessible, it can be hammered out gently from behind, using a mallet with a wooden or plastic head. Whilst doing this, hold a suitable block of wood firmly against the impact from the hammer blows and thus prevent a large area of bodywork from being 'belled-out'.

Should the dent be in a section of the bodywork which has a double skin or some other factor making it inaccessible from behind, a different technique is called for. Drill several small holes through the metal inside the dent area - particularly in the deeper sections. Then screw long self-tapping screws into the holes just sufficiently for them to gain a good purchase in the metal. Now the dent can be pulled out by pulling on the protruding heads of the screws with a pair of pliers.

The next stage of the repair is the removal of the paint from the damaged area, and from an inch or so of the surrounding 'sound' bodywork. This is accomplished most easily by using a wire brush or abrasive pad on a power drill, although it can be done just as effectively by hand using sheets of abrasive paper. To complete the preparations for filling, score the surface of the bare metal with a screwdriver or the tang of a file, or alternatively, drill small holes in the affected area. This will provide a really good 'key' for the filler paste.

To complete the repair see the Section on filling and respraying.

Repair of rust holes or gashes in the vehicle's bodywork

Remove all paint from the affected area and from an inch or so of the surrounding 'sound' bodywork, using an abrasive pad or a wire brush on a power drill. If these are not available a few sheets of abrasive paper will do the job just as effectively. With the paint removed you will be able to gauge the severity of the corrosion and therefore decide whether to replace the whole panel (if this is possible) or to repair the affected area. Replacement body panels are not as expensive as most people think and it is often quicker and more satisfactory to fit a new panel than to attempt to repair large areas of corrosion.

Remove all fittings from the affected area except those which will act as a guide to the original shape of the damaged bodywork (eg; headlamp shells etc.) Then, using tin snips or a hacksaw blade, remove all loose metal and any other metal badly affected by corrosion. Hammer the edges of the hole inwards in order to create a slight depression for the filler paste.

Wire brush the affected area to remove the powdery rust from the surface of the remaining metal. Paint the affected area with rust inhibiting paint, if the back of the rusted area is accessible treat this also.

Before filling can take place it will be necessary to block the hole in some way. This can be achieved by the use of one of the following materials: Zinc gauze, Aluminum tape or Polyurethane foam.

Zinc gauze is probably the best material to use for a large hole. Cut a piece to the approximate size and shape of the hole to be filled, then position it in the hole so that its edges are below the level of the surrounding bodywork. It can be retained in position by several blobs of filler paste around its periphery.

Aluminum tape should be used for small or very narrow holes. Pull a piece off the roll and trim it to the approximate size and shape required, then pull off the backing paper (if used) and stick the tape over the hole; it can be overlapped if the thickness of one piece is insufficient. Burnish down the edges of the tape with the handle of a screwdriver or similar, to ensure that the tape is securely attached to the metal underneath.

Polyurethane foam is best used where the hole is situated in a section of bodywork of complex shape, backed by a small box section (eg; where the sill panel meets the rear wheel arch - most vehicles). The unusual mixing procedure for this foam is as follows: Put equal amounts of fluid from each of the two cans provided in the kit, into one container. Stir until the mixture begins to thicken, then quickly pour this mixture into the hole, and hold a piece of cardboard over the larger apertures. Almost immediately the polyurethane will begin to expand, gushing frantically out of any small holes left unblocked. When the foam hardens it can be cut back to just below the level of the surrounding bodywork with a hacksaw blade.

Bodywork repairs - filling and re-spraying

Before using this Section, see the Sections on dent, deep scratch, rust hole and gash repairs.

Many types of bodyfiller are available, but generally speaking those proprietary kits which contain a tin of filler paste and a tube of resin hardener are best for this type of repair. A wide, flexible plastic or nylon applicator will be found invaluable for imparting a smooth and well contoured finish to the surface of the filler.

Mix up a little filler on a clean piece of card or board - use the hardener sparingly (follow the maker's instructions on the packet) otherwise the filler will set very rapidly.

Using the applicator, apply the filler paste to the prepared area; draw the applicator across the surface of the filler to achieve the correct contour and to level the filler surface. As soon as a contour that approximates the correct one is achieved, stop working the paste - if you carry on too long the paste will become sticky and begin to 'pick-up' on the applicator. Continue to add thin layers of filler paste at twenty-minute intervals until the level of the filler is just 'proud' of the surrounding bodywork.

Once the filler has hardened, excess can be removed using a metal plane or file. From then on, progressively finer grades of abrasive paper should be used, starting with a 40 grade production paper and finishing with 400 grade 'wet-and-dry' paper.

Always wrap the abrasive paper around a flat rubber, cork, or wooden block - otherwise the surface of the filler will not be completely flat. During the smoothing of the filler surface the 'wet-and-dry' paper should be periodically rinsed in water. This will ensure that a very smooth finish is imparted to the filler at the final stage.

At this stage the 'dent' should be surrounded by a ring of bare metal, which in turn should be encircled by the finely 'feathered' edge of the good paintwork. Rinse the repair area with clean water, until all of the dust produced by the rubbing-down operation is gone.

Spray the whole repair area with a light coat of grey primer - this will show up any imperfections in the surface of the filler. Repair these imperfections with fresh filler paste or bodystopper, and once more smooth the surface with abrasive paper. If bodystopper is used, it can be mixed with cellulose thinners to form a really thin paste which is ideal for filling small holes. Repeat this spray and repair procedure until you are satisfied that the surface of the filler, and the feathered edge of the paintwork are perfect. Clean the repair area with clean water and allow to dry fully.

The repair area is now ready for spraying. Paint spraying must be carried out in a warm, dry, windless and dust free atmosphere. This condition can be created artificially if you have access to a large indoor working area, but if you are forced to work in the open, you will have to pick your day very carefully. If you are working indoors, dousing the floor in the work area with water will 'lay' the dust which would otherwise be in the atmosphere. If the repair area is confined to one body panel, mask off the surrounding panels; this will help to minimise the effects of a slight mis-match in paint colors. Bodywork fitting (eg; chrome strips, door handles etc), will also need to be masked off. Use genuine masking tape and several thicknesses of newspaper for the masking operation.

Before commencing to spray, agitate the aerosol can thoroughly, then spray a test area (an old tin, or similar) until the technique is mastered. Cover the repair area with a thick coat of primer; the thickness should be built up using several thin layers of paint rather than one thick one. Using 400 grade 'wet-and-dry' paper, rub down the surface of the primer until it is really smooth. While doing this, the work area should be thoroughly doused with water, and the 'wet-and-dry' paper periodically rinsed in water. Allow to dry before spraying on more paint.

Spray on the top coat, again building up the thickness by using several thin layers of paint. Start spraying in the center of the repair area and then, using a circular motion, work outwards until the whole repair area and about 2 inches of the surrounding original paintwork is covered. Remove all masking material 10 to 15 minutes after spraying on the final coat of paint.

Allow the new paint at least 2 weeks to harden fully; then, using a paintwork renovator or a very fine cutting paste, blend the edges of the new paint into the existing paintwork. Finally, apply wax polish.

5 Major body damage - repair

Where serious damage has occurred or large areas need renewal due to neglect, it means certainly that completely new sections or panels will need welding in and this is best left to professionals. If the damage is due to impact it will also be necessary to completely check the alignment of the chassis structure. Due to the principle of construction the strength and shape of the whole can be affected by damage to a part. In such instances the services of the official agent with specialist checking jigs are essential. If a chassis is left misaligned it is first of all dangerous as the truck will not handle properly and secondly, uneven stresses will be imposed on the steering, engine and transmission, causing abnormal wear or complete failure. Tire wear may also be excessive.

6 Maintenance - hinges and locks

1 Periodically lubricate the hinges of the doors, hood and tailgate with a few drops of light oil.
2 Similarly lubricate the door catches, hood release mechanism and the release mechanism of the spare wheel.
3 Apply a smear of general purpose grease to the lock strikers, and striker plates.

7 Doors - rattles and their rectification

1 Check first that the door is not loose at the hinges and that the

Chapter 12/Bodywork and fittings

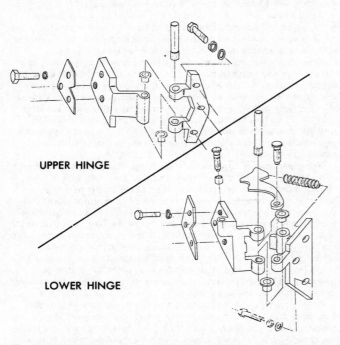

Fig. 12.1. The component parts of the upper and lower door hinges

latch is holding the door firmly in position. Check also that the door lines up with the aperture in the body.
2   If the hinges are loose, or the door is out of alignment, it will be necessary to reset the hinge positions. This is a straightforward matter after slackening the hinge retaining screws slightly, following which the door can be repositioned in/out or up/down.
3   If the latch is holding the door properly it should hold the door tightly when fully latched and the door should line up with the body. If adjustment is required, slacken the striker plate screws slightly and reposition the plate in/out or up/down as necessary. Fore and aft adjustment of the striker plate is effected by the use of shims.
4   Other rattles from the door could be caused by wear or looseness in the window winder, or the glass channels, seal strips and interior lock mechanism.

8 Doors - removal and installation

1   To remove a door, open and support it with a jack or preferably get an assistant to support it.
2   To aid refitting, mark around the hinge with a soft lead pencil and then unscrew the hinge retaining bolts, and lift the door clear.
3   If a replacement door is being fitted, remove the door trim panel, lock assembly, the window and its regulator assemblies (Sections 9 to 16) and fit them to the new door.
4   Installation of the door is a direct reversal of removal but adjustment will probably be necessary as described in Section 14.

9 Door trim - removal and installation

1   Press the door trim around the window regulator and withdraw the regulator handle retaining clip, then detach the handle (photo).

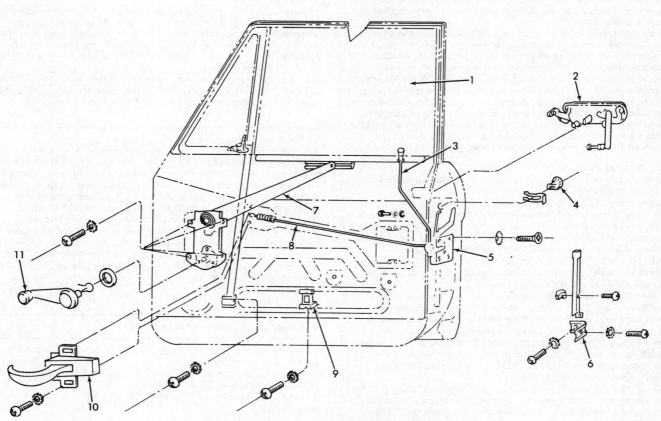

Fig. 12.2. The component parts of the door assembly

1 Window	4 Lock cylinder	7 Window regulator
2 Outside handle	5 Lock assembly	8 Inner lever to lock link
3 Lock rod	6 Support bracket	9 Glass stopper

10 Inner lever
11 Window regulator

Chapter 12/Bodywork and fittings

9.1 Remove the regulator handle

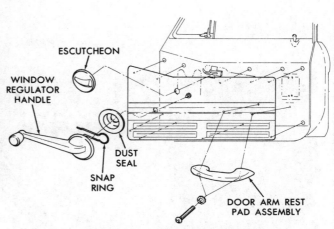
Fig. 12.3. The door trim and associate parts

2 Raise the latch lever and withdraw the escutcheon (photo). Undo the latch lever retaining screws to remove the latch lever.
3 Remove the armrest which is retained by two screws, and then carefully prise away the trim panel from the door. The trim is retained by nine nylon 'bullet type' fasteners.
4 Installation is the reversal of the removal procedure.

10 Door ventilation - removal and installation

1 Remove the inner door trim panel as described in Section 9.
2 Peel back the waterproof sheet from the check hole, and the glass stopper which is retained by a crosshead screw. Lower the door glass, withdraw it from the regulator channel and remove it from the door.
3 Remove the door inner weatherstrip and seal unit.
4 The door ventilation division bar is now detached from the support bracket, then undo the door ventilator to door frame retaining screw and withdraw the ventilator assembly from the door.
5 Installation is the reversal of removal but note the following:

   a) When installing the waterproof sheet to the check hole, ensure that there is a continuous even bead of adhesive round the opening and stretch the vinyl to avoid wrinkles. A firm finger pressure is sufficient to ensure good adhesion.
   b) Having installed the division bar and door frame, the upper clearance should be fitted with cemedine sealer or a similar sealant.

9.2 Remove the latch lever escutcheon

11 Door ventilator seal and inner frame - removal and installation

1 First remove the pivot by filing (Fig. 12.4).
2 Unscrew the nut and remove with the spring from the lower frame.
3 Detach the inner frame and the seal can then be removed.
4 Refit in the reverse order but note the following:

   a) When inserting the pin into position from the upper face, caulk it with a suitable drift.
   b) Ensure that the inner frame and vent seal are in good contact.
   c) Take care when installing the vent seal to avoid damaging it - soapy water will ease fitting.
   d) Ensure that the stop washer is correctly located.

12 Door lock - removal and installation

1 Remove the inner trim panel from the door as described in Section 9.

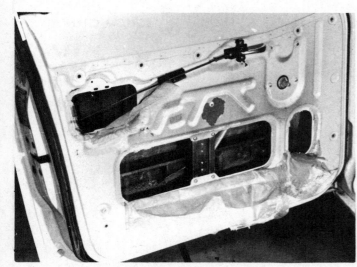
9.3 The inner door panel with trim removed

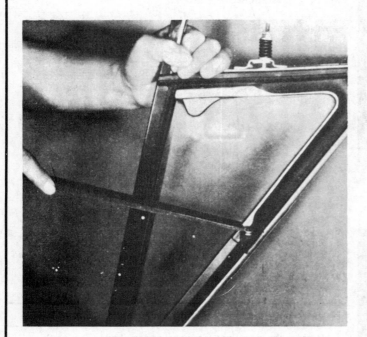

Fig. 12.4. File the pivot to remove

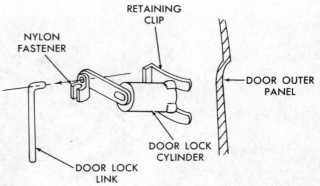

Fig. 12.5. The door lock cylinder

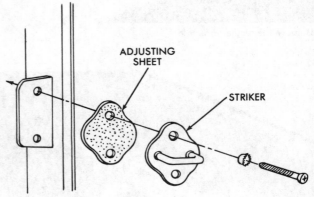

Fig. 12.6. The striker and adjustment sheet

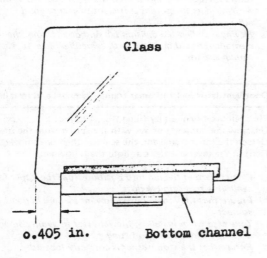

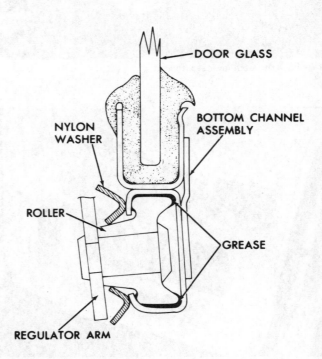

Fig. 12.7. The door glass and channel showing the correct relative position

2 Peel back the water deflectors from the check holes and extract the retaining clip from the lock side of the inner lever assembly.
3 Detach the cylinder link from the lock arm nylon fastener and remove the glass rim channel.
4 The relay lever bracket is now removed from the door lock side, and the door lock button from the lock rod.
5 Disconnect the outer link from the handle link (outside) and the retaining screws from the door lock unit. Withdraw the lock from the door.
6 To install the door lock assembly, first grease the respective linkages, then locate the lock into position, and the door lock knob and reconnect the outside handle to the lock link. Refit the securing bolts, but do not tighten.
7 Adjust the linkage so that the maximum handle free play is ¼ inch, and then tighten the adjustment nut with an open ended wrench. The securing bolts can now be tightened.
8 Reconnect the lock link to the cylinder and relocate the glass run channel and support bracket.
9 Fit the inner lever unit into position reconnecting it to the lock link. Check the lock action and if satisfactory, refit the trim panel and handles. The water deflectors must be refitted using a suitable adhesive.

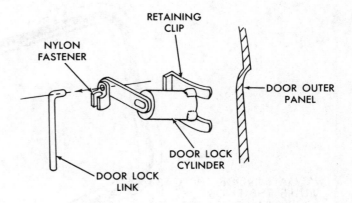

Fig. 12.8. The external door handle and door lock

13 Door lock cylinder - removal and installation

1 Remove the inner trim panel from the door as described in Section 9 and partially peel back the water deflector panel. With the window in the raised position, disconnect the door lock link.
2 Remove the lock cylinder retaining clip using a screwdriver to prise it with and withdraw the cylinder from the door.
3 Installation is a reversal of the above procedure.

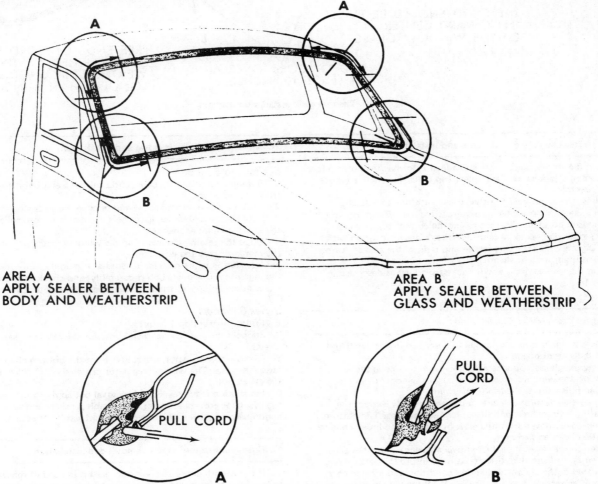

Fig. 12.9. The windshield upper and lower sealant points 'A' and 'B' together with the upper and lower weatherstrip fitting (insets)

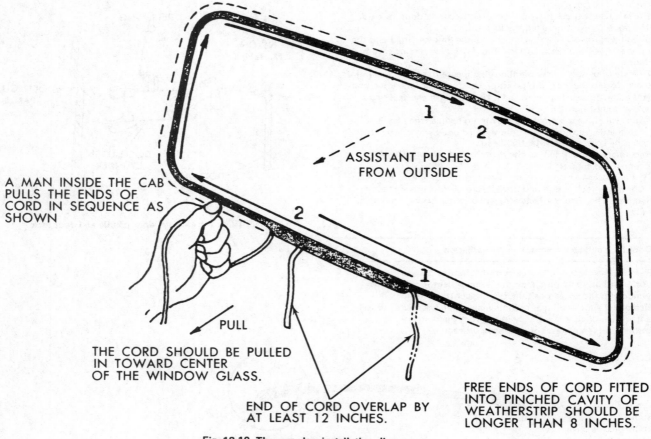

Fig. 12.10. The rear glass installation diagram

14 Door striker - adjustment, removal and installation

1 Slacken the striker plate screw and adjust it so that the bottom face of the dovetail is parallel to the striker. Tap the striker with a wooden drift and hammer.
2 To adjust the engagement of the striker with the door latch, adjustment sheets are available as required (see Fig. 12.7). Two to three sheets are normally required to get the correct adjustment.
3 If the striker plate is to be removed, first mark its location with a pencil. Unscrew the retaining screws and remove the striker noting the number of adjustment sheets.
4 Refit in the reverse order and if necessary adjust as described in paragraph 1.

15 Door glass - removal and installation

1 Remove the inner door trim panel as described in Section 9 and peel back the water deflectors.
2 Unscrew the glass stopper retaining screw and remove the stopper. With the window lowered, extract the window from the regulator and disconnect the inside lever unit.
3 Withdraw the glass run channel and then carefully extract the glass lowering it to its limit and tilting its side upwards. The bottom channel may be removed from the glass by tapping the lower channel lightly with a wooden hammer.
4 Installation is a reversal of the removal sequence but if the glass has been removed from the bottom channel, ensure that the glazing rubber located in the channel is in good condition and located correctly before reinserting the glass. See Fig. 12.8 for the correct bottom channel location to the glass.

16 Window regulator - removal and installation

Series 1 to 5 models
1 Remove the door trim panel and peel the water deflector sheets from the inner door.
2 Unscrew the retaining screw and remove the glass stopper.
3 With the glass lowered disconnect the regulator arm rollers from the bottom window channel.
4 Undo the retaining screw from the window regulator and extract the regulator thru the check hole.
5 Installation is the reversal of removal but apply some grease to the regulator arm rollers before assembly, and check the action of the window before refitting the inner panel and water deflectors.

Series 6 models
6 Follow instruction 1 above.
7 Support the glass by hand and unscrew the regulator securing screws.
8 Remove the regulator arm rollers from the bottom channel and lower the glass. The window regulator can now be withdrawn thru the check hole.
9 Installation is the reversal of removal but apply some grease to the regulator arm rollers prior to assembly, and check the window action before refitting the inner panel and water deflectors.

17 Windshield/rear window - removal and installation

1 If the windshield or rear window glass is in need of renewal due to breakage or the sealing rubber is to be renewed, it is recommended that the job be entrusted to a properly equipped repair shop. Special

Chapter 12/Bodywork and fittings

skills and tools are required to correctly refit the windshield and rear window glass and in the hands of a novice there is always the danger of breaking the new glass. However in the event of no local workshop having the required facilities the task may be very carefully undertaken as follows. The procedure is identical for the windshield and rear window (with the exception of paragraph 2 for the rear window).
2 Remove the windshield wipers, rear view mirror and sun visor.
3 Insert a flat blade tool between the inner face and flange of the rubber channel and separate the weatherstrip from the body taking great care. Do not use a sharp implement and avoid damaging the rubber strip if possible. The upper and lower corners of the windshield are filled with a sealer solution which is applied between the rubber channel and body flange.
4 If possible wear a pair of gloves and with the aid of an assistant, push on a corner of the windshield and gradually remove it from the body opening. Remove the weatherstrip and clean off all traces of old sealant from the glass and/or weatherstrip and body.
5 Commence installation by locating the weatherstrip round the windshield. Soapy water applied to the weatherstrip will ease fitting.
6 Locate a length of strong cord into and around the outer channel of the weatherstrip and overlap it in the middle at the lower edge by a minimum of 12 inches. Temporarily retain the cord in position by tapping the ends to the inner surface of the glass.
7 The upper and lower corners of the rubber channel should be smeared with a suitable water repellent sealer as shown in positions 'A' and 'B' in Fig. 12.9.
8 With the aid of an assistant, locate the windshield into position from the outside. Whilst the assistant applies pressure to the outside of the glass, carefully pull the ends of the cord inwards so that the inner lip of the rubber channel seats over the body flange, as in Fig. 12.10 (inset).
9 With the glass fully located, remove the cord. Refit the windshield wipers, interior mirrors and sun visor. Wipe away any surplus sealant that may have smeared onto the glass.

18 Rear view mirrors - removal and installation

Interior mirror
1 The rear view mirror is retained by a bracket which also supports the sun visor.
2 To remove the mirror, press it in the direction shown in Fig. 12.11.
3 To install the mirror, locate the lower part of the mirror assembly clip into the bracket and tap portion 'B' with a soft faced hammer to drive it into position.

External mirror
4 The external mirror is retained by two self-tapping screws which are simply unscrewed to remove the mirror complete with gasket.
5 When refitting, renew the gasket if the old one has perished or worn, align the screw holes and tighten the screws.

19 Seat - removal and installation

1 Hinge forward the driver and passenger seat back and disconnect the seat belts by removing their retaining bolts and spring washers.
2 With the belts removed, undo the four nuts with spring and plain washers which secure the seat adjusters (Fig. 12.12) to the floor. The nuts are located under the floor, and when all are unscrewed the seat can be removed.
3 To install the seat, reverse the removal instructions but tighten the seat belt retaining bolts to a torque of 50 lb f ft (6.9 kg f m).

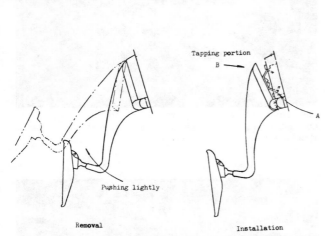

Fig. 12.11. The interior mirror removal and installation

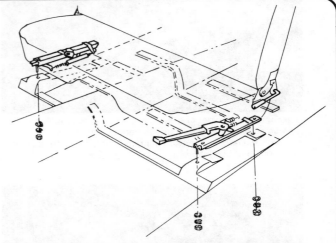

Fig. 12.12. The seat runner location nuts and washers

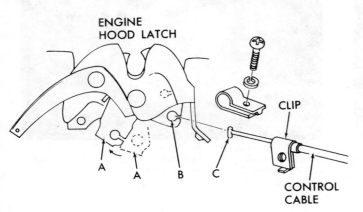

Fig. 12.13. The hood latch and control cable

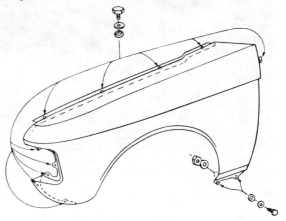

Fig. 12.14. The fender retaining bolt positions

This photo sequence illustrates the repair of a dent and damaged paintwork. The procedure for the repair of a hole is similar. Refer to the text for more complete instructions

After removing any adjacent body trim, hammer the dent out. The damaged area should then be made slightly concave

Use coarse sandpaper or a sanding disc on a drill motor to remove all paint from the damaged area. Feather the sanded area into the edges of the surrounding paint, using progressively finer grades of sandpaper

The damaged area should be treated with rust remover prior to application of the body filler. In the case of a rust hole, all rusted sheet metal should be cut away

Carefully follow manufacturer's instructions when mixing the body filler so as to have the longest possible working time during application. Rust holes should be covered with fiberglass screen held in place with dabs of body filler prior to repair

Apply the filler with a flexible applicator in thin layers at 20 minute intervals. Use an applicator such as a wood spatula for confined areas. The filler should protrude slightly above the surrounding area

Shape the filler with a surform-type plane. Then, use water and progressively finer grades of sandpaper and a sanding block to wet-sand the area until it is smooth. Feather the edges of the repair area into the surrounding paint.

Use spray or brush applied primer to cover the entire repair area so that slight imperfections in the surface will be filled in. Prime at least one inch into the area surrounding the repair. Be careful of over-spray when using spray-type primer

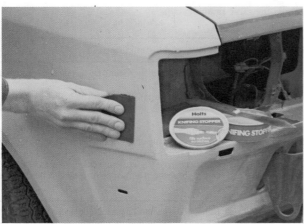

Wet-sand the primer with fine (approximately 400 grade) sandpaper until the area is smooth to the touch and blended into the surrounding paint. Use filler paste on minor imperfections

After the filler paste has dried, use rubbing compound to ensure that the surface of the primer is smooth. Prior to painting, the surface should be wiped down with a tack rag or lint-free cloth soaked in lacquer thinner

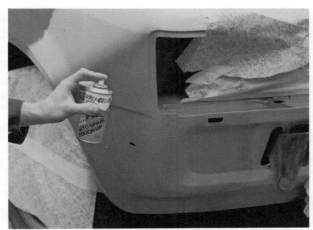

Choose a dry, warm, breeze-free area in which to paint and make sure that adjacent areas are protected from over-spray. Shake the spray paint can thoroughly and apply the top coat to the repair area, building it up by applying several coats, working from the center

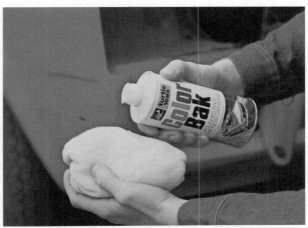

After allowing at least two weeks for the paint to harden, use fine rubbing compound to blend the area into the original paint. Wax can now be applied

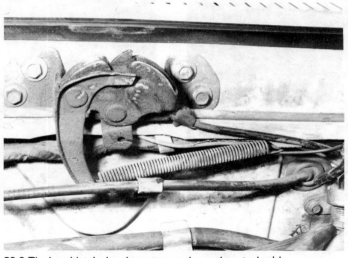

20.3 The hood latch showing return spring and control cable

20 Engine hood/latch/control cable - removal and installation

Engine hood
1 Raise the hood, and support. Undo the hinge retaining bolts and with the aid of an assistant, carefully lift the hood clear.
2 Installation is the reverse of removal, but to adjust the hood, position it correctly by adjustment of the hinge position to the body for vertical adjustment and hinge to hood for horizontal adjustment.

Hood latch
3 Disconnect the return spring and hood control cable and then remove the latch. Refit in the reverse order.

Hood control cable
4 Unscrew the cable clip retaining screw on the release lever side and disconnect the cable. Then remove the cable from the dashpanel clip.
5 Disconnect the cable from the latch and withdraw the cable thru the engine compartment.
6 To install, reverse the above procedure, inserting the cable end into the latch side clip, and then hinge the latch striker lever to position 'A' in Fig. 12.13. Locate the end of the cable 'C' into hole 'B' and tighten the cable retaining clip.

21 Front fender - removal and installation

1 The front fenders are retained by means of screws and bolts as shown in Fig. 12.14. Remove as follows.
2 Disconnect the windshield wiper arms and the grille.
3 Detach the headlight assembly and the side marker light (see Chapter 10).
4 Detach the front bumper.
5 Unscrew the fender retaining bolts/screws from these positions shown in Fig. 12.14 and remove the fender.
6 To install the fender, reverse the above procedure and check for alignment before fully tightening the fender retaining screws/bolts.

22 Front and rear bumpers - removal and installation

1 The front and rear bumpers may be removed with or without their brackets by unscrewing the bumper to bracket bolts, two on each side, or alternatively by unscrewing the respective bracket retaining bolts.
2 Installation is a direct reversal of the above procedure.

Chapter 13 Supplement: Revisions and information on later models

Contents

Introduction	1
Specifications	2
Routine maintenance	3
Engine	4
Valve clearances — adjustment	
Oil pan (4 x 4 Models) — removal and installation	
Engine — removal and installation	
Fuel system	5
Idle adjustment (Series 9 and 10)	
Idle adjustment (Series 11 and 12)	
Idle mixture adjustment (Series 11 and 12)	
Carburetor adjustments (Series 9 thru 12)	
Carburetor diaphragm chamber — disassembly	
Carburetor (Series 12) — general notes	
Mechanical diaphragm fuel pump (Series 12) — removal and installation	
Emission control systems	6
General description	
Air Injection Reactor (AIR) system (Series 12)	
Closed Loop Emission Control system (Series 12, California models)	
Coasting Fuel Cut-off system (Series 12)	
Coasting Richer system (Series 9 thru 11)	
Dash pot (Series 9 thru 11) — adjustment	
Early Fuel Evaporation (EFE) system (Series 12)	
Engine speed sensor (Series 10 thru 12) — testing	
Evaporative Emission Control system (Series 11 and 12)	
Exhaust Gas Recirculation (EGR) system (Series 9 thru 11)	
Exhaust Gas Recirculation (EGR) system (Series 12)	
Ignition system	7
General description	
Distributor (Series 11 and 12) — removal and installation	
Distributor (Series 11 and 12) — disassembly and reassembly	
Charging system	8
Generator (Series 11 and 12) — disassembly and reassembly	
Clutch	9
General description	
Clutch adjustment	
Clutch pedal — removal and installation	
Clutch — removal and installation	
Clutch release bearing — removal and installation	
Clutch cable — removal and installation	
Manual transmission	10
Removal and installation	
Disassembly and reassembly	
Automatic transmission	11
Fluid level checking	
Transmission/transfer case (4 x 4 models)	12
Removal and installation	
Transmission — disassembly	
Mainshaft — disassembly	
Transfer case — disassembly	
Transfer side case — disassembly	
Cleaning and inspection	
Mainshaft — reassembly	
Transfer case — reassembly	
Transfer side case — reassembly	
Transmission — reassembly	
Driveshaft	13
General description	
Driveshaft (long wheelbase models) — removal and installation	
Center bearing (long wheelbase models) — replacement	
Rear axle	14
Axleshaft — installation	
Braking system	15
Parking brake adjustment (Series 11 and 12)	
Brake pedal adjustment (Series 11 and 12)	
Drum brake adjustment (Series 11 and 12)	
Front disc brake (Series 11 and 12) — shoe replacement	
Disc brake caliper (Series 11 and 12) — removal, overhaul and installation	
Front hub/rotor, shield and adapter (Series 11 and 12) — removal and installation	
Rear brake shoes (Series 11 and 12) — removal and installation	
Rear wheel cylinder (Series 11 and 12) — removal, overhaul and installation	
Master cylinder — overhaul	
Power cylinder — overhaul	
Brake pedal (Series 8 thru 10) — removal and installation	
Brake pedal (Series 11 and 12) — removal and installation	
Parking brake handle (Series 11 and 12) — removal and installation	
Parking brake rear cables (Series 11 and 12) — removal and installation	
Parking brake first intermediate cable (Series 11 and 12) — removal and installation	
Parking brake second intermediate cable (Series 11 and 12) — removal and installation	
Fail indicator (combination valve) assembly	
Fail indicator — removal and installation	
Electrical system	16
Fuses	
Headlamps (Series 8 thru 10) — removal and installation	
Headlamps (Series 11 and 12) — removal and installation	
Combination switch (Series 11 and 12) — general description	
Combination switch (Series 11 and 12) — removal and installation	
Radio (Series 11 and 12) — removal and installation	
Speaker (Series 11 and 12) — removal and installation	
Radio interference suppression capacitor (Series 11 and 12) — general description	
Instrument panel (Series 8 thru 10) — removal and installation	

Chapter 13/Supplement: Revisions and information on later models

Instrument panel (Series 8 thru 10) — disassembly and reassembly
Instrument meter assembly (Series 11 and 12) — removal and installation
Instrument meter assembly (Series 11 and 12) — disassembly and reassembly
Wiper motor and linkage (Series 11 and 12) — removal and installation
Heater control cables (Series 9 thru 12) — adjustment
Heater control cable (Series 8) — adjustment
Air inlet door cable (Series 8) — adjustment
Heater unit (Series 8) — removal and installation
Dash outlet grilles (Series 8 thru 10) — removal and installation
Blower motor (Series 9 and 10) — removal and installation
Blower motor (Series 11 and 12) — removal and installation
A/C thermostatic switch (Series 11 and 12) — removal and installation
Front-wheel drive system (4 x 4 models)................ 17
Free wheeling hub — removal and installation
Free wheeling hub — disassembly and reassembly
Front axle assembly — removal and installation
Front axle shaft — removal and installation
Front axle shaft — overhaul
Front differential — removal and installation
Front suspension.. 18
Front wheel bearing adjustment
Front wheel bearing adjustment (4 x 4 models)
Front wheel hub (4 x 4) — removal and installation
Front wheel bearings — removal and installation
Lower balljoint (4 x 4) — removal and installation
Steering knuckle (4 x 4) — removal and installation
Upper control arm and balljoint assembly (Series 11 and 12) — removal and installation
Stabilizer bar (Series 11 and 12) — removal and installation
Ride height checking

Steering system (manual)................................ 19
Caster angle (Series 11 and 12) — adjustment
Steering wheel (Series 11 and 12) — removal and installation
Steering shaft flexible coupling (Series 11 and 12) — removal and installation
Steering column and shaft assembly (Series 11 and 12) — removal and installation
Steering gear (Series 8 thru 10) — reassembly
Steering gear (Series 11 and 12) — disassembly and reassembly
Intermediate rod and tie-rods (Series 11 and 12) — removal and installation
Steering damper (Series 11 and 12) — removal and installation
Idler arm assembly (Series 11 and 12) — removal and installation
Power steering system (Series 12)...................... 20
Power steering system — general description
Power steering pump drivebelt — checking and adjustment
Power steering system — bleeding
Power steering system — draining and refilling
Power steering pump — removal and installation
Power steering gear — removal and installation
Power steering gear — disassembly, inspection and reassembly
Body.. 21
Grille (Series 11 and 12) — removal and installation
Hood latch (Series 11 and 12) — removal and installation
Hood control cable (Series 11 and 12) — removal and installation
Front bumper (Series 11 and 12) — removal and installation
Front fender (Series 11 and 12) — removal and installation
Door trim panel (Series 11 and 12) — removal and installation
Door lock cylinder (Series 11 and 12) — removal and installation
Window regulator — removal and installation
Door glass — removal and installation
Windshield glass — removal

1 Introduction

This supplementary Chapter covers changes made in the Chevy Luv truck in the Series 8 thru 12 models (1978 – 1982), and the procedures affected by those changes.

Operations that are not included in this Chapter are the same as or similar to those described for the latest series model found in the first 12 Chapters of this manual.

There have been considerable changes made to the models included here; most notably in Series 11 and 12. The major changes are outlined below.

Beginning with Series 9, a 4-wheel drive version was offered, which includes a combination manual transmission and a 4WD transfer case, as well as a separate front-wheel drive system.

The distributor on Series 11 and 12 models is the electronic-type which contains no contact points.

The Exhaust Gas Recirculation (EGR) system and the Evaporative Emission Control system have received major alterations.

Beginning with the Series 8 models, the clutch has been changed from hydraulic operation to cable operation.

Many components of the braking system, in Series 11 and 12 models particularly, have been redesigned.

In the Series 12 model year, a power steering system was introduced, the electric fuel pump was replaced by a mechanical pump, use of the automatic transmission was discontinued and there were several changes to the emissions control systems.

The dash and body have received many alterations. In addition to these, there are many smaller changes made to these models, as noted in the following pages.

The recommended way of using this Supplement is, prior to any operation, check here first for any relevant information pertaining to your model. After noting any model differences, particularly in the Specifications section, you can then follow the appropriate procedure, either listed in this Chapter of one of the preceding 12.

2 Specifications

The specifications listed here are revised from or supplementary to the main specifications given at the beginning of each Chapter. Specifications not listed here are the same as those listed for the latest series model in Chapters 1 through 12.

Engine

Cylinder head
Intake valve closes at .... 65° ABDC
Intake valve head diameter (Series 10 and 11)......... 1.59 in
Valve spring tension (Series 11 and 12)
    Inner spring compressed to 1.614 in (standard) ... 19 to 22 lbs
    Limit for use..................................... 17 lbs
Camshaft journals, maximum variance between lobes .... 0.0020 in
Camshaft bearing inside diameter .... 1.3394 to 1.3406 in

Pistons (Series 12)
1st compression ring gap
    When fitted in bore .... 0.012 to 0.018 in
    When free .... 0.295 in

Chapter 13/Supplement: Revisions and information on later models

2nd compression ring gap
   When fitted in bore . . . . 0.012 to 0.018 in
   When free . . . . 0.472 in

Torque specifications
Connecting rod bearing cap nuts . . . . 43 ft-lb

Cooling system (Series 12)

Coolant capacity . . . . 7.4 qts (7.0 liters)

Pulley ratio 1.18

Fuel system

Carburetor (Series 9 thru 12)
Type . . . . DCH340 (Different versions used depending on year, locale and type of transmission)
Jets . . . . See Chevrolet dealer
Fast idle adjustment
   Throttle valve opening
      Series 9 thru 11 models . . . . 17° ± 1°
      Series 12 models . . . . 16° ± 1°
   Fast idle speed
      Series 9 and 10 manual transmission models. . . . . 3400 rpm
      Series 9 and 10 automatic transmission models and all
         Series 11 and 12 models . . . . 3200 rpm

Fuel pump (Series 12)
Type . . . . Mechanical diaphragm

Emission control system

Over Temperature Control System
Warning system operating temperature (Series 11) . . . . over 350°F (at converter)
Engine Speed sensor . . . . "Off" over engine speed of 2000 rpm

Exhaust Gas Recirculation (EGR) System
EGR valve operating vacuum
   Starting . . . . 1.9 in Hg
   Fully open . . . . 4.7 in Hg (Federal version with manual transmission)
                                                                                                   3.9 in Hg (all other versions)

Ignition system

Spark plugs (Series 11 and 12)
Type . . . . NGK BPR6ES11
Gap . . . . 0.040 in

Ignition timing (Series 12)
California models . . . . 6° BTDC @ 900 rpm
Federal models . . . . 6° BTDC @ 800 rpm

Ignition coil
Model
   Series 10 . . . . 029700—4980
   Series 11 and 12 . . . . 101311—2560
Secondary voltage (Series 11 and 12) . . . . 25 000 V at 3450 to 3550 rpm
Primary resistance (Series 11 and 12) . . . . 0.828 to 1.022 ohms
Secondary resistance (Series 11 and 12) . . . . 12.15 to 14.85 K ohms

Distributor
Model
   Series 9 . . . . 029100—3281
                                                                                                     —3940
                                                                                                     —4190
   Series 10 . . . . 029100—5970
                                                                                                     —5980
                                                                                                      —5990
                                                                                                      —6000
   Series 11 . . . . 029100—6900
                                                                               —6910
                                                                               —7260
   Series 12 . . . . 029100—6900
                                                                               —5930

Distributor (continued)
Ignition interval (Series 11 and 12)..................... 89.2° to 90.8°
Air gap (Series 11 and 12)............................ 0.008 to 0.016 in

Clutch
Driven plate
Facing material...................................... Woven SF105 M
Dimension (outer diameter x inner diameter x thickness)
  2-wheel drive..................................... 7.87 x 5.12 x 0.141 in
  4-wheel drive..................................... 8.46 x 5.90 x 9/64 in

Adjustments
Clutch pedal height
  Series 8 thru 10.................................. 5.9 to 6.3 in
  Series 11 and 12.................................. 6.40 to 6.85 in
Clutch pedal play.................................... 0.781 in
Shift fork play....................................... 0.2 in (5.0 mm)

Manual transmission
Gear ratios (Series 11 and 12)
1st.. 4.13 : 1
2nd.. 2.50 : 1
3rd.. 1.48 : 1
4th.. 1.00 : 1
Reverse.. 3.83 : 1

Speedometer gear ratio (Series 11 and 12)............. 6/17

Standard gear backlash (Series 9 and 10)
1st.. 0.0043 to 0.0075 in (0.11 to 0.19 mm)
2nd.. 0.0047 to 0.0075 in (0.12 to 0.19 mm)
3rd.. 0.0047 to 0.0075 in (0.12 to 0.19 mm)
4th.. 0.0020 to 0.0051 in (0.05 to 0.13 mm)
Reverse gear — main................................. 0.0026 to 0.0055 in (0.07 to 0.14 mm)
Reverse idler gear................................... 0.0026 to 0.0059 in (0.07 to 0.15 mm)
Maximum limit allowance............................. 0.016 in (0.4 mm)

Standard gear backlash (Series 11 and 12)
1st.. 0.0012 to 0.0043 in (0.03 to 0.11 mm)
2nd.. 0.0012 to 0.0043 in (0.03 to 0.11 mm)
3rd.. 0.0012 to 0.0039 in (0.03 to 0.10 mm)
4th.. 0.0012 to 0.0043 in (0.03 to 0.11 mm)
Reverse gear — main (4 x 2).......................... 0.0028 to 0.0055 in (0.07 to 0.14 mm)
Reverse gear — main (4 x 4).......................... 0.0020 to 0.0051 in (0.05 to 0.13 mm)
Reverse idler gear (4 x 2)............................ 0.0012 to 0.0043 in (0.03 to 0.11 mm)
Reverse idler gear (4 x 4)............................ 0.0020 to 0.0051 in (0.05 to 0.13 mm)
Maximum limit allowance............................. 0.014 in (0.35 mm)

Transmission/transfer case (4 x 4 models)
Type
Transmission.. MSG model; fully synchronized 4-forward gears with constant mesh-type Reverse gear
Transfer case.. Constant mesh-type gears with shifting of 2-speed gears while in 4-wheel drive and 2-to-4-wheel drive shifter
Control lever method................................ Transmission and transfer case each are direct control with a shift lever on the floor

Gear ratio
Transmission
  1st.. 3.79 : 1
  2nd.. 2.18 : 1
  3rd.. 1.42 : 1
  4th.. 1.00 : 1
  Reverse.. 3.73 : 1
Transfer case
  High... 1.00 : 1
  Low.. 1.87 : 1
Speedometer gear ratio............................... 6/20
Oil capacity... 2-2/3 qt

Transmission/transfer case (continued)

Torque specifications — **Ft-lb**

Output shaft nut	108
Output shaft cover bolt	20
Shift rod screw plugs	36
Countergear locknut	80
Mainshaft locknut	94
Lock plate bolt	14
Transmission case-to-center support bolts and nuts	27
Speedometer driven gear bolt	14
Bearing retainer bolts	14
Clutch fork ball stud	30
Shifter cover bolts	14

Front differential (4 x 4)

Oil capacity (Series 9 thru 11)	1-2/3 pints (0.8 L)
Oil capacity (Series 12)	1 qt (1.0 liter)

Front axle assembly

Torque specifications — **Ft-lb**

Front axle case mounting bracket bolts	15
Pitman arm-to-steering sector shaft nut	90
Idler arm-to-pivot shaft nut	87
Lower control arm-to-frame bracket nuts	92
Front axle shaft bracket-to-axle case bolts	45

Braking system

Rear wheel cylinder bore diameter	0.75 in

Torque specifications — **Ft-lb**

Flexible hose-to-caliper	29
Master cylinder end plug	47
Power cylinder-to-dash panel nuts (Series 12)	19

Electrical system

Wiring harness color coding

Circuit	Base color
Starting motor circuit	Black (B)
Charging circuit	White (W)
Light circuit	Red (R)
Signal circuit	Green (G)
Instrument circuit	Yellow (Y)
Wiper circuit	Blue (L)
Misc. circuits	Light green (Lg) or Brown (Br)

Suspension system

Alignment

Series 11
- Caster ... +30'
- Toe-in ... 0.080 in

Series 12
- Camber ... +30' ± 30'
- Caster ... +30' ± 30'
- Toe-in ... 0.080 in (2 mm) ± 0.080 in (2 mm)
- Kingpin inclination ... 7° 30' ± 30'

4 x 4 models (Series 9 thru 11)
- Camber ... +35'
- Caster ... +20'
- Toe-in ... 0
- Kingpin inclination ... 7° 25'

4 x 4 models (Series 12)
- Camber ... +35' ± 30'
- Caster ... +20' ± 30'
- Toe-in ... 0 ± 0.08 in
- Kingpin inclination ... 7° 25' ± 30'

Torsion bar spring
Series 8 thru 10 (long wheelbase and 4 x 4 models)
    Length x diameter . . . . 38.1 in x 0.906 in
    Wheel rate/side . . . . 122.6 lbs/in
    Spring constant . . . . 27.8 ft-lbs/degree
Series 11 and 12 (standard wheelbase models)
    Length x diameter . . . . 38.4 x 0.787 in
    Wheel rate/side . . . . 74.5 lbs/in
    Spring constant . . . . 15.9 ft-lbs/degree
Series 11 and 12 (long wheelbase models)
    Length x diameter . . . . 38.4 x 0.787 in
    Wheel rate/side . . . . 113.7 lbs/in
    Spring constant . . . . 25.3 ft-lbs/degree
Series 11 and 12 (4 x 4 models)
    Length x diameter . . . . 36.5 in x 0.894 in
    Wheel rate/side . . . . 126.6 lb/in
    Spring constant . . . . 27.6 ft

Shock absorbers
Series 11 and 12
    Maximum length . . . . 12.8 in
    Minimum length. . . . . 8.27 in
    Stroke . . . . 4.53 in
4 x 4 models
    Maximum length . . . . 14.0 in
    Minimum length. . . . . 8.86 in
    Stroke . . . . 5.14 in

Steering system
Manual steering gear oil capacity
    Series 9 and 10 . . . . 8 fl oz
    Series 11 and 12. . . . . 6.7 fl oz
Power steering system (Series 12)
    Steering gear type. . . . . Integral, ball screw
    Gear ratio . . . . 23.9 : 1
    Power steering pump type . . . . Vane type
    Fluid type . . . . Dexron II automatic transmission fluid
    Fluid capacity . . . . 1.1 qt (1 liter)
    Steering gear preload . . . . 5.2 to 7.8 in-lb

Torque specifications   Ft-lb
Idler arm mounting bolts (Series 11 and 12). 29
Idler arm/intermediate rod nut (Series 11 and 12) 50.5
Balljoint-to-lower arm nut (Series 11 and 12) 30
Control arm pivot-to-frame bolts (Series 11 and 12. 75
Upper control arm pivot shaft bushings (Series 11 and 12) 87
Stabilizer bar-to-frame bolts (Series 11 and 12) 55
Lower control arm-to-crossmember bolt (Series 11 and 12) 90
Adapter and dust shield-to-steering knuckle bolts (Series 11 and 12). 29
Strut bar-to-frame nuts 9
Strut bar-to-frame locknut. 30
Rear spring shackle pins 70
Power steering gear mounting bolts. 29
Steering column pinch bolt 18
Hose nut 33
Valve housing-to-gearbox bolts 35
Hose fitting locknut 30
Top cover-to-gearbox bolts 35
Adjusting screw locknut 30

Body

Overall length
Long wheelbase models . . . . 190.9 in
4 x 4 models. . . . . 174.5 in
Series 11 and 12 standard models. . . . . 174.5 in

Overall height
Long wheelbase models . . . . 59.3 in
4 x 4 models. . . . . 61.0 in
Series 11 and 12 standard models. . . . . 59.3 in

Wheelbase
Long wheelbase models . . . . 117.9 in
4 x 4 models. . . . . 104.3 in

3 Routine maintenance

The following routine maintenance items apply specifically to one or more model years within the Series 8 through 12 range. These items should be performed in addition to those listed for Series 5 in the Routine maintenance Section at the front of this manual.

Note that the clutch master cylinder fluid level check is no longer necessary, since beginning with Series 8, the clutch is no longer hydraulically operated, but rather uses a cable actuation system.

Also in Series 11 and 12 models, the distributor points and condenser were eliminated, making the related maintenance no longer necessary.

After initial 5000 miles
Check and adjust the engine idle speed using the specifications listed on the Emission Control Information Label on the underside of the hood.

After initial 7500 miles (4 x 4 models)
Replace the oil in the transmission, transfer case and front and rear differentials.

Every 7500 miles or 12 months, whichever comes first
4 x 4 models: Check the fluid levels in the transfer case and front and rear differentials. If necessary, add the correct type and grade of oil to bring the level up to the bottom of the filler plug hole. Check the flange bolt torque on the front and rear propeller shafts. Also lubricate the universal joints of the front and rear shafts and the sliding yoke of the front shaft.
Series 12 models: Check the fluid level in the power steering system, if equipped, and fill it to the specified mark if necessary. At first oil change, and every other oil change thereafter, replace the engine oil filter.

Every 15000 miles or 12 months, whichever comes first
Lubricate the clutch pedal springs, bushings and clevis pin. Also lubricate the clutch control cable end and check the clutch pedal free play.

Every 15 000 miles
Check and adjust the engine idle speed using the specifications listed on the Emission Control Information Label on the underside of the hood.
4 x 4 models only: If the truck is operated under severe conditions, replace the front and rear differential oil.

Every 22 500 miles (Series 12 models)
Change the fluid in the power steering system.

Every 30000 miles or 24 months, whichever comes first
Check the drivebelts that operate the air pump, generator, air-conditioning compressor (if equipped) and power steering pump (if equipped) for cracking, fraying or deterioration. Check the tension and adjust it, if required.
Check the operation of the distributor advance mechanism and lubricate it.

Every 30 000 miles
Replace the fuel line filter.
Check the ignition timing and adjust it as necessary. Also inspect and clean the distributor cap and rotor.
4 x 4 models only: Replace the oil in the transmission, transfer case and front and rear differentials. If necessary, add the correct type and grade of oil to bring the level up to the bottom of the filler plug hole.

Every 45 000 miles (Series 12 models)
Replace the rubber hoses of the power steering system, if equipped.

4 Engine

Valve clearances — adjustment
1 Remove the rocker arm cover as described in Chapter 1.
2 Prior to adjusting the valves, make sure that the rocker arm shaft bracket nuts are tight and at their proper torque.
3 Rotate the crankshaft so that either the number 1 or 4 piston is at top dead center (TDC) on the compression stroke. In this position the timing marks on the gear case and crankshaft pulley are in alignment.
4 When the number 1 piston is at TDC, the number 1 cylinder intake and exhaust valves, the number 2 intake valve and the number 3 exhaust valve can all be adjusted. When the number 4 piston is at TDC, the number 2 exhaust valve, the number 3 intake valve and the number 4 intake and exhaust valves can be adjusted.
5 Use a feeler gauge to measure the clearance between the rocker arm and the valve stem. There should be a slight drag on the gauge.
6 If the valve needs adjusting, loosen the locknut and turn the adjusting screw until the proper clearance is obtained. Then retighten the locknut and recheck the clearance to be sure it hasn't changed.
7 Reinstall the rocker arm cover.

Oil pan (4 x 4 models) — removal and installation
8 To remove the oil pan on 4 x 4 models, the engine must first be removed from the truck. Follow the procedure in Chapter 1.
9 With the engine removed, next remove the bellhousing braces.
10 Remove the oil pan bolts and lift off the oil pan. If the oil pan will not lift easily off, tap it lightly with a soft-headed hammer.
11 Prior to installation, clean the mating surfaces of the oil pan and engine block. Then apply a silicone-type sealant to the oil pan sealing surface.
12 Fit a new gasket into position and reinstall the oil pan onto the engine, tightening the mounting bolts to the proper torque.
13 The remainder of the installation procedure is the reverse of the removal procedure.

Engine — removal and installation
14 Although the removal and installation procedures remain basically unchanged from those described in Chapter 1, the changes and increases in emission control equipment make it necessary to also disconnect any vacuum line hoses or wiring leading from an engine mounted component to a non-engine mounted component.
15 Prior to disconnecting any hoses or wiring connectors, always label them with pieces of tape marked as to the properly installed location.

5 Fuel system

Refer to Figures 13.1 thru 13.9

Idle adjustment (Series 9 and 10)
1 The procedure for adjusting the engine idle speed is basically the same as described in Chapter 1, Section 15, with the following exceptions:
  a) In paragraph 4, on California models, the idle mixture screw should be unscrewed 1 1/2 turns.
  b) The rpm speeds specified in paragraphs 5 and 6 should be reduced 50 rpm for Series 9 Federal models and Series 10 Federal models with manual transmissions. The speeds should be increased by 50 rpm for Series 10 Federal models with automatic transmissions.

Idle adjustment (Series 11 and 12)
2 The parking brake should be set, the transmission in Neutral, and the engine should be at normal operating temperature.
3 The carburetor choke should be open with the air cleaner installed and the air-conditioner, if equipped, should be off.
4 The distributor vacuum line, the charcoal canister purge line and the EGR vacuum line should be disconnected and plugged. Also the idle compensator vacuum line on the air cleaner should be closed by bending the rubber hose.
5 Connect a tachometer according to the manufacturer's instructions and turn the throttle adjustment screw until the engine is idling at 900 ± 50 rpm (all models except Federal manual transmission) or 800 ± 50 rpm (Federal manual transmission models).
6 Models equipped with air-conditioning should also have the following steps performed:
  a) Turn the A/C controls to Maximum cold and High fan blower positions.

b) Open the throttle to about 1/3 and allow the throttle to close. This will allow the speed-up solenoid to reach full travel.
   c) Then adjust the speed-up controller adjusting screw so the engine idles at 900 ± 50 rpm.

Idle mixture adjustment (Series 11 and 12)
Note: *The idle mixture on these models was set at the factory and the adjustment screw is covered with a metal plug to prevent casual changing of the setting. Because of Federal emission laws, as well as the critical nature of later model emissions systems, altering the idle mixture adjustment from its original setting is not only illegal, but can adversely affect the performance of the engine. The idle mixture adjustment should not be changed unless absolutely necessary, as in the case of carburetor replacement.*

7 With the transmission in Neutral, set the parking brake and block the rear wheels.
8 In order to remove the idle mixture screw plug, the carburetor must be removed from the engine. Refer to Chapter 3 for this procedure.
9 With the carburetor removed, use a screwdriver and hammer, inserted in the slit on the lower carburetor flange (shown in the accompanying Figure), to tap the plug out.
10 Reinstall the carburetor on the engine.

All Series 11 models and Series 12 Federal models
11 Warm the engine up to normal operating temperature. The air-conditioner, if equipped, should be Off and the choke should be open. Also disconnect the distributor vacuum line, EGR valve vacuum line and charcoal canister purge line and plug them. The idle compensator vacuum line should also be closed by bending and clamping the rubber hose.
12 Connect a tachometer according to the manufacturer's instructions.
13 Turn the throttle adjusting screw to bring the engine speed to 800 ± 50 rpm (for Series 11 Federal models with manual transmission, and Series 12 models not equipped with air-conditioning) or 900 ± 50 rpm (for all other models).
14 Turn the idle mixture adjusting screw all the way in and then back it out two turns (for Series 11 Federal models and all Series 12 models) or one turn (for Series 11 California models).
15 Readjust the throttle adjusting screw to bring the engine speed to 850 rpm (for Series 11 Federal models with manual transmission), 950 rpm (for all other Series 11 models) or 800 rpm (for all Series 12 models).
16 Turn the idle mixture adjustment screw to achieve the maximum engine speed.
17 Once more, reset the throttle adjusting screw to 850 rpm (for Series 11 Federal models with manual transmission and all Series 12 models) or 950 rpm (for all other Series 11 models).
18 Finally turn the idle mixture adjustment screw clockwise (in the lean direction) until the engine speed is down to 800 ± 50 rpm (for Series 11 Federal models with manual transmission and all Series 12 models) or 900 ± 50 rpm (for all other Series 11 models).
19 Reinstall the idle mixture screw plug.
20 On models equipped with air-conditioning, complete the procedure by following the steps outlined in Paragraph 6 under Idle adjustment (Series 11 and 12).
21 Unplug and reconnect all hoses to their original positions.

Series 12 California models
22 Warm the engine up to normal operating temperature. The air-conditioner, if equipped, should be Off and the choke should be open. Also disconnect the distributor vacuum line, EGR valve vacuum line and charcoal canister purge line and plug them. The idle compensator vacuum line should also be closed by bending and clamping the rubber hose.
23 Connect a tachometer and a dwell meter according to the manufacturer's instructions.
24 Turn the idle mixture adjustment screw all the way in and then back it out 1/2 turn.
25 Turn the throttle adjusting screw to bring the engine speed to 900 ± 50 rpm.
26 Adjust the idle mixture adjustment screw to obtain an average dwell reading of 36°.
27 Reset the throttle adjusting screw so the engine speed is again 900 ± 50 rpm.
28 Reinstall the idle mixture screw plug.
29 Unplug and reconnect all hoses to their original positions.

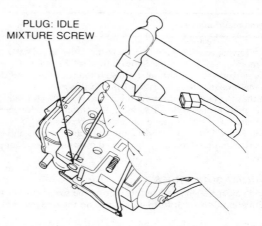

Fig. 13.1 Prior to idle mixture adjustment on Series 11 and 12 models, the screw plug must be removed as shown (Sec 5)

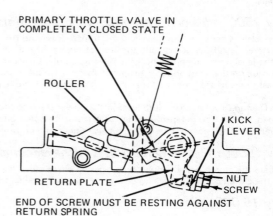

Fig. 13.2 Kick lever adjustment procedure for manual transmission models (Sec 5)

Carburetor adjustments (Series 9 thru 12)
Primary throttle valve opening adjustment
30 The procedure for this adjustment is the same as described in Chapter 3, Section 6, except that the specifications have changed. On Series 9 and 10 models, and Series 11 with automatic transmissions, the throttle valve should be at an angle of 17° when the choke valve is fully closed. On Series 11 models with manual transmissions, the angle should be 16°.
31 The standard clearance on Series 9 and 10 models is 0.047 to 0.051 in (1.2 to 1.3 mm). On Series 11 with automatic transmissions the standard clearance is 0.055 to 0.064 in (1.40 to 1.63 mm), while on manual transmissions it is 0.050 to 0.059 in (1.28 to 1.51 mm).

Linkage adjustment
32 This procedure is the same as described in Chapter 3, Section 7, except that the angle of the primary throttle valve mentioned in paragraph 1 is not 47°. Also, the standard clearance mentioned in paragraph 2 is now 0.24 to 0.30 in (6.1 to 7.6 mm).

Fast idle adjustment
33 The fast idle adjustment on the DCH-340 carburetor is set by adjusting the throttle valve opening. When set on the first step of the fast idle cam, the opening should be as shown in the Specifications.
34 To check for the correct setting, warm the engine up to normal operating temperature. Then disconnect the vacuum lines from the distributor, idle compensator and EGR valve and plug them. In this condition the rpm should be as shown in the Specifications.

Kick lever adjustment
35 Turn the throttle adjustment screw to bring the primary side throttle valve into a completely closed position.

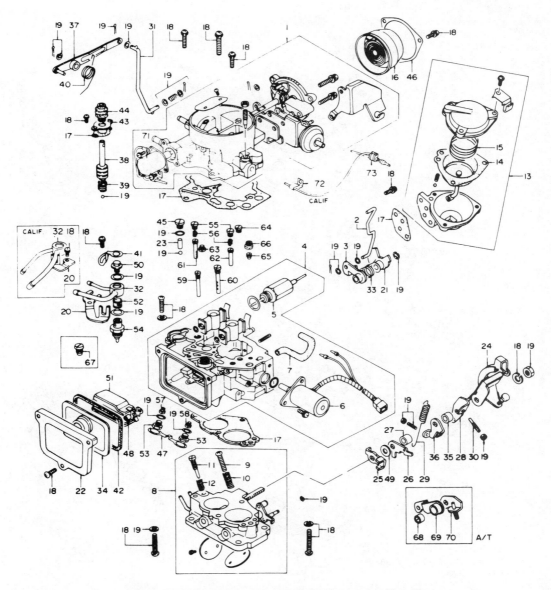

Fig. 13.3 Exploded view of a typical DCH340 carburetor (Sec 5)

1 Choke chamber assembly
2 Choke connecting rod
3 Counter lever
4 Float chamber assembly
5 Anti-dieseling or slow cut solenoid
6 Coasting richer solenoid (if equipped)
7 Hose
8 Throttle chamber assembly
9 Throttle chamber assembly
10 Adjusting screw spring
11 Idle adjusting screw
12 Adjusting screw spring
13 Diaphragm chamber assembly
14 Diaphragm
15 Diaphragm spring
16 Thermostat cover assembly
17 Gasket kit
18 Screw & washer kit, A
19 Screw & washer kit, B
20 Nipple stop plate
21 Fast idle cam
22 Cover
23 Injector weight
24 Throttle lever (Primary)
25 Throttle adjusting lever
26 Throttle return plate
27 Kick lever sleeve
28 Fast idle lever
29 Throttle return spring
30 Fast idle screw
31 Accelerator pump rod
32 Nipple
33 Fast idle cam spring
34 Fuel level gauge
35 Throttle lever sleeve
36 Kick lever
37 Accelerator pump lever
38 Accelerator pump piston
39 Piston return spring
40 Pump lever return spring
41 Lock lever
42 Rubber seal
43 Plate
44 Dust cover
45 Injector weight plug
46 Thermostat cover fixing plate
47 Main jet plug lock plate
48 Collar
49 Thrust washer
50 Nipple set screw
51 Float
52 Strainer
53 Main jet plug
54 Needle valve
55 Slow jet plug
56 Slow jet spring
57 Main jet (Primary)
58 Main jet (Secondary)
59 Main air bleed (Primary)
60 Main air bleed (Secondary)
61 Slow jet (Primary)
62 Slow jet (Secondary)
63 Slow air bleed (Primary)
64 Slow air bleed (Secondary)
65 Coasting jet
66 Coasting air bleed
67 Power valve
68 Collar A
69 Down shift lever
70 Connecting lever
71 Vent switching valve solenoid
72 Harness clip
73 Harness clip

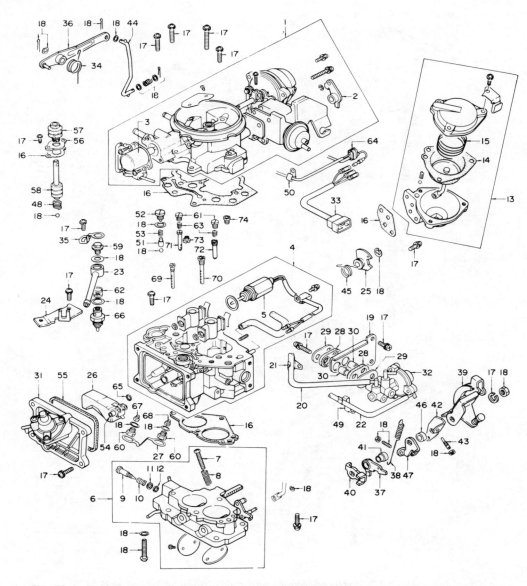

Fig. 13.4 Exploded view of the carburetor used on Series 12 California models (Sec 5)

1 Chamber assembly	20 Rubber hose	38 Secondary throttle spring	57 Dust cover
2 Counter lever	21 Hose clamp	39 Throttle lever	58 Pump piston
3 Vent switching valve	22 Rubber hose	40 Throttle adjust lever	59 Nipple set screw
4 Float chamber assembly	23 Fuel nipple	41 Throttle sleeve (A)	60 Drain plug
5 Slow cut valve	24 Stopping plate	42 Fast adjust lever	61 Slow jet taper plug
6 Throttle chamber assembly	25 Fast idle cam	43 Fast idle screw	62 Needle valve filter
7 Throttle adjust screw	26 Float	44 Pump rod	63 Slow jet spring
8 Throttle adjust spring	27 Drain plug lock plate	45 Fast idle cam spring	64 Lead wire connector
9 Idle adjust screw	28 Mounting rubber	46 Throttle sleeve (B)	65 O-ring
10 Idle adjust spring	29 Plate	47 Throttle kick lever	66 Needle valve
11 Idle adjust washer	30 Collar	48 Piston return spring	67 Primary main jet
12 Idle adjust rubber seal	31 Main actuator	49 Hose clamp (B)	68 Secondary main jet
13 Diaphragm chamber assembly	32 Slow actuator	50 Lead wire holder	69 Primary main air bleed
14 Diaphragm	33 With connector harness assembly	51 Injector weight	70 Secondary main air bleed
15 Diaphragm spring		52 Pump set screw	71 Primary slow jet
16 Overhaul gasket kit	34 Pump lever spring	53 Injector spring	72 Secondary slow jet
17 Screw & washer kit (A)	35 Lock lever	54 Float set collar (C)	73 Primary slow air bleed
18 Screw & washer kit (B)	36 Pump lever	55 Level gauge rubber seal	74 Secondary slow air bleed
19 Actuator bracket	37 Throttle return plate	56 Cylinder plate	

36 *Manual transmission models:* With the throttle valve completely closed, loosen the locknut on the kick lever screw and turn the screw just until it comes into contact with the return plate. Then retighten the locknut.
37 *Automatic transmission models:* With the throttle valve completely closed, bend the end of the kick lever just until it comes into contact with the return plate.

Carburetor diaphragm chamber — disassembly

38 Remove the screws that attach the diaphragm cover to the assembly. Then carefully separate the cover, spring and diaphragm, being careful not to lose the ball and spring.
39 Remove the jets in the upper part of the float chamber.
40 Remove the injector weight plug, then invert the float chamber to remove the injector with weight and check ball.
41 Remove the power jet. Be sure the screwdriver is properly seated in the slot to prevent damage.
42 Remove the two main jet plugs. Then remove the primary and secondary main jets.
43 Remove the coasting air bleed (if equipped) and the primary slow air bleed from the choke chamber.
44 Since the screws retaining the primary throttle valve, the secondary throttle valve and the choke valve are sealed with an adhesive compound to prevent air leaks, these parts should not be removed unless absolutely necessary.

Carburetor (Series 12) — general notes

45 On both Federal and California models, a slow cut solenoid, used with the Coasting Fuel Cut-off system (Section 6), has replaced the coasting richer solenoid used on previous years.
46 The carburetor used on Series 12 California models is basically the same as that used on Federal models and previous model years, but is equipped with fuel actuators that work in conjunction with the Loop Emissions Control system (described in Section 6).
47 These actuators, the main actuator and the slow actuator shown in the accompanying exploded view diagram, control the air/fuel mixture at a level optimum for catalytic converter functioning, through signals received from the Closed Loop system Electronic Control Module.
48 When disassembling the carburetor, the slow actuator can be removed by disconnecting the rubber vacuum hose and removing the two mounting screws.
49 The main actuator is held to the carburetor with four mounting screws and must be removed in order to gain access to the float chamber.
50 Both the slow and main actuators are factory adjusted and assembled and are not designed to be disassembled.
51 Prior to installing the main actuator, lithium grease should be applied to the O-ring. Tighten the mounting screws carefully to avoid cracking the O-ring.

Mechanical diaphragm fuel pump (Series 12) — removal and installation

52 On Series 12 models, the electric fuel pump was replaced with a mechanical diaphragm type. Located at the front right side of the cylinder head, it is driven by an eccentric cam mounted to the front of the camshaft.

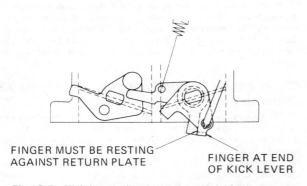

Fig. 13.5 Kick lever adjustment procedure for automatic transmission models (Sec 5)

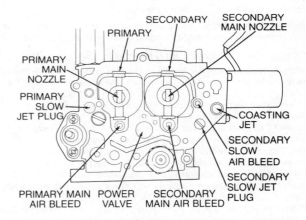

Fig. 13.6 Location of the float chamber components (Sec 5)

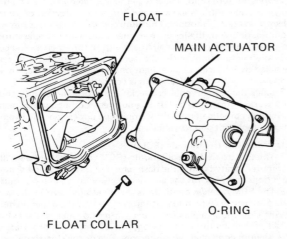

Fig. 13.7 Location of the main actuator O-ring on Series 12 California models (Sec 5)

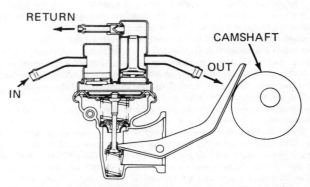

Fig. 13.8 The mechanical fuel pump used on Series 12 models is run off of the front of the camshaft (Sec 5)

53 The mechanical fuel pump is not designed to be disassembled and, if defective, must be replaced as a unit.
54 Use the following procedure for removal and installation of the pump:
  a) Remove the distributor assembly. Use the procedure described in Section 7.
  b) Remove the spark plug wires.
  c) Disconnect the fuel hoses from the fuel pump.
  d) Remove the engine hanger.
  e) Remove the fuel pump attaching bolt and lift off the pump.
  f) Installation is the reverse of the removal procedure.

6 Emission control systems

Refer to Figures 13.10 thru 13.17

General description

1 The emission control systems found on Series 9, 10, 11 and 12 models include the following:

Air Injector Reactor system
Catalytic converter
Closed Loop Emission Control system
Coasting Fuel Cut-off system
Coasting Richer system
Crankcase Emission Control system (PCV system)
Dash pot
Early Fuel Evaporation (EFE) system
Evaporative Emission Control system
Exhaust Gas Recirculation (EGR) system
Over Temperature Control system
Thermostatically-controlled air cleaner

2 Some of these systems overlap others in function and component use. Systems or components not included here are described in Chapter 3.

Air Injection Reactor (AIR) system (Series 12)

3 While function of the Air Injection Reactor system remains basically the same, there are minor design changes in the system for Series 12 models.
4 The mixture control valve, previously used only on California models, is used on all Series 12 models.
5 Series 12 California models use the vacuum switching valve, which was previously used as part of the Over Temperature Control system. The vacuum switching valve is described in Chapter 3.
6 An air switching valve, similar to the one used previously on California models, was integrated into the system on all models, although the valves used on Federal and California models are slightly different.
7 The air switching valve on Federal models is a simple design which uses intake manifold vacuum to divert the air from the air pump into the atmosphere during periods of low manifold vacuum. During periods of high manifold vacuum, the air switching valve routes the air through the check valve and into the exhaust manifold.
8 The air switching valve on California models also serves to divert air pump air flow from the exhaust manifold during certain times, but it depends on both intake manifold vacuum and air pump pressure and is regulated by the vacuum switching valve. During certain conditions the vacuum switching valve is electrically activated to allow intake manifold vacuum to open the air switching valve, allowing air pump flow to the check valve and exhaust manifold. When turned off, the vacuum switching valve shuts off the intake manifold vacuum. During this time, excessive air pump pressure will cause the air switching valve to close the air passage and divert the air flow to the atmosphere.
9 To check for proper operation of the vacuum switching valve, disconnect the valve wiring connector and use jumper wires to connect the connector terminals directly to the battery. Connect and disconnect the valve to the battery several times, listening carefully for the sound of the internal plunger being activated. If this sound is heard, the plunger is operating properly.
10 *To check the air switching valve on California models,* first be sure all vacuum hoses are in good condition. Connect the vacuum switching valve to the battery as described above. Disconnect the hose from the check valve and start the engine. With the vacuum switching valve

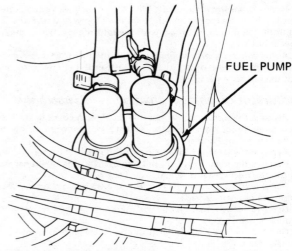

Fig. 13.9 The mechanical fuel pump is mounted at the right front of the cylinder head (Sec 5)

energized by the battery, the air switching valve should be allowing air flow from the air pump to pass through. This can be felt at the end of the disconnected hose. If no air is felt, disconnect the vacuum line closest to the end of the valve and check for suction. If vacuum is felt at the end of the hose, signifying the vacuum switching valve is operating properly, the air switching valve is suspect.
11 *To check the air switching valve on Federal models*, initally, check the condition of the vacuum hoses. Disconnect the hose from the check valve and allow the engine to idle. You should be able to feel air flow at the end of the hose. If not, disconnect the vacuum line from the valve and check for suction. If vacuum is present, the air switching valve is suspect.

Closed Loop Emission Control system (Series 12, California models)

12 The Closed Loop Emission Control system was introduced to Series 12 California models in order to increase the efficiency of the catalytic converter by monitoring and adjusting the fuel/air mixture ratio being burned.
13 The principal components of this system are an oxygen sensor, an Electronic Control Module (ECM), a vacuum controller, air switching valve, vacuum switches and a temperature switch, as well as the catalytic converter and a special controlled air/fuel ratio carburetor.
14 The operation of the catalytic converter is most efficient at a specific mixture ratio. The oxygen sensor measures the amount of oxygen present in the exhaust and relays this information to the electronic control module. The control module uses this information in combination with the input from the vacuum and temperature switches to determine the optimum air/fuel mixture for all operating conditions.
15 Any malfunctions in the Closed Loop Emission Control system are signaled by a 'Check Engine' light on the dash which goes on and remains lit until the malfunction is corrected.
16 Since the system requires special tools for maintenance and repair, any work on it should be left to your dealer or a qualified technician. Although complicated, the system can be understood by examining each component and its function.

Electronic Control Module (ECM)
17 The Electronic Control Module (ECM) is essentially a small onboard computer located under the dash which monitors several engine/vehicle functions and controls the air/fuel mixture ratio at an optimum level.
18 The ECM receives continuous electrical signals from the various switches/sensors which feed it information on the engine temperature, throttle position, engine speed and ratio of oxygen in the exhaust gases. After processing this information, the ECM sends appropriate signals to the vacuum controller, air switching valve or the Coasting Fuel Cut-off system, which in turn regulate the carburetor's air/fuel mixture.

Oxygen sensor
19 The oxygen sensor is mounted in the exhaust pipe, upstream of the catalytic converter. It monitors the exhaust stream and sends infor-

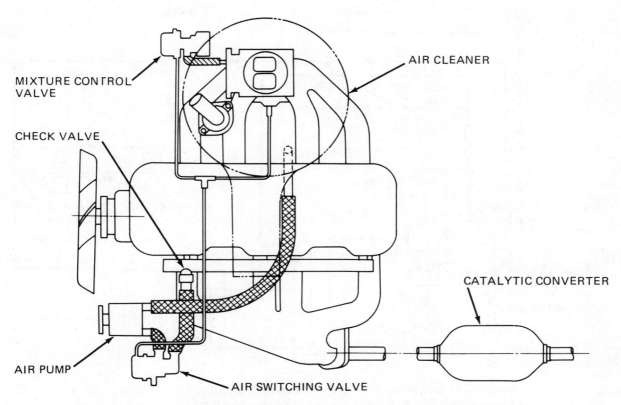

Fig. 13.10 Diagram of the Air Injection Reactor system for Series 12 Federal models (Sec 6)

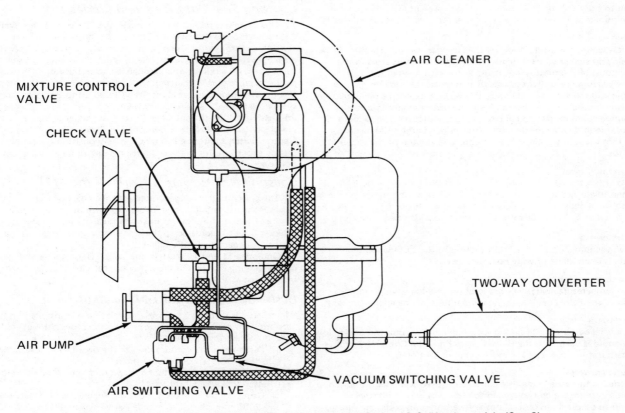

Fig. 13.11 Diagram of the Air Injection Reactor system for Series 12 California models (Sec 6)

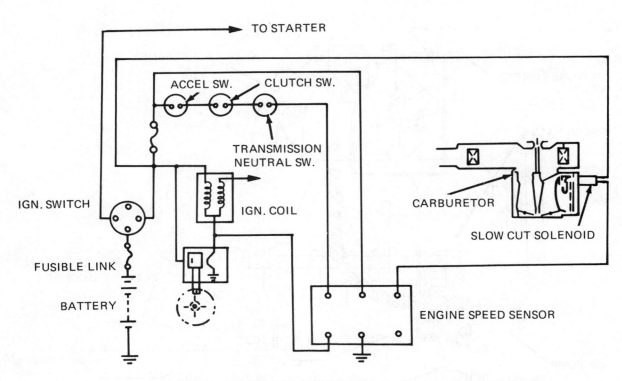

Fig. 13.12 Diagram of the Coasting Fuel Cut-off system used on Series 12 models (Sec 6)

mation to the ECM on how much oxygen is present. The oxygen level is determined by how rich or lean the fuel mixture in the carburetor is.

Vacuum controller

20 The vacuum controller converts the electrical signals from the ECM into vacuum signals which directly control the fuel actuators in the carburetor; these actuators adjust the air/fuel mixture ratio.
21 The vacuum controller consists of two parts; the vacuum control solenoid and the vacuum regulator. The vacuum regulator regulates the inconstant vacuum signal coming from the intake manifold into the constant vacuum needed to control the fuel actuators. The vacuum control solenoid is the mechanism which converts the ECM electrical signal into vacuum signals, using the constant vacuum of the vacuum regulator

Air switching valve

22 The air switching valve, which is controlled directly by the ECM, is designed to allow the injection of air into the exhaust ports either when the coolant temperature is below a certain level, or during a certain time period after the accelerator is moved to a wide open position.

Vacuum switch

23 Two vacuum switches feed information to the ECM when the throttle is in an idling or wide open position.

Temperature switch

24 The temperature switch is a bi-metal type which supplies information to the ECM concerning the temperature of the engine coolant.

Carburetor

25 A special carburetor equipped with fuel control actuators and a slow cut solenoid is used with the Closed Loop Emission Control system. This carburetor is discussed in more detail in Section 5.

Catalytic converter

26 A special three-way catalytic converter is used with the Closed Loop Emission system, which is designed to reduce oxides of nitrogen in addition to oxidizing hydrocarbons and carbon monoxide. For this to be achieved at maximum efficiency, the fuel/air mixture ratio must be kept at an optimum point. This essentially is the job of the entire Closed Loop Emission Control system.

Coasting Fuel Cut-off system (Series 12)

27 This system is basically the same as the Coasting Richer system for California models described in Chapter 3. The system detects coasting conditions through the use of a speed sensor, transmission Neutral switch, accelerator switch and clutch switch. The slow cut solenoid valve opens an auxiliary fuel passage in the carburetor to provide a richer mixture during coasting. On California models, the Coasting Fuel Cut-off system is integrated into the Closed Loop Emission Control system, described elsewhere in this Section.
28 Testing for the transmission Neutral switch, accelerator switch and clutch switch is the same as described in Chapter 3. Note that the Neutral switch should be turned On in any position except Neutral. The testing procedure for the speed sensor is described elsewhere in this Section.

Coasting Richer system (Series 9 thru 11)

29 The clutch switch incorporated into the coasting richer system on Series 8 thru 10 models is relocated as shown in the accompanying Figure. The switch functions in an identical manner to that described in Chapter 3, but is adjusted so that the clutch pedal height is the same as the brake pedal when fully released.
30 On Series 11 and 12 models, the switch is located as shown in the accompanying Figure. On these models, the clutch pedal height should be adjusted as shown.

Dash Pot (Series 9 thru 11) — adjustment

31 In Series 9 through 11 models, the dash pot is installed on Federal versions with manual transmissions. To adjust the dash pot, follow the procedure described below:
   a) Disconnect the vacuum lines at the distributor, idle compensator and EGR valve and plug them.
   b) Loosen the dash pot locknut and then screw the dash pot counterclockwise all the way.
   c) Using the throttle lever, hold the engine speed at between 2200 and 2600 rpm (Series 10 and 11) or between 2600 and 3000 (Series 9). Then turn the dash pot clockwise until the end of the dash pot contacts the throttle lever. Then retighten the locknut.
   d) Following adjustment, reconnect all vacuum lines.

Chapter 13/Supplement: Revisions and information on later models

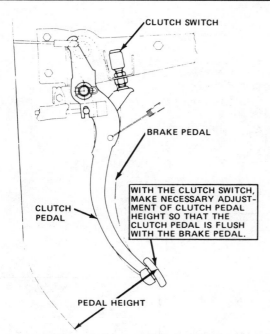

Fig. 13.13 Location of the Coasting Richer system clutch switch and correct clutch pedal height (Series 8 thru 10) (Sec 6)

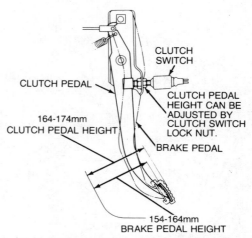

Fig. 13.14 Location of the clutch switch used in the Coasting Richer (Series 11) and Coasting Fuel Cut-off (Series 12) systems, and correct clutch pedal height (Sec 6)

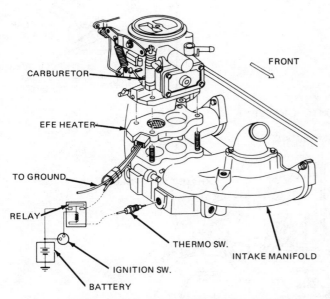

Fig. 13.15 Location of the heater unit and thermo switch used in the Early Fuel Evaporation system (Series 12) (Sec 6)

Early Fuel Evaporation (EFE) system (Series 12)

32 The Early Fuel Evaporation (EFE) system used on Series 12 models only, is designed to increase engine efficiency and lower hydrocarbon emissions levels during cold engine operation. This is done by providing a rapid source of heat to the intake air supply by way of a ceramic heater grid which is integral with the carburetor base gasket and located under the primary bore.

33 The system also uses a thermo switch, which measures coolant temperature in the intake manifold, and a relay. When the ignition switch is turned on, voltage is applied to the thermo switch. If the coolant is below a pre-determined level, the thermo switch is in an On position and the relay applies current directly to the EFE heater unit. The heater unit self-regulates itself at a certain temperature, until coolant temperature rises enough to turn off the thermo switch and, in turn, the heater unit.

34 If the EFE heater is not coming on, poor cold engine performance will be experienced. If the heater is not shutting off when the engine is fully warmed up, the engine will run as if it is out of tune due to the constant flow of hot air through the carburetor.

35 If the EFE system is suspected of malfunctioning while the engine is cold, first visually check all electrical wires and connectors to be sure they are clean, tight and in good condition.

36 With the ignition switch On, use a circuit tester or voltmeter to be sure current is reaching the relay. If not, there is a problem in the wiring leading to the relay.

37 Next, with the engine cold, but the ignition switch On, disconnect the heater unit wiring connector and use a circuit tester or voltmeter to check that current is reaching the heater unit. If so, use a continuity tester to check for continuity in the wiring connector attached to the heater unit. If continuity exists, the system is operating correctly in the cold engine mode.

38 If current was not reaching the heater unit in the above test, disconnect the thermo switch wiring connector and connect a continuity tester between the switch's terminal and a good ground (the intake manifold can be used if clean). With the engine cold, continuity should be shown. If not, the thermo switch should be replaced.

39 If continuity was shown at the thermo switch, but no current is reaching the heater unit, replace the relay.

40 To check that the system turns off at normal engine operating temperature, first allow the engine to warm up thoroughly. With the engine idling, disconnect the heater unit wiring connector and use a circuit tester or voltmeter to check if current is reaching the heater unit. If not, the system is okay.

41 If current is reaching the heater unit, disconnect the wiring connector from the thermo switch and connect a continuity tester between the switch's terminal and a good ground (the intake manifold can be used if clean). With the engine warmed up, no continuity should be shown. If there is, replace the thermo switch.

42 If there is no continuity at the thermo switch, but current is still reaching the heater unit, replace the relay.

43 The heater unit is removed by disconnecting the wiring connector, removing the carburetor and lifting it off.

44 The thermo switch is removed by disconnecting the wiring connector and unscrewing it from the intake manifold. When installing the thermo switch, wrap the threads with thread-sealing tape to prevent coolant leaks.

Engine speed sensor (Series 10 thru 12) — testing

45 Disconnect the wiring connector from the sensor, and with suitable jumper wires, connect the BY and LgW color-coded wiring terminals. Start the engine and check the voltage between the Lg and B terminals. The sensor is normal if the voltage is zero when the engine runs over 2100 to 2300 rpm.

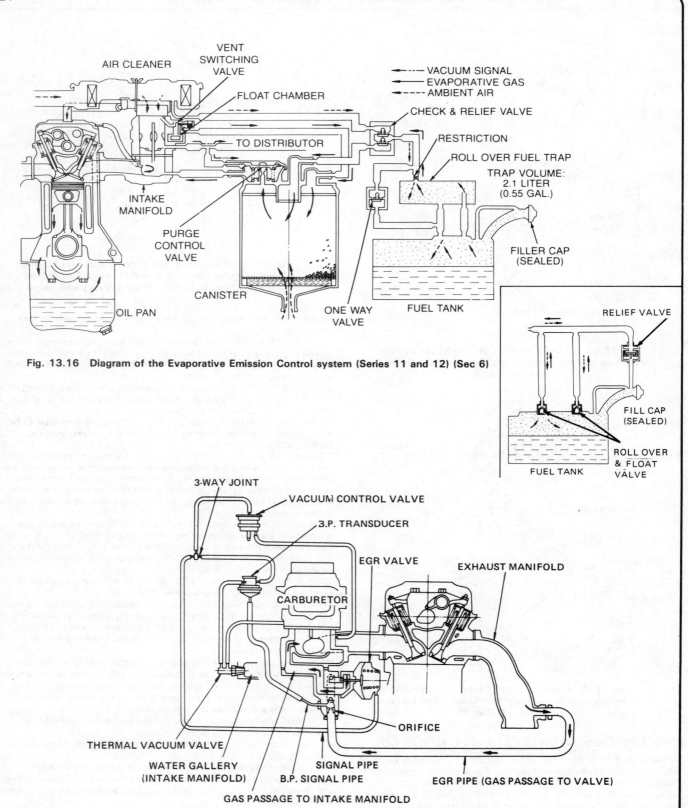

Fig. 13.16 Diagram of the Evaporative Emission Control system (Series 11 and 12) (Sec 6)

Fig. 13.17 Diagram of the EGR system on Federal models; California models similar, but without the vacuum control valve (Sec 6)

Evaporative Emission Control system (Series 11 and 12)

46 The Evaporative Emission Control system on Series 11 and 12 models has been redesigned to include a charcoal canister, in which the fuel vapors emitted by the fuel tank and carburetor are stored when the engine is not running. When the engine is running, the stored vapors in the canister are mixed with ambient air and fed into the engine for burning.

47 The system also incorporates a vent switching valve, which is an electrically-operated solenoid valve built into the carburetor. This valve remains open when the engine is not running, to allow fuel vapors to bleed into the charcoal canister. When the engine is running, the valve closes to prevent outside air from exiting the carburetor.

48 One minor difference in the system between Series 11 and 12 models is that on Series 12 models the rollover fuel trap was replaced by two individual rollover and float valves mounted in the top of the fuel tank. Also, the one way valve bypassing the rollover fuel trap has been eliminated and a pressure relief valve attached to the fuel filler tube has been added.

Exhaust Gas Recirculation (EGR) system (Series 9 thru 11)

49 The EGR system used in Series 9 through 11 models has been redesigned to provide greater control of the opening and closing of the EGR valve. On California models, a back pressure transducer has been added, which provides an air leak in the EGR valve vacuum signal line until the exhaust back pressure becomes great enough to close the transducer. On Federal models, both a back pressure transducer and a vacuum control valve are included in the system. The vacuum control valve also provides an air bleed in the EGR valve vacuum signal line until the vacuum in the intake manifold becomes great enough to close the vacuum control valve. The thermal vacuum valve is still part of the system and is intended to regulate the EGR valve in relation to the engine temperature, in a similar fashion as described in Chapter 3.

50 The individual components of the EGR system can be tested by using the following procedures:

  a) To test the EGR valve, connect an outside vacuum source to the vacuum supply tube at the top of the valve. Apply vacuum to the valve and note the movement of the diaphragm. It should move to the full up position at about 15.8 kPa (4.7 in Hg) on Federal models with manual transmissions, and at about 13.3 kPa (3.9 in Hg) on all other models. Hold the vacuum for several seconds to be sure the valve does not "leak down".
  b) To test the thermal vacuum valve, remove it from the engine with its vacuum tubes still attached. Place the sensing part (below the threads) in a pan of water and slowly heat the water. When the water temperature reaches about 115 to 129°F (46 to 54°C), you should be able to blow air through the vacuum tubes. If air will not pass through the valve at these temperatures, the valve must be replaced with a new one.
  c) To check the back pressure transducer, connect an air pump to the signal port and apply air pressure to the transducer. With a pressure of about 50 mm Ag (2 in Ag), the diaphragm should be fully closed and should not leak air.
  d) To check the vacuum control valve, connect a vacuum pump to the intake manifold signal port on the valve. Also attach a tube to the port connected to the EGR signal vacuum line. Apply about 13.2 to 14.4 in Hg (335 to 365 mm Hg) of vacuum to the valve and attempt to blow air through the tube. If air will not pass through the valve, it should be replaced.

Exhaust Gas Recirculation (EGR) system (Series 12)

51 The EGR system used in all Series 12 models is the same as the system used in Series 9 through 11 California models. Refer to the above sub-section and Chapter 3.

---

7 Ignition system

Refer to Figures 13.18 thru 13.20

General description

1 Series 11 and 12 models are equipped with a pointless-type distributor, in which a control module and magnetic pick-up assembly

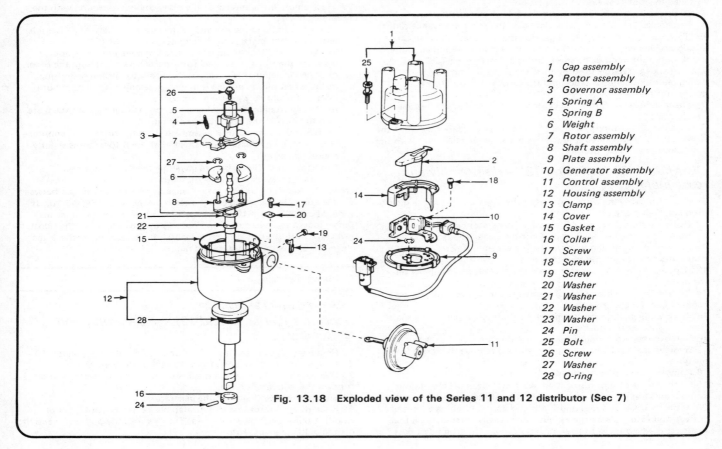

Fig. 13.18 Exploded view of the Series 11 and 12 distributor (Sec 7)

1 Cap assembly
2 Rotor assembly
3 Governor assembly
4 Spring A
5 Spring B
6 Weight
7 Rotor assembly
8 Shaft assembly
9 Plate assembly
10 Generator assembly
11 Control assembly
12 Housing assembly
13 Clamp
14 Cover
15 Gasket
16 Collar
17 Screw
18 Screw
19 Screw
20 Washer
21 Washer
22 Washer
23 Washer
24 Pin
25 Bolt
26 Screw
27 Washer
28 O-ring

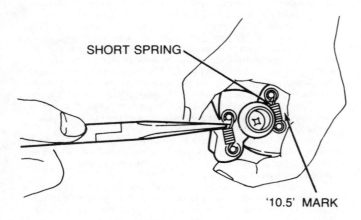

Fig. 13.19 When installing the distributor weights, the side with the '10.5' mark should be aligned with the stopper (Sec 7)

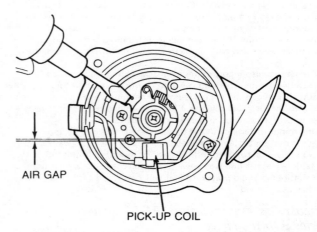

Fig. 13.20 On Series 11 and 12 distributors, the air gap between the pick-up coil and the trigger should be checked periodically (Sec 7)

replace the contact points of a conventional distributor. With this type of distributor, maintenance is greatly decreased, as there are no points to wear and regularly adjust.

Distributor (Series 11 and 12) — removal and installation

2 Disconnect the negative battery cable.
3 Disconnect the high-tension wires from the spark plugs and ignition coil, then remove them from the distributor cap. Be sure to note their installed locations for ease of installation.
4 Disconnect the wiring connector leading to the distributor.
5 Disconnect the vacuum hose leading to the vacuum advance canister.
6 To simplify reinstallation of the distributor, rotate the crankshaft (using a wrench on the pulley retaining nut) clockwise and align the crankshaft pulley and front cover timing pointer at the TDC position. Index mark the relative position of the distributor body with the fixing plate and cylinder block and the rotor in relation to the distributor body.
7 Remove the distributor bracket mounting bolt and lift out the distributor.
8 Installation is a direct reversal of removal, but ensure that when the distributor is fully located, the relative marks are in alignment with the crankshaft at the TDC position.

Distributor (Series 11 and 12) — disassembly and reassembly

9 After removing the distributor from the vehicle, remove the distributor cap, O-ring and rotor.
10 Using a screwdriver, pry off the signal generator cover.
11 Remove the two screws and pull out the signal generator.
12 Remove the snap-ring that secures the vacuum canister to the distributor body, then pull out the canister.
13 Remove the two screws that retain the breaker plate and lift out the plate.
14 Grind or file the head from the collar retaining pin. Then, using a hammer and punch, drive the pin out.
15 Remove the collar from the governor shaft and separate the shaft from the distributor housing.
16 Remove the two springs from the signal rotor.
17 Remove the cap from the signal rotor and remove the screw from the governor shaft.
18 Remove the snap-rings and lift off the two weights.
19 With the distributor dismantled, wash and clean off the various components but do not wash the inner face of the vacuum advancer.
20 Refer to Chapter 4, Section 2, for checks and cleaning details of the cap and rotor. Check the weights for excessive looseness on their pivots; it is advisable to fit new springs in view of their modest cost.

Suck (by mouth) on the vacuum diaphragm unit and check that the link is drawn in. When the vacuum is held by placing the tongue or a finger over the tube, the link should remain in; if it fails to do so, the vacuum unit must be replaced. Ensure that the driven gear is securely pinned to the cam spindle and that the gear is undamaged. Also check the shaft and the governor weights and springs for wear. Replace any suspect parts.
21 Use an ohmmeter to measure the resistance of the signal generator at the wiring connector. If it is not 140 to 180 ohms, replace the signal generator.
22 Reassembly is the reverse of the disassembly procedure, with the following notes:
   a) When installing the weights to the governor shaft, align the side with the '10.5' mark with the stopper (Figure 13.19).
   b) When installing the signal rotor to the governor shaft, apply grease to the top of the signal rotor before sealing it with the cover.
   c) When installing the governor shaft to the distributor housing, install a new pin by driving it in with a hammer. Then peen both sides of the pin by squeezing it in a vise.
   d) When installing the breaker plate, fit the four clips on the plate into the slots in the distributor housing.
   e) When the distributor is assembled, turn the rotor shaft counterclockwise, then release it and check that the rotor returns slightly counterclockwise.
   f) Install a new O-ring before installing the distributor cap.
23 Following reassembly of the distributor, the air gap must be adjusted. Do this by using a feeler gauge to measure the air gap between the pick-up coil projection and the teeth on the rotor (Fig. 13.20). It should be 0.008 to 0.016 in (0.2 to 0.4 mm). If it is not within this range, loosen the two screws and use a screwdriver to move the signal generator until the gap is correct. Tighten the screws and recheck that the gap is still within the specified range.

8 Charging system

Refer to Figures 13.21 and 13.22

Generator (Series 11 and 12) — disassembly and reassembly

1 Remove the generator from the truck, as described in Chapter 10.
2 Remove the through bolts from the rear of the generator.
3 Depress the locking tab on the electrical wiring connector and disconnect the generator wiring.
4 Separate the two halves of the generator.
5 Place the generator in a vise (preferably between two blocks of wood). Do not overtighten the vise. Then remove the pulley nut and lift off the pulley, fan and rotor.

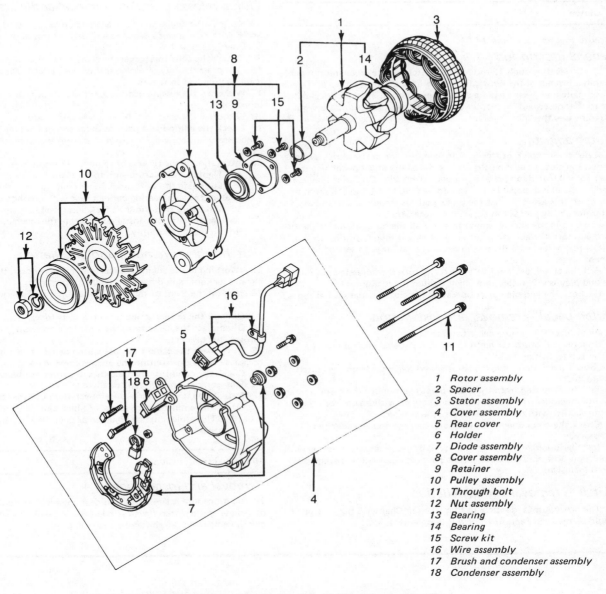

1 Rotor assembly
2 Spacer
3 Stator assembly
4 Cover assembly
5 Rear cover
6 Holder
7 Diode assembly
8 Cover assembly
9 Retainer
10 Pulley assembly
11 Through bolt
12 Nut assembly
13 Bearing
14 Bearing
15 Screw kit
16 Wire assembly
17 Brush and condenser assembly
18 Condenser assembly

Fig. 13.21 Exploded view of the Series 11 and 12 generator (Sec 8)

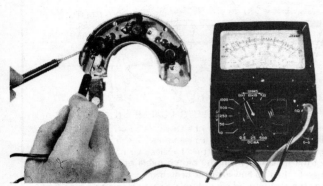

Fig. 13.22 When testing the generator diodes, the ohmmeter probes should be positioned as shown (Sec 8)

6   Remove the bearing retainer mounting screws and lift out the retainer.
7   Remove the rear side nuts. Then remove the stator from the rear cover together with the diodes, brushes and capacitor.
8   Unsolder the connections between the diode and the stator coil, then separate the diodes from the stator complete with brushes and capacitor.
9   Remove the screws that attach the brush holder and remove the diodes, brushes and capacitor.
10 The generator components can be tested in the same manner as described in Chapter 10, but note Fig. 13.22 in this Chapter for the testing of the diodes.
11 Reassembly is the reverse of the removal procedure, with the following note. When reassembling the front and rear sections, insert a thick piece of wire into the hole in the rear face of the rear cover from the outside to support the brush in the raised position. Then insert the front section, to which the rotor is assembled. Be careful not to scratch the face of the brushes that contact the wire. Also be sure all of the insulated components are positioned correctly and that they are free from oil or grease.

9 Clutch

Refer to Figures 13.23 and 13.24

General description

1 On Series 8 through 12 models the clutch release mechanism is cable operated instead of the hydraulic system used on earlier models. The driven plate and pressure plate assemblies function in an identical manner to that described in Chapter 5, although the dimensions are slightly different (see the Specifications).

Clutch adjustment

2 Refer to Section 5 of this Supplement and adjust the clutch switch so that the clutch pedal height is the same as the brake pedal height when fully released (on Series 8 through 10 models). On Series 11 and 12 models, the clutch pedal height should be 6.40 to 6.85 in (164 mm to 174 mm). It is assumed that the brake pedal is already adjusted but, if not, refer to Chapter 9 and adjust it as necessary.
3 Working in the engine compartment, pull the outer clutch cable forward and turn the adjusting nut until all play is taken up (Fig. 13.24).
4 Depress and release the clutch pedal several times to settle the component parts.
5 Pull the clutch cable fully forward and turn the adjusting nut until the end play is 0.2 in (5.0 mm) then tighten the locknut.
6 Recheck the end play after operating the clutch pedal several times.

Clutch pedal — removal and installation

7 Disconnect the ground cable from the battery.
8 Unhook the return spring from the clutch pedal and remove the starter safety switch.
9 Loosen the clutch cable adjustment and release it from the clutch pedal.
10 Pry the snap-ring from the pedal pivot and remove the plain washer and nylon washer; the clutch pedal can now be withdrawn from the pivot together with the bushing and nylon washer.
11 Check the components for wear and damage and replace them if necessary.
12 Installation is the reverse of removal, but lubricate the bushing with a multi-purpose grease and adjust the pedal as described in paragraphs 2 to 6 inclusive.

Clutch — removal and installation

13 The procedure is identical to that given in Chapter 5 but reference should be made to Section 10 of this Supplement.

Clutch release bearing — removal and installation

14 Remove the engine and/or transmission as described in Chapters 1 and 6 with reference to Sections 4 and 10 of this Supplement where necessary.
15 Detach the shift fork cover from the transmission casing.
16 Compress the two retaining springs and detach them from the release bearing and shift fork.
17 Withdraw the release bearing and shift block over the transmission input shaft.
18 Withdraw the shift fork from the fulcrum stud.
19 Check the component parts as described in Chapter 5, Section 10; if necessary, unscrew the fulcrum stud from the transmission front cover.
20 Installation is the reverse of removal, but note the following points:
    a) Tighten the fulcrum stud to 30 ft-lbs.
    b) Lubricate the shift block inner collar groove, shift fork ball seat and release bearing contact face with graphite grease.
    c) Make sure that the shift fork and release bearing retaining springs are correctly installed. Operate the mechanism and check that it moves smoothly.

Clutch cable — removal and installation

21 Working in the engine compartment, fully loosen the cable adjusting and locknuts and detach the cable support clips.
22 Jack up the front of the vehicle and support it adequately on axle-stands.
23 Unhook the return spring from the shift fork.
24 Disconnect the cable from the shift fork and pull it forward through the bracket.
25 Disconnect the cable from the clutch pedal, then withdraw the complete cable from within the engine compartment.
26 Check the condition of the cable, in particular the damper rubber, for signs of deterioration; replace them as necessary.
27 Installation is the reverse of removal, but note the following points:
    a) There must not be any sharp bends or kinks in the cable.
    b) The cable must be adjusted as described in paragraphs 2 to 6 inclusive.

10 Manual transmission

Removal and installation

1 The procedure is similar to that described in Chapter 6 but, instead of removing the clutch slave cylinder, it will be necessary to detach the clutch cable from the shift fork and casing.

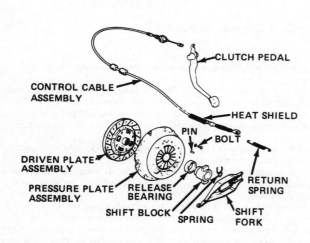

Fig. 13.23 Components of the clutch system (Sec 9)

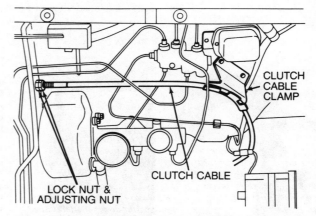

Fig. 13.24 Location of the clutch cable adjusting nut and locknut (Sec 9)

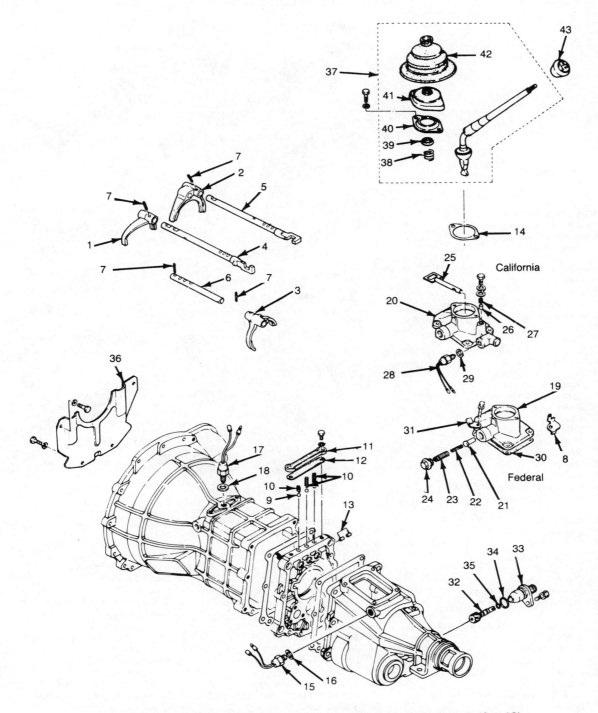

Fig. 13.25 Exploded view of the 4 x 4 manual transmission shift controls (Sec 12)

1 Fork, 3rd/4th shifter
2 Fork, 1st/2nd shifter
3 Fork, reverse shifter
4 Rod, gear shift, 3rd/4th
5 Rod, gear shift, 1st/2nd
6 Rod, gear shift, reverse
7 Pin, spring, shift fork
8 Bracket
9 Ball, detent, gear shift
10 Spring, detent ball
11 Plate, detent spring
12 Gasket
13 Pin, interlock
14 Plug, interlock
15 Switch asm., reverse lamp
16 Gasket
17 Switch asm., 3rd/4th (Federal Spec.)
18 Gasket (Federal Spec.)
19 Shifter cover (Federal Spec.)
20 Shifter cover (California Spec.)
21 Plunger, reverse, stop
22 Spring, inner
23 Spring, outer
24 Cap, reverse, stop
25 Rod, neutral switch
26 Damper pad
27 Spring
28 Switch asm., neutral (California Spec.)
29 Gasket
30 Gasket, shifter cover
31 Bracket
32 Gear, speed, driven
33 Bush w/O-ring
34 O-ring, bush outer
35 O-ring, bush inner
36 Cover, front, transmission case
37 Lever asm., gear-shift control
38 Spring, control lever
39 Seat, spring
40 Cover, control lever
41 Cover, dust, control lever
42 Grommet, lever, gear-shift
43 Knob, control lever

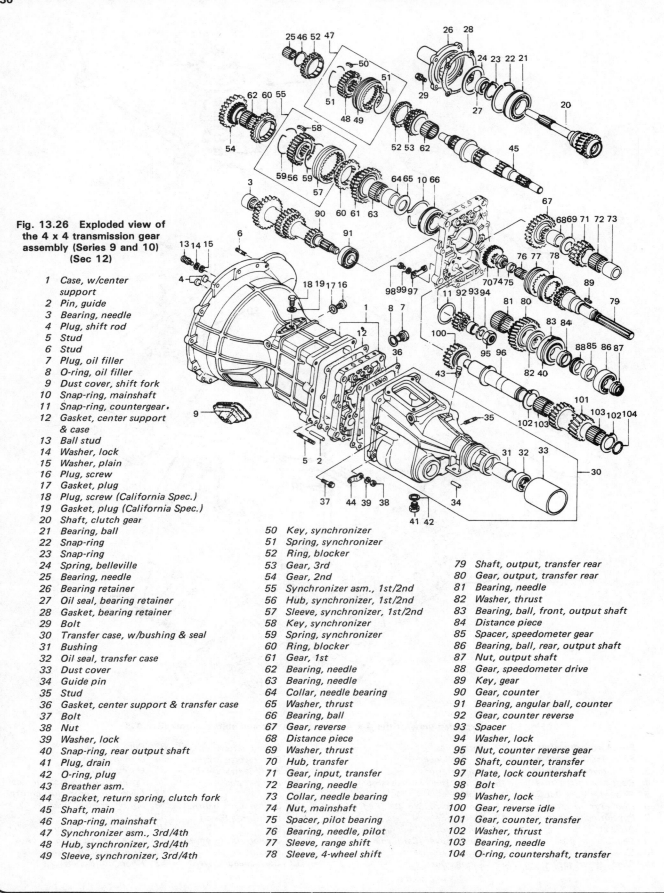

Fig. 13.26 Exploded view of the 4 x 4 transmission gear assembly (Series 9 and 10) (Sec 12)

1 Case, w/center support
2 Pin, guide
3 Bearing, needle
4 Plug, shift rod
5 Stud
6 Stud
7 Plug, oil filler
8 O-ring, oil filler
9 Dust cover, shift fork
10 Snap-ring, mainshaft
11 Snap-ring, countergear
12 Gasket, center support & case
13 Ball stud
14 Washer, lock
15 Washer, plain
16 Plug, screw
17 Gasket, plug
18 Plug, screw (California Spec.)
19 Gasket, plug (California Spec.)
20 Shaft, clutch gear
21 Bearing, ball
22 Snap-ring
23 Snap-ring
24 Spring, belleville
25 Bearing, needle
26 Bearing retainer
27 Oil seal, bearing retainer
28 Gasket, bearing retainer
29 Bolt
30 Transfer case, w/bushing & seal
31 Bushing
32 Oil seal, transfer case
33 Dust cover
34 Guide pin
35 Stud
36 Gasket, center support & transfer case
37 Bolt
38 Nut
39 Washer, lock
40 Snap-ring, rear output shaft
41 Plug, drain
42 O-ring, plug
43 Breather asm.
44 Bracket, return spring, clutch fork
45 Shaft, main
46 Snap-ring, mainshaft
47 Synchronizer asm., 3rd/4th
48 Hub, synchronizer, 3rd/4th
49 Sleeve, synchronizer, 3rd/4th
50 Key, synchronizer
51 Spring, synchronizer
52 Ring, blocker
53 Gear, 3rd
54 Gear, 2nd
55 Synchronizer asm., 1st/2nd
56 Hub, synchronizer, 1st/2nd
57 Sleeve, synchronizer, 1st/2nd
58 Key, synchronizer
59 Spring, synchronizer
60 Ring, blocker
61 Gear, 1st
62 Bearing, needle
63 Bearing, needle
64 Collar, needle bearing
65 Washer, thrust
66 Bearing, ball
67 Gear, reverse
68 Distance piece
69 Washer, thrust
70 Hub, transfer
71 Gear, input, transfer
72 Bearing, needle
73 Collar, needle bearing
74 Nut, mainshaft
75 Spacer, pilot bearing
76 Bearing, needle, pilot
77 Sleeve, range shift
78 Sleeve, 4-wheel shift
79 Shaft, output, transfer rear
80 Gear, output, transfer rear
81 Bearing, needle
82 Washer, thrust
83 Bearing, ball, front, output shaft
84 Distance piece
85 Spacer, speedometer gear
86 Bearing, ball, rear, output shaft
87 Nut, output shaft
88 Gear, speedometer drive
89 Key, gear
90 Gear, counter
91 Bearing, angular ball, counter
92 Gear, counter reverse
93 Spacer
94 Washer, lock
95 Nut, counter reverse gear
96 Shaft, counter, transfer
97 Plate, lock countershaft
98 Bolt
99 Washer, lock
100 Gear, reverse idle
101 Gear, counter, transfer
102 Washer, thrust
103 Bearing, needle
104 O-ring, countershaft, transfer

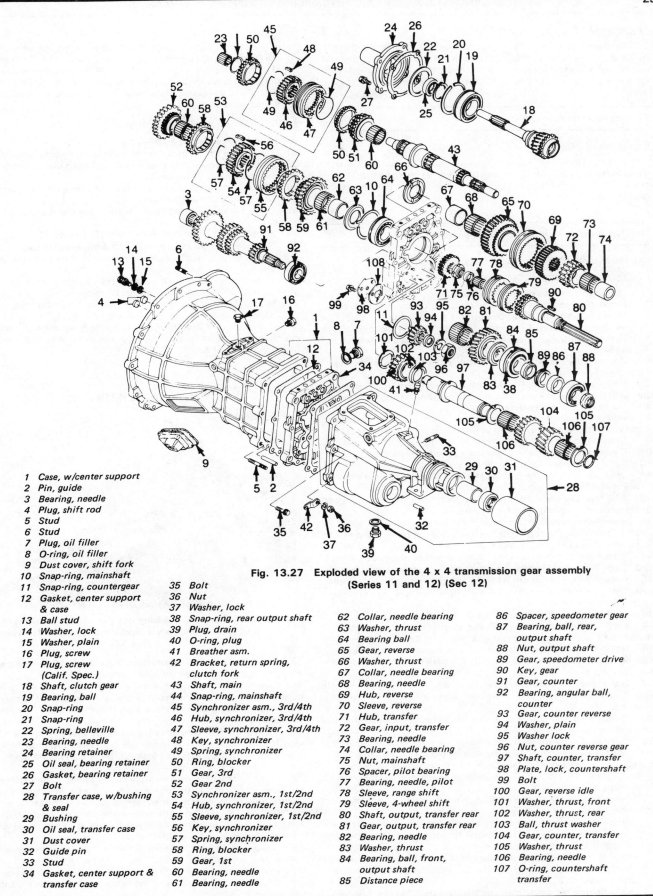

Fig. 13.27 Exploded view of the 4 x 4 transmission gear assembly (Series 11 and 12) (Sec 12)

1 Case, w/center support
2 Pin, guide
3 Bearing, needle
4 Plug, shift rod
5 Stud
6 Stud
7 Plug, oil filler
8 O-ring, oil filler
9 Dust cover, shift fork
10 Snap-ring, mainshaft
11 Snap-ring, countergear
12 Gasket, center support & case
13 Ball stud
14 Washer, lock
15 Washer, plain
16 Plug, screw
17 Plug, screw (Calif. Spec.)
18 Shaft, clutch gear
19 Bearing, ball
20 Snap-ring
21 Snap-ring
22 Spring, belleville
23 Bearing, needle
24 Bearing retainer
25 Oil seal, bearing retainer
26 Gasket, bearing retainer
27 Bolt
28 Transfer case, w/bushing & seal
29 Bushing
30 Oil seal, transfer case
31 Dust cover
32 Guide pin
33 Stud
34 Gasket, center support & transfer case
35 Bolt
36 Nut
37 Washer, lock
38 Snap-ring, rear output shaft
39 Plug, drain
40 O-ring, plug
41 Breather asm.
42 Bracket, return spring, clutch fork
43 Shaft, main
44 Snap-ring, mainshaft
45 Synchronizer asm., 3rd/4th
46 Hub, synchronizer, 3rd/4th
47 Sleeve, synchronizer, 3rd/4th
48 Key, synchronizer
49 Spring, synchronizer
50 Ring, blocker
51 Gear, 3rd
52 Gear 2nd
53 Synchronizer asm., 1st/2nd
54 Hub, synchronizer, 1st/2nd
55 Sleeve, synchronizer, 1st/2nd
56 Key, synchronizer
57 Spring, synchronizer
58 Ring, blocker
59 Gear, 1st
60 Bearing, needle
61 Bearing, needle
62 Collar, needle bearing
63 Washer, thrust
64 Bearing ball
65 Gear, reverse
66 Washer, thrust
67 Collar, needle bearing
68 Bearing, needle
69 Hub, reverse
70 Sleeve, reverse
71 Hub, transfer
72 Gear, input, transfer
73 Bearing, needle
74 Collar, needle bearing
75 Nut, mainshaft
76 Spacer, pilot bearing
77 Bearing, needle, pilot
78 Sleeve, range shift
79 Sleeve, 4-wheel shift
80 Shaft, output, transfer rear
81 Gear, output, transfer rear
82 Bearing, needle
83 Washer, thrust
84 Bearing, ball, front, output shaft
85 Distance piece
86 Spacer, speedometer gear
87 Bearing, ball, rear, output shaft
88 Nut, output shaft
89 Gear, speedometer drive
90 Key, gear
91 Gear, counter
92 Bearing, angular ball, counter
93 Gear, counter reverse
94 Washer, plain
95 Washer lock
96 Nut, counter reverse gear
97 Shaft, counter, transfer
98 Plate, lock, countershaft
99 Bolt
100 Gear, reverse idle
101 Washer, thrust, front
102 Washer, thrust, rear
103 Ball, thrust washer
104 Gear, counter, transfer
105 Washer, thrust
106 Bearing, needle
107 O-ring, countershaft transfer

Disassembly and reassembly

2 The procedure is identical to that described in Chapter 6 but the clutch shift fork fulcrum stud is located in the transmission casing and not in the front bearing retainer. Refer to Section 9 of this Supplement for the release bearing and shift fork removal procedure.

11 Automatic transmission

Fluid level checking

The fluid level indicator on Series 8 through 11 models may not incorporate two dimples. If not, the fluid level at an ambient temperature of 21°C (70°F) should be between 1/8 and 3/8 in (3.0 and 10.0 mm) below the Add mark.

12 Transmission/transfer case (4 x 4 models)

Refer to Figures 13.25 thru 13.37

Removal and installation

1 Disconnect the negative battery cable.
2 Remove the drain plug from the transmission and drain the oil.
3 Slide the rubber boots up on both the transmission and the transfer case shift levers. Then remove the shift lever attaching bolts.
4 Remove the transfer gearshift lever return spring and then lift out the levers.
5 Remove the starter motor attachment bolts and connecting cable. Withdraw the starter motor.
6 From inside the cab, lift the gearshift lever boot up the lever and with a suitable wrench, unscrew and remove the gearshift lever attachment bolts. Lift the lever clear.
7 Unless the vehicle is located over a work pit, jack it up and block it up sufficiently to enable work to be carried out underneath the vehicle.
8 Unscrew and remove the exhaust pipe-to-transmission hanger.
9 Detach the speedometer cable from the transmission unit.
10 Refer to Chapter 7 and remove the driveshaft.
11 Drain the oil from the transmission and plug the rear extension to prevent spillage of oil during removal.
12 After marking their positions, disconnect both ends of the front driveshaft and remove it from the vehicle.
13 Remove the clutch return spring from the clutch fork. Then disconnect the clutch cable from the fork's hooked portion and pull it forward through the stiffener bracket.
14 Disconnect the flywheel stone shield and undo the three (Series 9 and 10) or 2 (Series 11) bracket-to-transmission rear mounting bolts.
15 Place jacks under the engine and transmission and raise the units sufficiently to undo and remove the crossmember-to-frame bracket bolts.
16 Remove the transfer case rear mounting nuts.
17 Lower the engine and transmission assembly and support the rear of the engine with blocks or a jack.
18 Remove the nuts and bolts that retain the side case and lift the case off. Be careful not to lose the shift rod detent spring and ball when removing them.
19 Disconnect the electrical connectors from the CRS switch, reverse lamp switch, Neutral switch and 4-wheel drive indicator switch, as required.
20 Remove the four bolts that attach the shifter cover to the transfer case and remove the cover.
21 Support the transmission and remove the engine bellhousing bolts. Withdraw the transmission by turning the side case fitting face down, then pull the case straight back until it disengages from the clutch. Tip the front of the transmission down and then slide it out from under the truck.
22 Installation is the reverse of the removal procedure, with the following notes:
    a) When reinstalling the transmission, position it with the side case fitting face down. Then slide it forward, carefully guiding the clutch gear into the pilot bearing.
    b) When installing the side case to the transfer case, it should be in the 4-H position and the grooves in the shift arms and shift sleeves should be aligned.
    c) Be sure the marks made on the driveshaft flanges match up when installed.
    d) After reconnecting the clutch cable and return spring, be sure the clutch cable is properly adjusted, as described in Section 9.
    e) When installing the shift levers, the transmission should be in the Neutral position and the transfer case in the 4L or 2H position.
    f) Check the clutch pedal height and adjust it, if necessary, as described in Section 9.

Transmission — disassembly

23 Drain any remaining oil from the casing.
24 Detach the clutch fork, rubber boot and throwout bearing.
25 Undo and remove the bearing retainer bolts and withdraw the retainer, gasket and spring washer.
26 Unscrew the speedometer gear bush retaining bolt and withdraw the speedometer-driven gear assembly.
27 Undo the retaining bolts and remove the gear shift cover and gasket.
28 Remove the reverse light switch and CRS switches.
29 Remove the eight bolts that hold the transmission case to the center support and transfer case and remove tthe transmission case and gasket.
30 Remove the four bolts that retain the transfer countershaft lock plate and lift off the lock plate and distance piece.
31 Withdraw the center support unit from the transfer case.
32 Support the center support unit and drive the spring pins from the reverse, 1st/2nd and 3rd/4th shifter forks. Support the shift rods during removal to prevent the rods from bending and to ease removal of the pins (Figure 13.28).

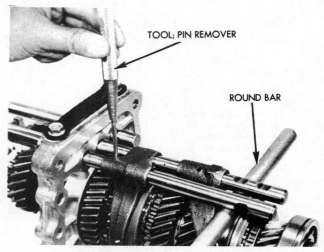

Fig. 13.28 When driving out the pins, the shift rods should be adequately supported to prevent damage (Sec 12)

33 Undo the detent spring retaining plate bolts and remove the plate and gasket. Extract the detent springs and balls.
34 Now remove the respective shifter rods, taking care not to lose the detent interlock plugs in the rods in the center support. The reverse shifter rod is fitted with a stop pin and must therefore be removed forwards.
35 To prevent the mainshaft rotating, move the synchronizers to the rear. If they are reluctant to move, tap them with a hammer handle.
36 Flatten the mainshaft nut lock washer tab, and undo the locknut. Remove the nut and washer.
37 From the rear of the mainshaft on Series 11 and 12 models, remove the transfer clutch hub, transfer input gear, needle bearings, collars, reverse clutch hub, reverse sleeve, reverse gear and thrust washer.
38 On Series 9 and 10 models, remove the transfer clutch hub, transfer input gear, needle bearing, collar, reverse gear, shim and thrust washer from the rear of the mainshaft.
39 Undo the locknut from the countershaft and remove the washer, reverse gear and collar.
40 Disengage the snap-ring from the center support using a pair of snap-ring pliers. Tap the front of the center support while expanding the snap-ring.

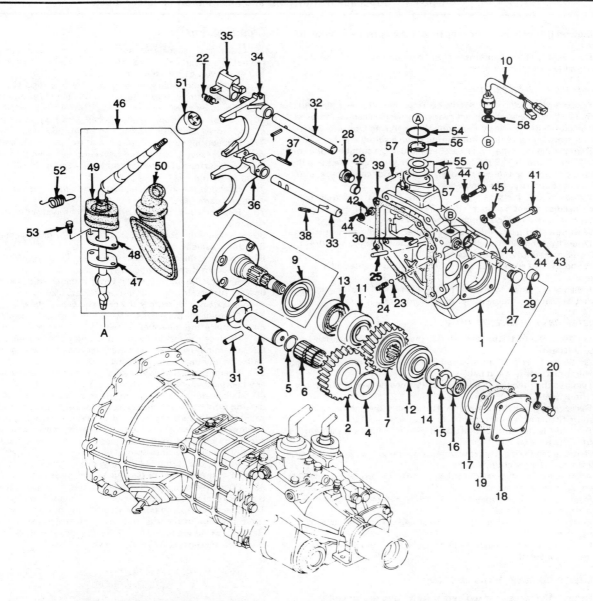

Fig. 13.29 Exploded view of the 4 x 4 transfer case gear assembly (Sec 12)

1 Case, side, transfer
2 Gear, idler
3 Gear, idler
4 Washer, thrust
5 O-ring, idler shaft
6 Bearing, needle
7 Gear, transfer output front
8 Shaft, output, front
9 Cover, dust
10 Switch, 4-wheel drive, indicator
11 Bearing, ball, front, output shaft
12 Bearing, ball, rear, output shaft
13 Oil seal, front output shaft
14 Washer, plain
15 Washer, lock
16 Nut, front output shaft
17 Distance piece
18 Cover, front output shaft
19 Gasket, cover
20 Bolt
21 Washer, lock
22 Spring, gear lock release
23 Ball, detent
24 Spring, detent
25 Pin, interlock
26 Plug, shift rod
27 Plug, screw
28 Plug, screw
29 Plug, shift rod
30 Pin, dowel
31 Pin, dowel
32 Rod, shift, 4-wheel
33 Rod, shift range
34 Arm, shift, 4-wheel
35 Block, shift, 4-wheel
36 Arm, shift, range
37 Pin, spring, select stop
38 Pin, spring, shift arm
39 Gasket, side case & transfer case
40 Bolt
41 Bolt
42 Bolt
43 Bolt
44 Washer, lock
45 Nut
46 Lever, gear shift, transfer
47 Cover, ball seat
48 Retainer, gear shift lever
49 Cover, dust
50 Grommet, gearshift lever
51 Knob, gearshift lever
52 Spring, return, gearshift lever
53 Bolt
54 O-ring, gearshift lever
55 Spring, ball seat
56 Seat, ball
57 Pin, straight
58 Gasket

41 Expand the mainshaft retaining snap-ring and remove the center support.
42 If necessary, the countergear front bearing can be removed by using suitable extracting tools.

Mainshaft — disassembly

43 Separate the clutch gear, needle bearings and blocker ring from the mainshaft. If necessary, the clutch gear front bearing can be removed from the clutch gear shaft by removing the snap-ring from the shaft retaining groove, and remove the bearing using a suitable puller.
44 Remove the mainshaft rear ball bearing using a suitable puller or press. If special tool J-22912 is available, this should be used together with an arbor press to remove the bearing. This tool can also be used to remove the countershaft bearing.
45 Withdraw the 1st speed gear assembly complete with thrust washer, needler rooler bearings, collar and blocker ring.
46 Withdraw the 1st/2nd synchronizer assembly.
47 Withdraw 2nd gear with blocker ring and needle roller bearing.
48 Disengage and remove the snap-ring, then withdraw 3rd/4th synchronizer assembly and blocker ring.
49 Withdraw 3rd speed gear from the shaft with needle roller bearings.
50 If necessary, the countershaft reverse gear and bearing can be removed by using a suitable puller or an arbor press.

Transfer case — disassembly

51 On Series 9 and 10 models, remove the reverse idler gear and reverse shift arm.
52 On Series 11 and 12 models, remove the reverse idler gear along with the thrust washers and ball.
53 Remove the range shift sleeve and pilot needle bearing from the rear output shaft.
54 Remove the transfer countershaft assembly by lightly tapping it from its fitting hole in the case.
55 Expand the rear output shaft front bearing snap-ring and remove the output shaft assembly.
56 Remove the O-ring, thrust washers, countergear and needle bearings from the transfer countershaft.
57 After removing the rear output shaft nut, the rear bearing can be pressed from the output shaft.
58 Remove the spacer, speedometer drive gear and key distance piece from the output shaft.
59 Press the front bearing, thrust washer and rear output gear from the output shaft.
60 Remove the output gear needle roller bearing and 4WD shift sleeve from the output shaft.

Transfer side case — disassembly

61 Remove the range shift rod detent spring and detent ball.
62 Remove the 4WD indicator light switch.
63 Remove the spring pin from the 4-wheel shift arm.
64 With the range shift rod held in the high range position, drive the 4-wheel shift rod from the rear side, along with the plug. Then remove the shift rod, shift arm and shift block. Also from the case remove the detent ball, spring and interlock pin. If the detent ball did not snap out of position when the shift rod was removed, remove it now, along with the spring and interlock pin.
65 Using a suitable punch, remove the spring pin from the range shift arm.
66 From the front, extract the range shift rod with the plug. Then remove the shift rod and shift arm.
67 Remove the dowel pin from the idler shaft.
68 The idler shaft can now be removed by screwing an 8x1.25 mm bolt into the threaded hole at the end of the shaft. Then remove the thrust washers, idler gear and needle bearings.
69 Remove the front output shaft cover and distance piece or shim.
70 Using a suitable tool, remove the front output shaft nut, spring washer and plain washer.
71 Remove the output gear from the case.
72 Using a suitable tool, drive the output shaft front and rear bearings and oil seal from the case.
73 Remove the gear lock release spring from the 4-wheel shift arm. Disconnect the shift arm and shift block.
74 Drive the spring pin from the range shift arm.

Cleaning and inspection

75 All parts, except seals and ball bearing assemblies, should be soaked in a suitable cleaning solvent, brushed or scraped as necessary, and dried with compressed air.
76 Ball bearing assemblies may be carefully dipped into cleaning solvent and spun with the fingers. While being prevented from turning, they should be dried with compressed air. After inspection (paragraphs 86 thru 88), they should be lubricated with SAE 90EP transmission oil and stored carefully in clean conditions until ready for use.
77 Inspect the transmission case and extension housing for cracks, worn or damaged bearings and bores and damaged threads and machined surfaces. Small nicks and burrs can be locally dressed out using a fine file.
78 Examine the shift levers, forks, shift rods and associated parts for wear and damage.
79 Replace roller bearings which are chipped, corroded or rough running.
80 Examine the countershaft (layshaft) for damage and wear; replace it if this is evident.
81 Examine the reverse idler and sliding gears for damage and wear. Replace them if this is evident.
82 Replace the input shaft and gear if damaged or worn. If the roller bearing surface in the counterbore is damaged or worn, or the cone surface is damaged, replace the gear and gear rollers.
83 Examine all the gears for wear and damage, replacing as necessary.
84 Replace the mainshaft if bent, or if the splines are damaged.
85 Replace the seal in the transmission front bearing cover.
86 Examine the inner and outer raceways for pitting and corrosion, replacing any bearings where this is found.
87 Examine the ball cage and races for signs of cracking, replacing any bearings where this is found.
88 Lubricate the raceways with a small quantity of SAE 90EP transmission oil then rotate the outer race slowly until the balls are lubricated. Spin the bearing by hand in various attitudes, checking for roughness. If any is found after the bearings have been cleaned they should be replaced. If they are satisfactory they should be stored carefully while awaiting assembly into the transmission.
89 Inspect the clutch hub, sliding sleeve, blocker rings and inserts for signs of excessive wear or damage — replace as required.
90 Check the hub taper and the gear teeth. Inspect fitting grooves of the blocker rings for excessive wear or damage and replace as required. To inspect the tapered portions, locate the blocker ring against the conical section of the gear and, using a feeler gauge, measure the clearance as shown in Figure 13.30. Replace the blocker ring if the clearance exceeds 0.032 in (0.8 mm).

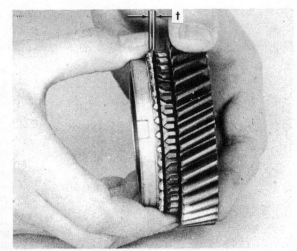

Fig. 13.30 Checking the blocker ring clearance (Sec 12)

91 Measure the clearance of the clutch hub splines and mainshaft splines and replace if the clearance exceeds 0.008 in (0.2 mm).
92 Place the mainshaft on two 'V' blocks and check for runout using a dial test indicator. If the runout exceeds 0.002 in (0.5 mm), replace the mainshaft.

93 Measure the outside diameter of the reverse idle gear shaft and the inside diameter of the idle gear bushing. Then compare and note the difference between these measurements. The outside diameter of the idle shaft should be 1.379 in (35 mm) for Series 9 and 10 models, and 1.181 in (30 mm) for Series 11 and 12 models.
94 Check the condition of the shift mechanism, checking for wear, bending or other damage.
95 Check the shift arm and shift block grooves for wear or damage. Also check the thickness at the end of the shift arm. The arm should be replaced if the thickness is less than 0.256 in (6.5 mm) on the 3rd/4th arm, 0.276 in (7.0 mm) on 1st/2nd arms, 0.276 in (7.0 mm) on reverse (Series 9 and 10 models), 0.256 in (6.5 mm) on reverse (Series 11 and 12 models), 0.236 in (6.0 mm) on transfer range shift arm and transfer 4-wheel shift arm.
96 Check the shift rods, detent balls and springs for wear or damage. Measure the free length of the detent springs and replace them if they are less than 1.083 in (27.5 mm) for 1st/2nd and 3rd/4th, 1.051 in (26.7 mm) for reverse, or 1.615 in (41 mm) for transfer range shift and transfer 4-wheel shift.

Mainshaft — reassembly

Note: During reassembly of the transmission/transfer case assembly, all oil seals, O-rings and bushings should be replaced with new ones.

97 Smear the needle roller bearing with grease and fit it to the mainshaft together with the third speed gear. The gear is fitted with its tapered side to the front.
98 Fit a blocker ring with clutching teeth up over the third speed gear synchronizing surface.
99 To reassemble the 3rd/4th synchronizer unit, locate the synchronizer hub face with its heavy boss to the sleeve face. Locate the keys into the groove and insert the springs into the hole in the side face of the synchronizer hub. Check that the hub and sleeve slide smoothly.
100 Fit the 3rd/4th synchronizer into position on the mainshaft with the small chamfer on the sleeve face to the rear of the shaft (Figure 13.31). Retain with the snap-ring.

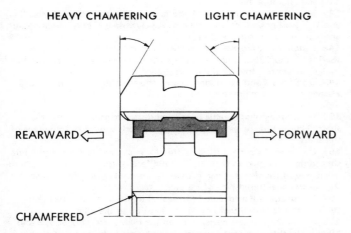

Fig. 13.31 The synchronizer units should be assembled as shown above (Sec 12)

101 Invert the mainshaft and fit the 2nd speed gear and needle bearing onto the shaft — grease the bearing before assembly and ensure that the 2nd gear taper face is to the rear of the shaft.
102 Fit the blocker ring with its clutching teeth down over the 2nd gear synchronizer surface.
103 To reassemble the 1st/2nd synchronizer, position the hub with the chamfer on the inner edge (Series 9 and 10) or with the oil grooves (Series 11 and 12) facing to the heavy chamfer on the outer edge of the sleeve. Locate the keys in their grooves and the synchronizer spring in the hole in the synchronizer hub side face. Check that the hub and sleeve slide smoothly.
104 Fit the 1st/2nd synchronizer unit into position with its sleeve chamfer facing forwards.

105 Fit the blocker ring with its clutching teeth to the rear.
106 Smear the 1st speed gear roller bearing with grease and fit it together with its collar and gear onto the mainshaft. The gear is fitted with the tapered side facing forwards.
107 Fit the 1st speed gear thrust washer with its grooved side against the gear.
108 Press the rear bearing into position on the shaft with the snap-ring groove offset to the front of the mainshaft.
109 Fit the bearing to the clutch gear shaft if it has been removed. The snap-ring groove is positioned offset to the front of the shaft. Retain the bearing with the snap-ring.
110 Smear the clutch gear pilot needle roller bearing with grease and fit it together with the blocker ring and clutch gear to the mainshaft.

Transfer case — reassembly

111 Install a new oil seal into the transfer case, using a suitably sized socket for a drift.
112 Install the output shaft front bearing snap-ring into the transfer case.
113 Install the 4-wheel shift sleeve onto the rear output shaft, so its heavily chamfered side is toward the output gear. Then install the needle bearing, output gear and thrust washer onto the shaft. The side of the thrust washer with the oil groove should face the output gear.
114 Using an appropriate press, install the output shaft front bearing onto the output shaft, with its snap-ring groove facing to the rear.
115 Install the shim, with its oil grooved side facing the front bearing. Then install the speedometer drive gear key, drive gear and spacer.
116 Again using a press, install the output shaft rear bearing onto the shaft, so its sealed face is turned toward the rear.
117 Tighten the output shaft nut to the proper torque. Then, using a punch, stake the nut so it engages with the groove in the shaft.
118 Apply grease to the countershaft needle bearings and to both faces of the thrust washers. Then install the thrust washer on the countershaft so that its oil grooved side faces the gear. Then install the needle bearings, counter gear, other thrust washer and O-ring in their proper order.
119 Expand the front bearing snap-ring and insert the output shaft assembly into the case until the snap-ring fits into the bearing ring groove.
120 Install the countershaft assembly into the case so that its cutaway side, at the front end, is positioned as shown in Figure 13.32. The finger on the thrust washer fitted to the rear of the shaft should be aligned with the groove in the case.
121 After applying grease to the output shaft pilot bearing, install it and the range shift sleeve onto the output shaft. The heavily chamfered end of the sleeve should face away from the gear assembly.
122 On Series 9 and 10 models, install the reverse idler gear and reverse shift arm, with the gear shift arm fitting groove facing the rear.
123 On Series 11 and 12 models install the ball, thrust washer, reverse idler gear and thrust washer. The oil grooved side of the washer should face the reverse idler gear.

Fig. 13.32 When installing the countershaft assembly into the transfer case, the cutaway side should be positioned as shown (Sec 12)

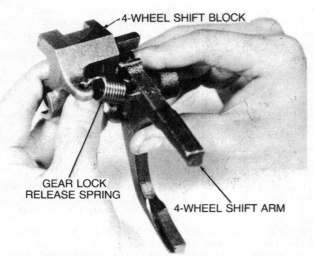

Fig. 13.33 Assembling the 4WD shift arm and block (Sec 12)

Fig. 13.34 The interlock pin and detent ball and spring should be installed in the holes shown above (Sec 12)

Transfer side case — reassembly

124 Assemble the 4-wheel shift arm with the shift block. Then install the gear lock release spring as shown in Figure 13.33.
125 Using a suitable-sized socket, install the front output shaft rear bearing into the case until it contacts the snap-ring.
126 Install the output gear so that the end with the splines not cut to the edge is facing forwards.
127 Using a suitably-sized socket, install the output shaft front bearing just until it contacts the output gear. Do not drive the bearing beyond that point as it may dislodge the rear bearing from its position. After installing the front bearing, check that the rear bearing snap-ring is still in contact with the case.
128 Using a suitably-sized socket, install a new output shaft oil seal.
129 Install the output shaft, along with its washer, spring washer and nut. Keep the shaft from turning, and torque the nut to the proper specifications.
130 Install a shim on the case. Then install the output shaft cover, torquing the cover bolts.
131 Apply a coat of grease to both sides of the idle gear thrust washer. Then install the thrust washer by aligning the stopper finger with the case groove. The oil grooved side of the thrust washer should face the gear.
132 Apply grease to the idler gear needle bearing and install the bearing to the gear. Insert the bearing into the case so that the heavier bossed end is turned toward the front.
133 Apply a thin coat of grease to the idler shaft O-ring. Then insert the idler shaft into the case until the groove lines up with the dowel fitting hole.
134 Install the dowel into the case so that its projection is 0.355 to 0.433 in (9 to 11 mm).
135 Install the range shift arm and shaft shift rod and secure them in position with the spring pin.
136 Install the interlock pin, the 4-wheel shift rod detent spring and the detent ball into the case.
137 With the range shift rod held in the high range position, install the 4-wheel shift arm and shift block, with the shift rod, into the case. Secure them with the spring pin.
138 Install the shift rod plugs and screw plugs, and tighten the screw plugs to the proper torque.
139 Install the range shift rod detent ball and spring into position.
140 On Series 11 and 12 models, install the 4-wheel drive indicator light switch.

Transmission — reassembly

141 Press the countershaft reverse gear and the countergear ball bearing into position with the snap-ring groove offset to the rear.
142 Install the snap-rings into their grooves in the mainshaft and countergear shaft locations in the center support.
143 Using suitable installing tools, install the countergear front bearing onto the shaft if it has been removed.
144 The mainshaft assembly and countergear shaft unit are meshed together in parallel and fitted to the center support. If special tool No. J-26545-5 is available, this should be used to retain the two gear shaft assemblies in parallel mesh when fitting the center support.
145 Expand the countergear bearing and mainshaft bearing snap-rings in the center support and get an assistant to press the center support over the bearings so that the snap-rings are fitted in their respective grooves.
146 Push the synchronizers to the rear to prevent the mainshaft rotating. If the synchronizers are reluctant to move, tap them with a wooden hammer handle.
147 Now locate on the rear of the countergear the collar (Series 9 and 10), washer and self-locking nut. Always use a new self-locking nut, and tighten to a torque of 80 ft-lb.
148 Install the reverse gear shim (Series 9 and 10) or thrust washer (Series 11 and 12) on the mainshaft so that the side with the groove or under-cut teeth is toward the reverse gear.
149 On Series 9 and 10 models, install the transfer input gear thrust washer onto the mainshaft. The grooved side should be toward the input gear.
150 Install the collar, needle bearing and reverse gear onto the mainshaft. The side of the gear with the clutch teeth should face the rear of the shaft.
151 On Series 11 and 12 models, the sleeve and the reverse clutch hub should be assembled so that the stepped face of the hub is on the chamfered side of the sleeve. Then install this assembly onto the mainshaft so that the chamfered face of the sleeve faces forward.
152 Install the collar, needle bearing and input gear onto the mainshaft, with the clutch teeth of the gear facing to the rear.
153 Install the transfer clutch hub on the mainshaft. On Series 9 and 10 models, the oil grooved side of the hub should face the input gear. On Series 11 and 12 models, the side with the outer edge chamfer should face the rear of the shaft.
154 Install the mainshaft locknut and torque it to specifications.
155 Install the spacer onto the mainshaft. Then grease the detent plugs and install them into their location holes from the middle hole of the center support.
156 Position the 1st/2nd and 3rd/4th shifter forks into their location grooves in the synchronizers (Figure 13.35).
157 Insert the 3rd/4th shifter rod through the rear of the center support, the middle hole, and into the respective shifter forks. The 3rd/4th shifter rod has two detent grooves on the side of the rod.
158 Insert the 1st/2nd shifter rod through the rear of the center support and into the 1st/2nd shifter fork.
159 Align the shifter forks and shaft spring pin holes.
160 On Series 9 and 10 models, install the reverse shifter rod from the rear of the center support.

Chapter 13/Supplement: Revisions and information on later models

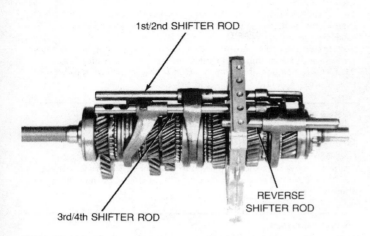

Fig. 13.35 Correct installed positions of the shift rods and forks (Sec 12)

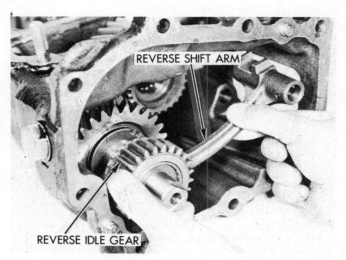

Fig. 13.36 On series 9 and 10 models the reverse shift arm should be installed as shown (Sec 12)

161 On Series 11 and 12 models, install the reverse shifter fork into its groove in the reverse sliding sleeve. Then install the reverse shifter rod from the front of the center support and through the reverse shifter fork. Be sure the hole in the rod is aligned with the hole in the shifter fork.
162 Install the two (Series 9 and 10) or three (Series 11 and 12) spring pins into the 1st/2nd shifter forks by tapping them in. Be sure the forks are well supported.
163 Install the detent balls, detent spring, gasket and retainer onto the center support and torque the bolts to specs. The reverse detent spring is the shortest one.
164 Support the transfer case in an upright position on wooden blocks, and apply a coat of grease to the end of the reverse shift rod, mainshaft and transfer countershaft.
165 Install the center support assembly and gasket into the transfer case, using the following sequence.
   a) Install the reverse shift rod to the reverse shift arm on Series 9 and 10 models (Figure 13.36), or the reverse shift rod hole fitting on Series 11 and 12 models.
   b) Install the mainshaft to the output shaft fitting.
   c) Install the transfer countershaft to the countershaft hole fitting.
   d) On Series 11 and 12 models, engage the countershaft reverse gear with the reverse idler gear.
   e) Engage the input gear with the transfer counter gear.
   f) Engage the range shift sleeve with the clutch hub.
   g) Fit the dowel pins into the center support pin holes.
166 On Series 9 and 10 models, support the reverse shifter fork and drive the spring pin into place.
167 Measure the distance between the front end of the center support and the end of the transfer countershaft, as shown in Figure 13.37. After finding this measurement, refer to the chart below to determine the correct shim to be used. The correct countergear thrust clearance is 0.004 to 0.014 in (0.1 to 0.35 mm).

Measurement	Shim thickness
0.106 to 0.114 in. (2.7 to 2.9 mm)	0.118 in. (3.0 mm)
0.114 to 0.122 in. (2.9 to 3.1 mm)	0.126 in. (3.2 mm)
0.122 to 0.130 in. (3.1 to 3.3 mm)	0.134 in. (3.4 mm)
0.130 to 0.138 in. (3.3 to 3.5 mm)	0.142 in. (3.6 mm)
0.138 to 0.142 in. (3.5 to 3.6 mm)	0.150 in. (3.8 mm)

168 Remove the bolts from the transfer case and center support once more, and install the lock plate and selected shim. Then reinstall the bolts, torquing them to the proper specifications.
169 Install the transmission case to the center support/transfer case assembly, making sure the dowel pins are aligned with the dowel pin holes. Torque the bolts.
170 Reinstall the reverse lamp switch and CRS switch.
171 Apply some oil to the O-ring on the speedometer driven gear unit and fit it into position in the rear extension. Tighten the retaining bolt.
172 Fit the new front bearing retainer seal into position.

Fig. 13.37 To determine proper shimming, measure the distance between the center support and the transfer countershaft end (Sec 12)

173 Fit the snap-ring to the outer clutch gear bearing race. Grease the retainer spring washer and position it in the bearing retainer. The dished face of the washer must face the bearing outer race.
174 Fit the bearing retainer into position in the front of the gearcase and retain with bolts. The shorter bolts are fitted to the countergear front bearing side of the retainer. Tighten the bolts to 14 ft-lb (1.9 kg f m).
175 If it has been removed, reinstall the ball stud and tighten to a torque of 30 ft-lb (4.1 kg f m).
176 Reinstall the clutch fork, dust boot and throwout bearing, and install the retaining spring. Grease the shift fork support with molybdenum disulphide grease, and ensure that the clutch fork hook is correctly located to the support.
177 If not already done, reinstall the drain plug.

13 Driveshaft

Refer to Figure 13.38

General description

1 Long wheelbase models are fitted with a two section driveshaft (propeller shaft) incorporating a center support bearing.

Driveshaft (long wheelbase models) — removal and installation

2   Remove the bolts that connect the flange yokes of the first and second driveshafts.
3   Mark the rear driveshaft flanges in relation to the rear axle pinion flange and the front driveshaft flange.
4   Unscrew and remove the rear driveshaft flange bolts and withdraw the rear driveshaft.
5   Unscrew and remove the bolts securing the center support bearing to the chassis crossmember, then carefully withdraw the front driveshaft assembly from the transmission housing taking care not to damage the rear oil seal.
6   Plug the transmission rear cover to prevent loss of lubricant.
7   Installation is the reverse of removal but observe the following points:

   a) Properly align all marks made during removal.
   b) Take care not to damage the oil seal in the transmission rear cover.
   c) Tighten the bolts to the recommended torque wrench settings.
   d) Top-up the transmission lubricant if necessary.

Center bearing (long wheelbase models) — replacement

8   Remove the first driveshaft, as previously described.
9   With the shaft held securely in a vise, relieve the staking on the locknut and remove the nut and washer.
10 Remove the pinion flange from the first driveshaft, using a suitable gear puller.
11 Remove the two bearing cushion retainer bolts, then remove the retainer.
12 Remove the rubber cushion support ring from the cushion.
13 Remove the rubber cushion from the bearing.
14 Using a suitable gear puller, remove the spacer ring bearing assembly from the shaft.
15 To begin installation, install the rubber support cushion and the support ring onto the shaft and push them onto the shaft so that they clear the bearing fitting face.
16 Install the cushion support ring, aligning the pin in the ring with the cushion's hole.
17 After repacking the bearing with grease, install it onto the shaft.
18 The spacer ring and pinion flange can now be installed by driving them on with a suitable installation tool.
19 Install the washer and locknut. Tighten the locknut to the specified torque and then stake the outer face of the flange so that it engages the slot in the shaft.
20 Install the rubber cushion. Be sure to align the stop groove in the inner face of the cushion with the projected bearing projection.
21 The remainder of the installation is the reverse of the removal procedure.

14 Rear axle

Axleshaft — installation

When calculating the axleshaft shim thickness on Series 8 through 12 models, follow the procedure given in Chapter 8, Section 2, but add 0.012 in (0.3 mm) to the clearance (without shims) to determine the shim requirement.

15 Braking system

Refer to Figures 13.39 thru 13.52

Parking brake adjustment (Series 11 and 12)

1   Raise the rear of the truck and support it on jackstands.
2   Release the parking brake all the way.
3   Locate the parking brake cable adjuster under the vehicle (Figure 13.39) and loosen the locknut.
4   Turn the adjuster in a clockwise direction until the actuating lever stopper at each rear wheel is lifted off the flange plate. Note Figure 13.40.
5   Next turn the adjuster in a counterclockwise direction until the actuating lever stopper at each wheel just makes contact with the flange plate.
6   Tighten the adjuster locknut.
7   Apply and release the parking brake three or four times.
8   Loosen the locknut once again and turn the adjuster clockwise once more until the actuating lever stoppers are lifted from their flanges.
9   Now turn the adjuster counterclockwise once more until all looseness in the actuating lever stoppers is removed. This can be checked by

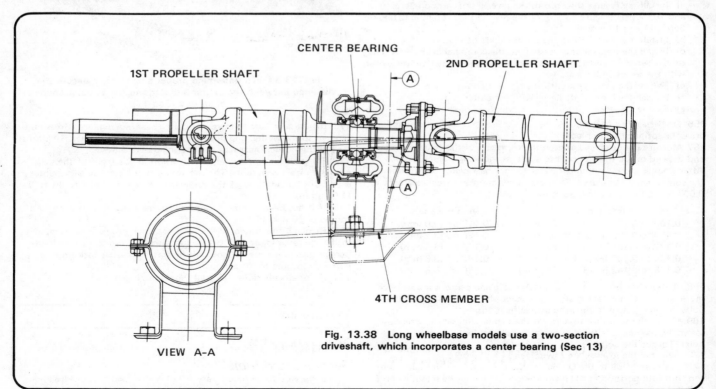

Fig. 13.38 Long wheelbase models use a two-section driveshaft, which incorporates a center bearing (Sec 13)

moving the stopper on the right side wheel with your fingers.
10 Turn the adjuster an additional 1-1/2 to 2 turns and then tighten the locknut again.
11 Check that the parking brake lever can be pulled out 12 to 13 notches and that the parking brake is fully set in this position.
12 Fully release the parking brake and rotate the rear wheels to ensure there is no drag.
13 Lower the truck to the ground.

Brake pedal adjustment (Series 11 and 12)

14 The correct brake pedal height for Series 11 and 12 models is 6.1 to 6.5 in (154 to 164 mm).

Drum brake adjustment (Series 11 and 12)

15 Normally, adjustment of the brakes is not necessary as they are self-adjusting. However, after the brake linings have been replaced or the adjuster position altered, an initial adjustment can be made as follows.
16 Raise the rear of the truck and support it firmly on jackstands.
17 Remove the wheel and tires.
18 Remove the brake drum.
19 Disconnect the rear parking brake cable from the actuator lever.
20 There are two methods of making the adjustment.

   a) With a brake shoe gauge, measure the inside diameter of the brake drum (Figure 13.41). Next invert the gauge and fit it over the brake shoes and rotate the star wheel with the parking brake actuating lever until the gauge just slides over the linings. Rotate the gauge over the linings to assure the proper clearance.

   b) With the brake drum installed, move the parking brake actuating lever until the clicks, which indicate the rotation of the star wheel, are no longer heard.

21 Reconnect the rear parking brake cable.
22 Reinstall the drum and wheel and lower the vehicle to the ground.
23 The final adjustment is made with the self-adjusting feature, by operating the parking brake repeatedly while depressing the brake pedal until a sufficient pedal stroke is obtained.

Front disc brake (Series 11 and 12) — shoe replacement

24 Each front disc brake has a spring steel wear sensor riveted to the rear edge of each brake shoe. When the shoe lining has worn to within 0.067 in (1.7 mm) of the back side of the shoe base, the sensor will rub against the rotor, emitting a high-pitched squeal. This is an indication that the disc pads need immediate replacement.
25 Raise the front of the vehicle and support it on jackstands. All disc pads should be replaced at the same interval; however, it is a good idea to complete one assembly at a time, using the other side for reference if necessary.
26 Remove the wheels and tires.
27 Remove the caliper lock bolt (Figure 13.42).
28 Rotate the caliper up. Use heavy wire to suspend the caliper while work is done. Do not allow the caliper to hang by the hose.
29 Lift out the shoes and shims, noting their locations if the same ones are to be reinstalled.
30 Remove the shoe clips from the support.
31 The shoes should be replaced if the lining is worn to within

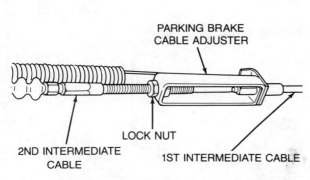

Fig. 13.39 The parking brake adjuster on Series 11 and 12 models is at the junction of the first and second intermediate cables (Sec 15)

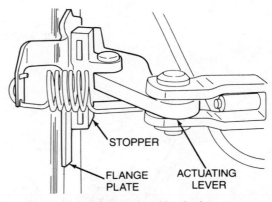

Fig. 13.40 Rear wheel parking brake components (Series 11 and 12) (Sec 15)

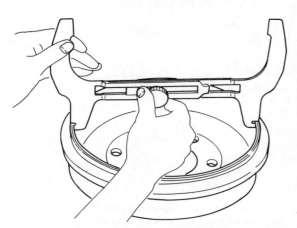

Fig. 13.41 Using a brake shoe gauge is the most accurate way of adjusting the drum brakes (Sec 15)

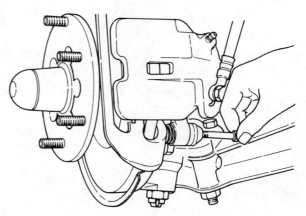

Fig. 13.42 Location of the caliper lock bolt (Series 11 and 12) (Sec 15)

0.067 in (1.7 mm) of the base.

32 Apply a suitable lubricant, such as Delco Brake Lube 5450032 to the shims.
33 Install new shoe clips into the support.
34 Install the shims and shoes into the support. Be sure the wear indicators on the shoes are positioned to the lower side of the support.
35 Lower the caliper and install the lock bolt.
36 Reinstall the wheels and tires and lower the vehicle to the ground.

Disc brake caliper (Series 11 and 12) — removal, overhaul and installation

37 Raise the front of the vehicle and support it on jackstands.
38 Remove the wheel and tire.
39 Remove the caliper lock bolt (Figure 13.42).
40 Disconnect the flexible brake hose from the caliper and plug both openings to prevent the entry of foreign material.
41 Disengage the guide pin dust boot from the guide pin. Then raise the caliper and remove it from the guide pin built into the support.
42 Remove the sleeve, sleeve dust boot and guide pin dust boot.
43 Using a small screwdriver, pry the dust seal ring from the caliper. Then pry out the dust seal.
44 Remove the piston from the caliper by applying compressed air into the flexible hose connection or by using special pliers designed for this purpose. Do not catch the piston by hand, but place a suitable block of wood into the caliper to check it on exit.
45 Detach and discard the piston square ring seal.
46 Clean all components in clean brake fluid and blow dry with compressed air. Check the cylinder bore and pistons for signs of wear, corrosion or surface defects such as scoring and replace if necessary.
47 The dust seal and piston seal must always be replaced when the caliper is dismantled.
48 To reassemble the caliper, lubricate the caliper bore and the piston square ring seal with Delco Silicone Lube 5459912 or an equivalent before inserting the piston seal into the bore.
49 Carefully insert the piston into the caliper unit using only finger pressure.
50 Apply Delco Silicone Lube 5459912 or an equivalent to the piston and then fit the dust seal to the piston and caliper. Insert the seal ring into the dust seal.
51 Apply Delco Silicone Lube 5459912 or an equivalent into the sleeve and guide pin dust boots. Then install the sleeve dust boot to the caliper and insert the sleeve into the dust boot.
52 Apply Delco Brake Lube 5450032 or an equivalent into the guide pin fitting hole and then install the guide pin dust boot to the caliper.
53 Install the caliper on the guide pin and then fit the dust boot over

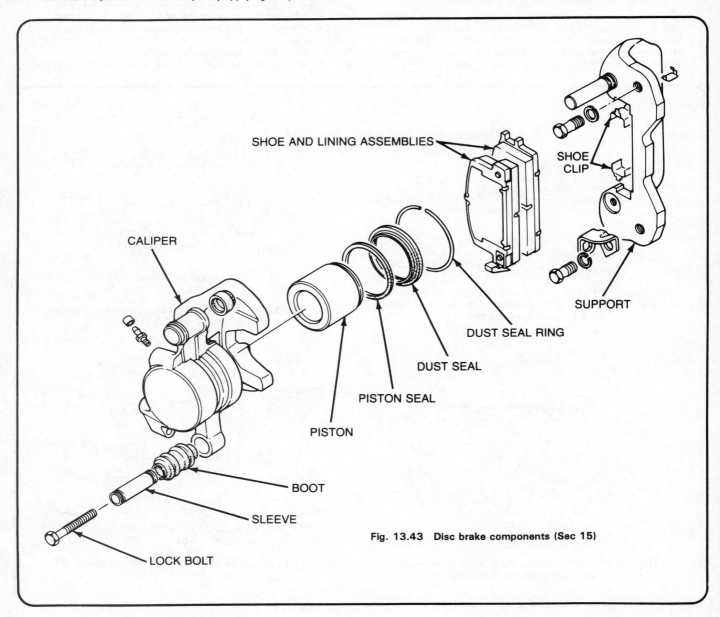

Fig. 13.43 Disc brake components (Sec 15)

the guide pin.
54 Install the caliper lock bolt and torque it to specifications.
55 Install the flexible brake hose to the caliper. Be sure the hose does not interfere with any suspension or steering components.
56 Install the wheel and tire.
57 Bleed the brake system as described in Chapter 9.
58 Lower the vehicle to the ground.

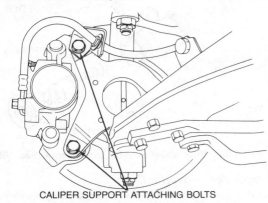

Fig. 13.44 Location of the disc caliper support mounting bolts (Sec 15)

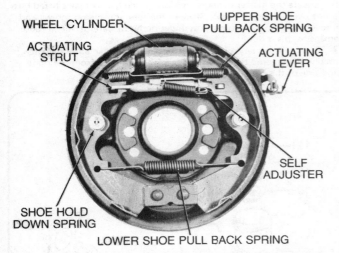

Fig. 13.45 Components of the rear drum brake (Series 11 and 12) (Sec 15)

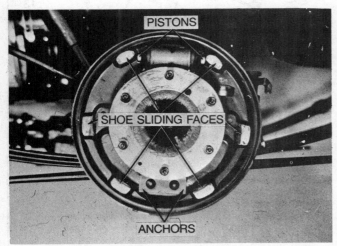

Fig. 13.46 Prior to installing the brake shoes, grease should be applied to the sliding surfaces shown above (Sec 15)

Front hub/rotor, shield and adapter (Series 11 and 12) — removal and installation

59 Raise the front of the vehicle and support it on jackstands.
60 Remove the wheel and tire.
61 Remove the bolts that attach the caliper support to the brake adapter (Figure 13.44) and suspend the assembly from the upper link or frame with wire. Do not allow it to hang by the brake hose.
62 Remove the front hub/rotor as described in Section 18 of this Chapter.
63 If the dust shield or adapter needs to be removed, disconnect the tie-rod from the adapter. Then remove the four attaching bolts and lift off the shield and adapter.
64 Installation is the reverse of the removal procedure with the following notes.

   a) When installing the hub/rotor assembly, be sure to follow the procedure for adjusting the front wheel bearings described in Section 18.
   b) Tighten all bolts to the specified torque settings.

Rear brake shoes (Series 11 and 12) — removal and installation

Note: *All shoes should be replaced at the same interval; however, it is a good idea to work on one side at a time, using the other (fully assembled) side for reference.*

65 Raise the rear of the vehicle and support it on jackstands.
66 Remove the wheel and tire. The linings can be initially inspected for wear at this point (see Step 73).
67 Disconnect the rear brake cable from the actuating lever.
68 Remove the screws that attach the drum to the hub and lift the drum off. If the drum is severely worn, it may be necessary to retract the brake shoes away from the inside of the drum. This is done by removing the stopper from the parking brake actuating lever and then bringing the actuating lever into contact with the flange plate. **Note**: *The brake pedal must not be depressed while the drum is removed.*
69 Using pliers, remove the brake shoe hold-down springs by depressing the retainer and rotating it 90° until the retainer slot lines up with the flanged end of the pin.
70 Using pliers, remove the lower pull-back spring.
71 Expand the brake shoes until they disengage from the actuator assembly. Then remove both shoes together with the upper pull-back spring. If the same shoes are to be reinstalled, note their positions so they will be installed the same way they were removed.
72 If it is necessary to remove the actuator assembly, disconnect the actuating spring, then remove the actuating strut and lever.
73 The shoes must be replaced if they are damaged, oil soaked or if the lining is worn to within 0.039 in (1.0 mm) of the base.
74 Prior to installing the shoes, apply a thin coat of a suitable lubricant such as Delco Brake Lube 5450032 to the 6 points on the flange plate over which the brake shoes slide (Figure 13.35). Also apply it to the slots in the wheel cylinder pistons where the brake shoes contact. Be careful you do not get any lubricant on the brake shoe surfaces.
75 Installation is the reverse of the removal procedure.
76 Prior to installing the brake drum, check the operation of the self-adjuster by moving the actuator lever.
77 Adjust the rear brakes and parking brake as described in this Section.

Rear wheel cylinder (Series 11 and 12) — removal, overhaul and installation

Removal
78 Raise the rear of the truck and support it on jackstands.
79 Remove the wheel and tire.
80 Mark the position of the brake drum and remove it from the hub. If necessary, refer to step 68.
81 Disconnect the brake line from the wheel cylinder and plug the end of it to prevent foreign material from entering the line.
82 Disconnect the rear parking brake cable from the actuator lever.
83 Move the actuator lever to move the self-adjuster and expand the brake shoes until they are disengaged from the wheel cylinder pistons.
84 Remove the screws that attach the wheel cylinder to the rear flange plate and lift it off.

Overhaul

85 Remove the dust boots from both ends of the cylinder (Figure 13.47).
86 Remove the piston assemblies and springs.
87 Remove the seal cups from the pistons.
88 Using clean brake fluid, clean the wheel cylinder components.

Caution: *Do not use gasoline, kerosene or any other type of cleaning solvent, as they will damage the rubber brake parts.*

89 Inspect the components carefully and if the cylinder bore and/or piston is scored or corroded, they will have to be replaced. The rubber cups and boots must always be replaced irrespective of their condition.
90 Assemble by first lubricating the cylinder bore with clean brake fluid. As each component is assembled it should be lubricated in a similar fashion.

Installation

91 Installation is the reverse of the removal procedure with the following notes.

   a) Prior to reassembly, the cylinder bore should be lubricated with clean brake fluid.
   b) The seal cups should be installed on the pistons so that the flared end is facing the inner end of the piston.
   c) Do not lubricate the pistons, cups, or boots prior to installation.
   d) After inserting the piston assembly into the cylinder, apply Delco Silicone Lube 5459912 or an equivalent to the pistons and inner face of the boots prior to attaching the boots to the cylinder.
   e) If there is trouble fitting the brake drum over the shoes, remove the actuator lever stopper and retract the shoes by moving the lever toward the flange plate.

92 Following installation, adjust the rear brakes and parking brake as described in this Section.
93 Bleed the brake system as described in Chapter 9.

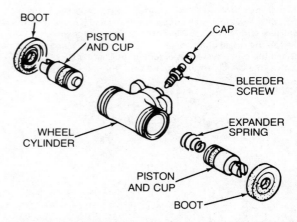

Fig. 13.47 Rear wheel cylinder components (Sec 15)

Master cylinder — overhaul

94 Dismantle the components of the master cylinder on a clean work surface and lay the parts out in the order of dismantling.
95 Remove the reservoir caps and filters. Drain any remaining fluid from the reservoirs.
96 Locate the master cylinder in a vise which has soft jaws fitted but do not overtighten the jaws. Slacken the reservoir clamp screws and remove the reservoirs from the cylinder body. Be careful not to cause damage as this is done.
97 Press the primary piston fully into the cylinder bore and then unscrew the stopper bolt and remove with the gasket from the right-hand side of the cylinder.

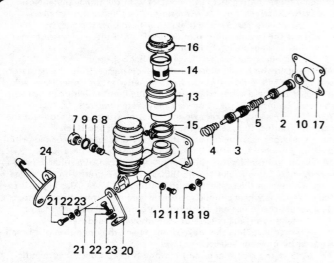

Fig. 13.48 Components of the master cylinder (Series 9 and 10) (Sec 15)

1	Master cylinder	13	Reservoir
2	Primary piston	14	Filter
3	Secondary piston	15	Clamp
4	Spring (Primary)	16	Clamp cap
5	Spring (Secondary)	17	Gasket
6	Check valve	18	Nut
7	Connector	19	Spring washer
8	Return spring	20	Bracket
9	Gasket	21	Bolt
10	Snap-ring	22	Spring washer
11	Stopper bolt	23	Washer
12	Gasket		

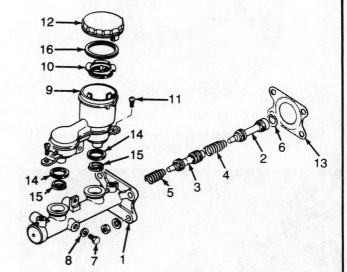

Fig. 13.49 Components of the master cylinder (Series 11 and 12) (Sec 15)

1	Master cylinder	9	Reservoir
2	Primary piston	10	Filter
3	Secondary piston	11	Clamp screw
4	Spring (Primary)	12	Clamp cap
5	Spring (Secondary)	13	Gasket
6	Snap-ring	14	Upper rubber seal
7	Stopper bolt	15	Lower rubber seal
8	Gasket	16	Cap seal

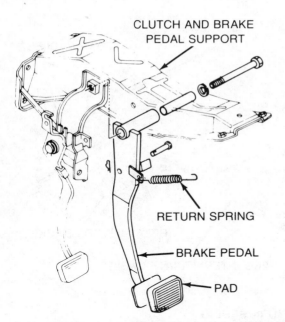

Fig. 13.50 Components of the brake pedal assembly (typical) (Sec 15)

98 With a pair of snap-ring pliers, extract the primary piston snap-ring from the cylinder and then withdraw the primary and secondary piston units from the cylinder. Make a special note of which way they are installed.
99 Remove the seals and again note which way they are located.
100 Clean all parts thoroughly in clean brake fluid. **Caution:** *Do not use gasoline, kerosene or any other type of solvent.* After drying the items with a lint free cloth, inspect the seals for signs of distortion, swelling, splitting, or hardening although it is normal practice and also recommended that new seals are always installed after dismantling.
101 Inspect the bore and piston for signs of deep scoring which, if evident, means a new cylinder should be purchased. Ensure that the cylinder bore parts are clean.
102 As the respective parts are installed into the cylinder bore ensure that they are thoroughly lubricated with clean brake fluid of the specified type.
103 Install the new seals to the pistons making sure that they are installed the correct way as noted during dismantling.
104 Take care not to scratch the piston cups when fitting the secondary and primary piston units. If damaged, they must be replaced.
105 The remainder of the reassembly procedure is the reverse of the disassembly.
106 Following reassembly, the master cylinder should be "bench bled" to remove all air from it.

  a) Plug up all of the outlet ports of the cylinder securely. Plastic plugs which are supplied with a replacement cylinder are well suited for this.
  b) Fill the reservoir with clean brake fluid.
  c) Using a smooth round rod or similar tool, depress the piston into the cylinder body. Then release it and watch for air bubbles in the reservoir fluid.
  d) Repeat this procedure until bubbles no longer appear in the reservoir.

Power cylinder — overhaul
107 Disassembly of the power cylinder used on later models requires that the special Spanner Wrench J-9504 be modified by drilling three 3/8 in diameter holes. The actual procedure is the same as that in Chapter 9.

Brake pedal (Series 8 thru 10) — removal and installation
108 Series 8 through 10 models are equipped with a brake pedal mounted on a separate pivot bolt and removal is as follows.

109 Detach the ground cable from the battery.
110 Disconnect the master cylinder operating rod from the brake pedal by removing the cotter pin, washer, and clevis pin.
111 Detach the return spring from the brake pedal.
112 Unscrew and remove the pivot bolt nut, spring washer and plain washer, then withdraw the pivot bolt and lower the pedal together with the bushing.
113 Installation is the reverse of removal, but lubricate the pedal bushing with Delco Brake Lube 5450032 or equivalent and tighten the pivot bolt to the specified torque wrench setting. Adjust the brake light switch on completion.

Brake pedal (Series 11 and 12) — removal and installation
114 Disconnect the negative battery cable.
115 Disconnect the brake pedal return spring from the pedal.
116 Remove the snap pin and clevis pin at the pedal.
117 Remove the pivot bolt and lock washer and lift out the pedal with the bushing.
118 Prior to installation, lubricate the pedal bushing with a suitable lube such as Delco Brake Lube 5450032.
119 Installation is the reverse of the removal procedure. Following installation, adjust the pedal height if necessary.

Parking brake handle (Series 11 and 12) — removal and installation
120 Disconnect the negative battery cable.
121 Disconnect the parking brake front cable at the relay lever, located on the left side of the engine compartment.
122 Disengage the dust boot at the lower end of the parking brake handle housing.
123 Remove the screws that attach the handle assembly to the instrument panel.
124 Disconnect the parking brake light switch harness connector.
125 Remove the parking brake handle assembly, complete with front cable.
126 Remove the parking brake light switch locknut and lift off the switch.
127 Manually release the lever ratchet and depress the handle until the point where the cable attaches to the handle is visible. Then disengage the cable from the handle.
128 Align the parking brake light switch actuating pin with the hole in the housing. Then using a flat-head punch, drive out the pin.
129 Separate the handle ratchet and spring.
130 Installation is the reverse of the removal procedure.

Parking brake rear cables (Series 11 and 12) — removal and installation
131 Remove the cotter pins, plain washers waved washers and pins that attach the rear cables to the actuating levers at each rear wheel.
132 Remove the same attaching components that secure the rear cables to the rear relay lever and lift out the left rear cable.
133 Raise the cable guides on the rear axle case to release them from the guide bracket, then disconnect the right rear cable.
134 Installation is the reverse of the removal procedure.
135 Following installation, adjust the parking brake as described in this Section.

Parking brake first intermediate cable (Series 11 and 12) — removal and installation
136 If necessary for clearance, raise the front of the truck and support it on jackstands.
137 Disconnect the front end of the front cable from the front relay lever.
138 Pry out the one (2-wheel drive models) or two (4-wheel drive models) cable guides from the frame crossmember located under the driver's compartment.
139 Remove the return spring from the adjuster at the rear end of the first intermediate cable.
140 Loosen the adjuster locknut. Then separate the first and second intermediate cables by turning the second cable out from the adjuster.
141 Installation is the reverse of the removal procedure.
142 Following installation, adjust the parking brake as described in this Section.

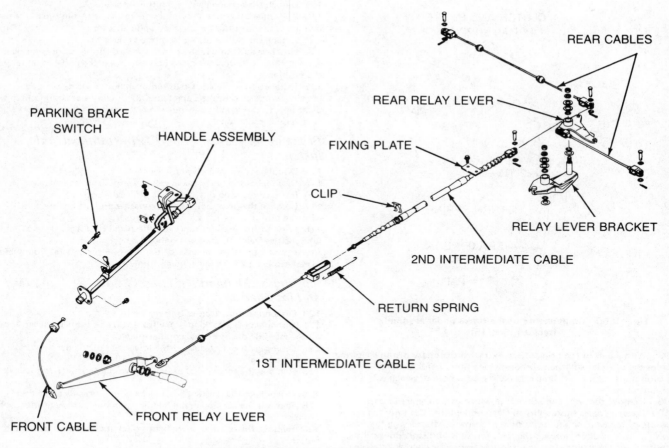

Fig. 13.51 Components of the parking brake system (Sec 15)

Parking brake second intermediate cable (Series 11 and 12 — removal and installation)

143 If necessary for clearance, raise the rear of the truck and support it on jackstands.
144 Remove the return spring from the adjuster located at the rear of the first intermediate cable.
145 Loosen the adjuster locknut. Then turn the second intermediate cable out of the adjuster until the two are separated.
146 Remove the clip attaching the front of the cable to the frame bracket and remove the cable from the bracket.
147 Remove the cotter pin, plain washer, waved washer and pin that attach the cable to the rear relay lever.
148 Remove the plate that attaches the cable to the rear axle case and lift out the cable.
149 Installation is the reverse of the removal procedure.
150 Following installation, adjust the parking brake as described in this Section.

Fail indicator (combination valve) assembly

151 Series 8 through 12 models are equipped with an hydraulic fluid pressure fail indicator (Figure 13.52) which illuminates a warning lamp on the instrument panel when the hydraulic fluid pressure fails in either the front or rear brake lines.

Fail indicator — removal and installation

152 Note that the fail indicator cannot be repaired and must be replaced if proven faulty. To prevent loss of hydraulic fluid, remove the reservoir filler cap and place a piece of polyethylene sheet over the top of the reservoir and refit the cap.
153 Note the location of the hydraulic lines, then disconnect them from the fail indicator assembly by unscrewing the unions; plug the line ends.
154 Pull the supply wire from the unit terminal.
155 Unscrew and remove the mounting bolts and withdraw the fail indicator assembly.
156 Installation is a reverse of removal, but it will be necessary to bleed the brake system as described in Chapter 9. Make sure that the plastic sheet is removed from the brake fluid reservoir filler cap before proceeding onto the road.

16 Electrical system

Refer to Figures 13.53 thru 13.58

Fuses

1 The fuse box for Series 8 through 12 vehicles was changed slightly in design and location of fuses. All fuses and their respective circuits are clearly marked on the box.

Headlamps (Series 8 thru 10) — removal and installation

2 Series 8, 9 and 10 models are equipped with single headlamp units. However, the removal, installation and adjustment procedures are identical to those given in Chapter 10.

Headlamps (Series 11 and 12) — removal and installation

3 The procedure for removing and installing the headlamps is the same as described in Chapter 10, except that on Series 11 and 12 models the grille must be removed initially, as described in Section 19, and there are four attaching screws retaining the sealed beam unit.

Chapter 13/Supplement: Revisions and information on later models

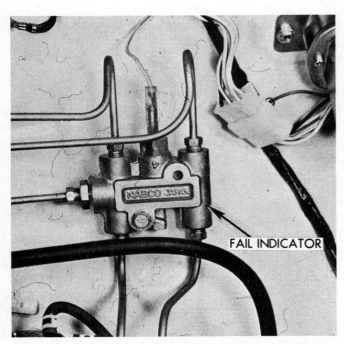

Fig. 13.52 Location of the brake fail indicator (combination valve) (Sec 15)

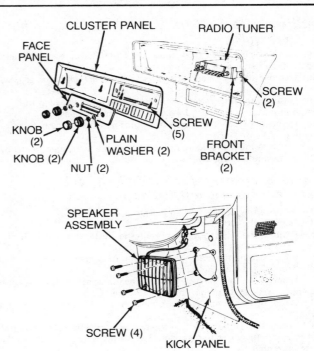

Fig. 13.53 Location of the radio and speaker components (Series 11 and 12) (Sec 16)

Combination switch (Series 11 and 12) — general description

4 On Series 11 and 12 models the turn signal/dimmer switch lever on the left side of the steering column not only controls those functions, but also incorporates the windshield wiper switch and the windshield washer switch. Also included in this combination switch, and located on the right side of the steering column, is the hazard warning switch.

5 The turn signals and dimmer are controlled as before. The windshield wiper is controlled by rotating the end of the lever and the windshield washer is controlled by depressing the knob at the end of the lever.

6 A continuity tester can still be used to check the operation of the turn signal and dimmer functions as described in Chapter 10.

Combination switch (Series 11 and 12) — removal and installation

7 Disconnect the negative battery cable.
8 Remove the five screws that attach the steering column cowling to the column.
9 Disconnect the combination switch wiring connectors from the harness.
10 Remove the steering wheel as previously described.
11 Remove the two screws that retain the combination switch to the column flange and lift off the switch.
12 Installation is the reverse of the removal procedure. Following installation, check the switch for proper functioning.

Radio (Series 11 and 12) — removal and installation

13 Disconnect the negative battery cable.
14 Remove the radio knobs, nuts, washers and face panel from the front of the radio.
15 Remove the five screws that retain the cluster panel and lift off the panel.
16 Remove the screws from the radio mounting brackets.
17 Disconnect the wiring connectors and antenna lead from the rear of the radio.
18 Lift out the radio.
19 Remove the mounting brackets from the radio.
20 Installation is the reverse of the removal procedure.

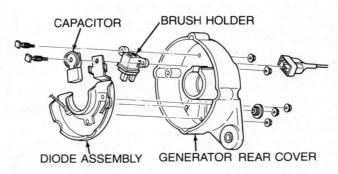

Fig. 13.54 On Series 11 and 12 models, the radio interference suppression capacitor is located inside the generator (Sec 16)

Speaker (Series 11 and 12) — removal and installation

21 The radio speaker on Series 11 and 12 models is located under the dash in the side trim panel in front of the door.
22 Remove the four screws retaining the speaker assembly to the panel.
23 Lift the speaker out and disconnect the wires from the rear.
24 Installation is the reverse of the removal procedure.

Radio interference suppression capacitor (Series 11 and 12) — general description

25 On Series 11 and 12 models the radio interference suppression capacitor is located in the generator, as shown in Figure 13.54. To remove the capacitor, refer to Section 8 for disassembling the generator.

Instrument panel (Series 8 thru 10) — removal and installation

26 A triple cluster instrument panel is installed on Series 8, 9 and 10 models. However, the removal and installation procedures are identical to those given in Chapter 10.

Instrument panel (Series 8 thru 10) — disassembly and reassembly

27 Remove the screws from the rear of the instrument panel and separate the meter case from the panel.
28 Remove the six retaining screws and withdraw the meter glass and surrounds.
29 Unscrew the nuts retaining the speedometer, fuel gauge and temperature gauge and withdraw each unit from the base.
30 Remove the panel and indicator bulbs by turning the holders counterclockwise.
31 Separate the printed circuit board from the instrument panel base.
32 Reassembly is the reverse of dismantling.

Instrument meter assembly (Series 11 and 12) — removal and installation

33 Remove the radio knobs and nuts.
34 Remove the radio bezel.
35 Remove the five screws that retain the cluster panel and lift off the cluster panel.
36 Reach behind the meter assembly and disconnect the speedometer cable from the rear of the assembly.
37 Remove the four screws retaining the meter assembly and pull the assembly part way out from the dash (Figure 13.55).
38 Disconnect the wiring at the connector and lift the meter assembly out.
39 Installation is the reverse of the removal procedure, but be careful not to catch the wiring harness between the instrument panel and the meter assembly.

Instrument meter assembly (Series 11 and 12) — Disassembly and reassembly

40 Disengage the six fingers that retain the meter glass to the assembly and lift off the glass.
41 Disengage the six fingers that retain the meter bezel to the assembly and lift the bezel off.
42 Remove the six nuts that retain the combination gauge to the assembly and lift off the gauge.
43 Remove the lens from the high beam indicator light.
44 Remove the two screws that retain the speedometer to the assembly and lift off the speedometer.
45 Remove the screw retaining the seat belt warning buzzer and lift off the buzzer.
46 Remove the illumination lights and indicator lights by turning the sockets counterclockwise.
47 Pull out the seat belt warning buzzer timer.
48 Disengage the six fingers that retain the printed circuit to the assembly and lift off the circuit.
49 Reassembly is the reverse of the disassembly procedure.

Wiper motor and linkage (Series 11 and 12) — removal and installation

50 The wiper motor can be removed either with or without the linkage.
51 To remove the motor without the linkage, use the following procedure.

   a) Remove the four wiper motor mounting bolts.
   b) Disconnect the wiring at the connector.
   c) Pull the motor part way out and remove the motor shaft nut. Then remove the wiper motor.

52 To remove the wiper motor with the linkage, use the following procedure.

   a) Remove the wiper arm cover and remove the nuts attaching the arms to the pivot shaft. Lift off the blade and arm assemblies.
   b) Remove the pivot covers and the pivot mounting nuts. Then push the pivots down into the panel.
   c) Remove the hole covers on the engine compartment side of the dash panel.
   d) Remove the four wiper motor mounting bolts.
   e) Disconnect the motor wiring at the connector.
   f) Remove the wiper motor together with the linkage.
   g) Disengage the finger on the link joint retainer from the hole in the link. Then disconnect the link from the pivot arms by turning the retainers clockwise.
   h) Installation is the reverse of the removal procedure, but be sure that the wiper linkage is not twisted or contacting any adjacent parts. This would load the motor with additional friction and cause poor wiping action.

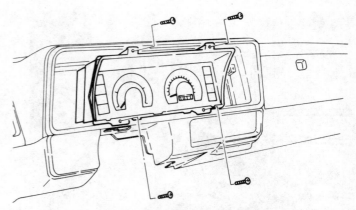

Fig. 13.55 Location of the instrument panel meter assembly screws (Sec 16)

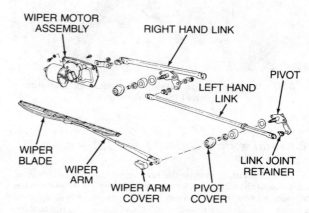

Fig. 13.56 Windshield wiper motor and linkage components (Sec 16)

Heater control cables (Series 9 thru 12) — adjustment

53 Check the positioning of each of the doors and the water valve. If they are not closing properly, loosen the cable clamp at the control

   a) On Series 9 and 10 models, the defroster door should be in the defroster position when the control lever is moved to the DEF position.
   b) On Series 9 and 10 models, the diverter door should be positioned so as to direct all air to the center and side outlets in the VENT position, and allow no air to these outlets in the HEATER and DEF positions.
   c) On Series 11 and 12 models, the defroster door should be in the defroster position when the control lever is in the DEF or HEAT position.
   d) On Series 11 and 12 models the vent door should direct all air to the center and side outlets when the lever is on the VENT or BI-LEVEL position, and should block the air in any other position.
   e) On Series 11 and 12 models, the heater door should direct air out the floor outlets when the lever is in the HEAT or BI-LEVEL position, with some air delivered to the defroster outlet in the HEAT position.
   f) On Series 11 and 12 models when the temperature control lever is in the HOT position, the water valve should be fully open and hot air should be delivered. In the COLD position, the water valve should be closed and cold air delivered.

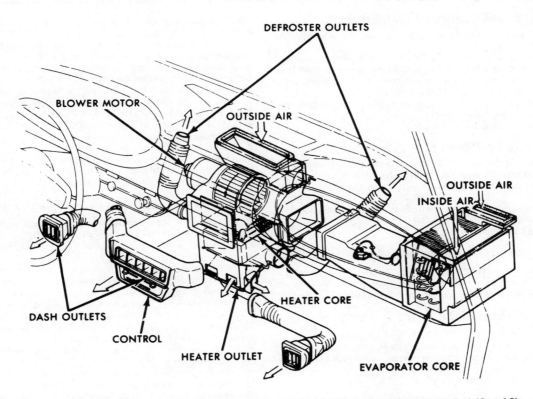

Fig. 13.57 Location of the heater system components (Series 8; Series 9 and 10 similar) (Sec 16)

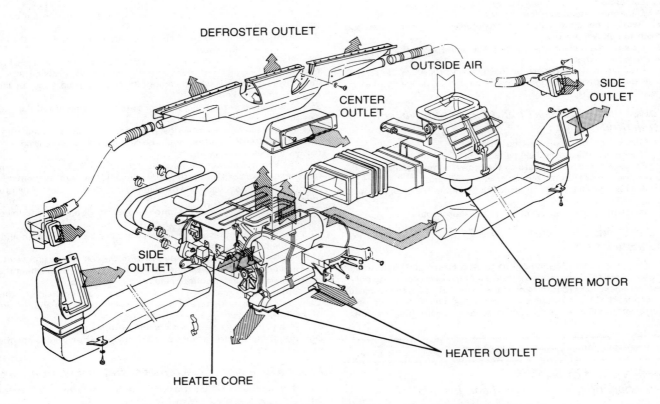

Fig. 13.58 Location of the heater system components (Series 11 and 12) (Sec 16)

Heater control cable (Series 8) — adjustment

54 If the control unit is already in position, loosen the retaining nuts to the instrument panel.
55 Position the selector lever to the DEF position.
56 Retain the vent door in the DEF (HEAT) position and move the control panel fore and aft until the loop on the end of the vent cable is aligned with the hook on the end of the vent lever.
57 Tighten the nuts retaining the control to the instrument panel, then install the cable to the vent lever.
58 Tighten the control cable clamp screws.

Air inlet door cable (Series 8) — adjustment

59 Position the selector lever to DEF, and connect the A/C cable to the link lever hook on the evaporator unit lower portion.
60 Tighten the cable clamp screw, then position the selector lever to A/C.
61 Close the heater outside door and then connect the cable to the outside door shaft.
62 Tighten the cable clamp screw.

Heater unit (Series 8) — removal and installation

63 The procedure is similar to that given in Chapter 10 for Series 3 to 6 models, but it will be necessary to remove the side outlet air hoses. Also note that nuts are installed instead of screws in some instances.

Dash outlet grilles (Series 8 thru 10 — removal and installation

64 Reach under the dash and disconnect the air hose.
65 Unscrew and remove the two wing nuts and withdraw the duct and grille.
66 Installation is the reverse of removal.

Blower motor (Series 9 and 10) — removal and installation

67 Disconnect the negative battery cable.
68 Disconnect the blower lead wires.
69 Remove the screws that attach the blower to the heater case and lift out the blower motor.
70 Remove the blower wheel nut, washer and spacer, then separate the blower and wheel.
71 Installation is the reverse of the removal procedure. Be sure to install the damper in position between the motor flange and heater case.

Blower motor (Series 11 and 12) — removal and installation

72 Disconnect the negative battery cable.
73 Disconnect the blower lead wires.
74 Remove the screws that attach the blower motor assembly to the blower case and lift out the blower assembly and motor.
75 Remove the snap-ring and spacer and separate the blower from the wheel.
76 Installation is the reverse of the removal procedure.

A/C thermostatic switch (Series 11 and 12) — removal and installation

77 On Series 11 and 12 models, the thermostatic switch is located on the upper section of the evaporator case and the evaporator assembly must be removed in order to remove the switch. Because this procedure necessitates the disconnection of refrigerant lines of the A/C system, it is better left to a GM dealer or an automotive air-conditioning specialist.

17 Front-wheel drive system (4 x 4 models)

Refer to Figures 13.59 thru 13.62

Free wheeling hub — removal and installation

1 Raise the front of the truck and support it firmly on jackstands. The 4WD shift lever should be in the 2H position.
2 With the free-wheeling hub in the Free position, remove the screws that retain the hub cover assembly.
3 Remove the snap-ring and shims from the end of the spindle.
4 Remove the hub body.
5 Upon installation, thoroughly lubricate the hub body, then reinstall the lock washer and hub body.
6 Temporarily reinstall the snap-ring onto the shaft. Then using a feeler gauge, measure the gap between the hub body and the snap-ring. While doing this, push the axle shaft out with your hand.
7 Choose the correct shim(s) so that this gap is less than 0.3 mm (0.01 in). Shims are available in the following thicknesses: 0.2 mm (0.008 in), 0.3 mm (0.01 in), 0.5 mm (0.02 in), and 1.0 mm (0.04 in).
8 Remove the snap-ring. Then install the selected shims and reinstall the snap-ring once again.
9 Reinstall the hub cover assembly along with a new gasket. Be sure the stopper nails are aligned with the grooves in the hub body. Tighten the hub cover bolts to the correct torque.

Free wheeling hub — disassembly and reassembly

10 Remove the hub assembly as previously described.
11 To disassemble the cover assembly, first depress the follower toward the knob. Then turn the clutch assembly clockwise and remove the clutch assembly.
12 Remove the snap-ring. Then remove the knob from the cover. Be careful you do not lose the detent ball when removing the knob.
13 Remove the detent ball and spring from the knob.
14 Using only your fingers, remove the X-ring from the knob.
15 Remove the compression spring from the follower.
16 Remove the retaining spring from the follower hanger, then remove the follower from the clutch assembly.
17 Turn the retaining spring counterclockwise and remove it from the clutch assembly.
18 To disassemble the hub body, first remove the snap-ring.
19 Remove the inner assembly from the body.
20 Remove the second snap-ring and separate the ring, inner assembly and spacer.
21 Clean and inspect all parts for wear or damage and replace as necessary.
22 To begin reassembly, apply grease to the X-ring and install it into its groove in the knob.
23 Apply grease to the outer circumference of the knob and the inner circumference of the cover.
24 Place the detent ball in position in the knob, then install the knob to the cover, aligning the detent ball to either groove of the cover.
25 Install the snap-ring to the knob. Be sure the smooth face of the ring is against the knob.
26 Apply grease to the cam portion of the knob and install the gasket.
27 With the knob set in the Free position, install the retaining ring spring into its groove in the clutch. Be sure the end of the spring correctly matches the end of the groove.
28 Install the follower to the clutch. When correctly installed, the follower nail contacted with the cam will come closer to the bent portion of the retaining spring by aligning the follower stipper nail to the outer clutch teeth.
29 Hook the retaining spring onto the upper portion of the follower hanger nails.
30 Install the compression spring. The smaller diameter side should be against the follower.
31 Align the follower nail contacted with the cam to the groove in the handle. Then assemble the clutch with the knob by pushing the clutch in and turning it counterclockwise to the knob.
32 Begin the assembly of the hub body by applying grease to the spacer and to the inside face of the ring. Install the spacer and the ring to the inner, and secure them with the snap-ring.
33 Apply grease to the splined portion of the body.
34 Install the inner assembly into the body and secure it with the snap-ring.

Front axle assembly — removal and installation

35 Remove the bolts that attach the front driveshaft to the front axle differential, and separate the two components.
36 Raise the front of the truck and support it firmly on jackstands placed under the forward frame members.
37 Remove the front wheels and tires.
38 Remove the skid plate.

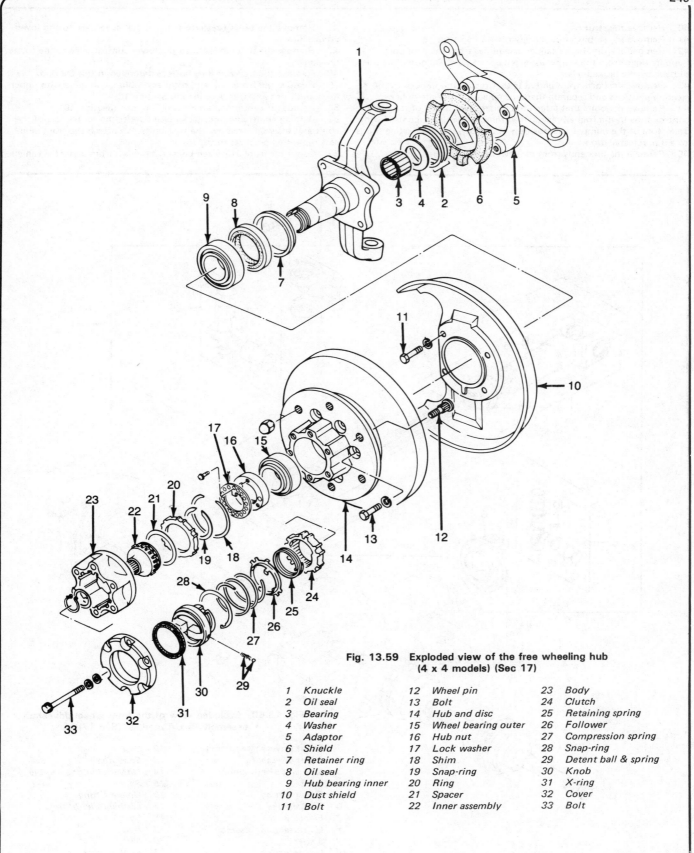

Fig. 13.59 Exploded view of the free wheeling hub (4 x 4 models) (Sec 17)

1 Knuckle
2 Oil seal
3 Bearing
4 Washer
5 Adaptor
6 Shield
7 Retainer ring
8 Oil seal
9 Hub bearing inner
10 Dust shield
11 Bolt
12 Wheel pin
13 Bolt
14 Hub and disc
15 Wheel bearing outer
16 Hub nut
17 Lock washer
18 Shim
19 Snap-ring
20 Ring
21 Spacer
22 Inner assembly
23 Body
24 Clutch
25 Retaining spring
26 Follower
27 Compression spring
28 Snap-ring
29 Detent ball & spring
30 Knob
31 X-ring
32 Cover
33 Bolt

40 Remove the strut bars.
41 Remove the stabilizer bar as described in Section 18.
42 Remove the disc brake caliper assemblies from their supports. Support them from the frame, using a piece of wire. Do not allow them to hang by the brake lines.
43 Remove the cotter pins and nuts that secure the tie-rods to the steering knuckles and separate them, using an appropriate puller.
44 Remove the bolts that attach the upper pivot shafts of the upper control arms to the frame brackets and remove the upper control arms. Take note of the number and position of camber/caster adjusting shims to aid in reinstallation.
45 Remove the link ends from the lower control arms.
46 Remove the bolts that attach the shock absorbers to the lower control arms.
47 Remove the bolts that attach the lower control arms to the frame brackets.
48 Remove the free-wheeling hubs as described in this Section.
49 Remove the front hub and rotor assemblies, complete with upper links and front axles, as described in Section 18.
50 Remove the intermediate rod and tie-rods (Section 19).
51 With an adjustable floor jack placed under the center part of the front axle assembly, remove the four bolts that attach the front axle case mounting brackets to the frame.
52 Lower the front axle assembly and remove it from under the vehicle.

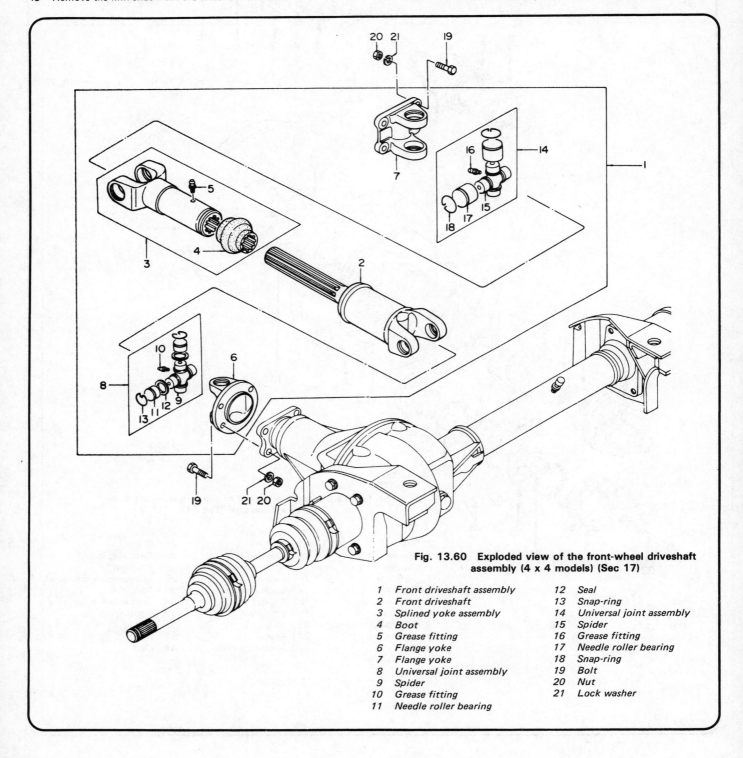

Fig. 13.60 Exploded view of the front-wheel driveshaft assembly (4 x 4 models) (Sec 17)

1 Front driveshaft assembly
2 Front driveshaft
3 Splined yoke assembly
4 Boot
5 Grease fitting
6 Flange yoke
7 Flange yoke
8 Universal joint assembly
9 Spider
10 Grease fitting
11 Needle roller bearing
12 Seal
13 Snap-ring
14 Universal joint assembly
15 Spider
16 Grease fitting
17 Needle roller bearing
18 Snap-ring
19 Bolt
20 Nut
21 Lock washer

53 Installation is the reverse of the removal procedure, with the following notes.

a) For proper installation of the individual components refer to the appropriate Sections in either this Chapter, Chapter 9 or Chapter 11.
b) Be sure to tighten all bolts and nuts to the proper torque.
c) Following installation, adjust the height control arm adjusting bolts as described in Chapter 11.

Front axle shaft — removal and installation

54 Remove the front axle assembly as previously described.
55 Support the front axle assembly on jackstands or similar supports and drain the oil from the front differential.
56 Remove the four bolts attaching the axle mounting bracket to the axle case. Then pull the shaft assembly out from the case.
57 For disassembly of the axle shaft assembly, refer to the following sub-section.
58 To install the axle, fit it into the case and tighten the four attaching bolts to the correct torque.
59 The remainder of the installation is the reverse of the removal procedure.

Front axle shaft — overhaul

60 Remove the front axle shaft from the vehicle as previously described.

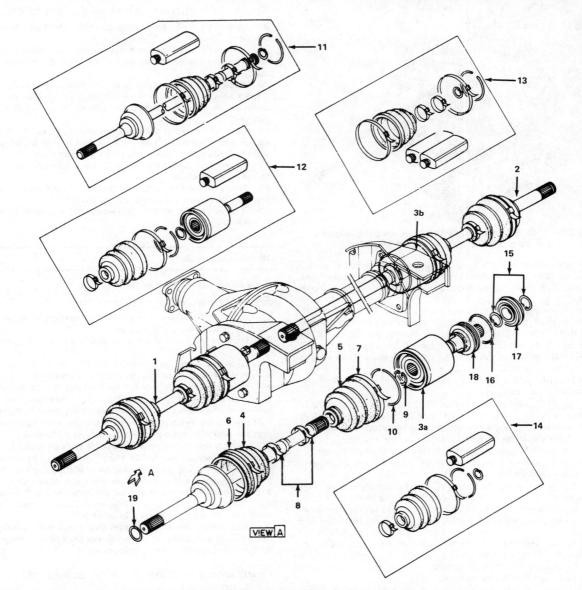

Fig. 13.61 Exploded view of the front axle shafts (4 x 4 models) (Sec 17)

1	BJ shaft assembly (R)	5	Bellows (B)	10	Circlip	15	Snap-ring
2	BJ shaft assembly (L)	6	Bellows band (A)	11	BJ shaft assembly	16	O-ring
3a	DOJ assembly (R)	7	Bellows band (B)	12	DOJ assembly	17	Ball bearing
3b	DOJ assembly (L)	8	Bellows band (C)	13	Bellows kit	18	Oil seal
4	Bellows (A)	9	Snap-ring	14	Bellows kit	19	Shim

61 Using a screwdriver or similar tool, loosen both bellows bands from the rubber boot covering the double offset joint (DOJ) and remove it from the DOJ assembly. Since bellows bands should not be reused, these should be discarded.
62 Slide the rubber bellows off of the DOJ joint.
63 Using a screwdriver or similar tool, pry out the spring clip from inside the DOJ assembly.
64 The axle shaft can now be withdrawn from the case.
65 Wipe off the grease with a rag and then use a screwdriver to remove the six balls from the joint assembly as shown in Figure 13.62.

Fig. 13.62 With the front axle shaft cage properly aligned, the balls can be pried out with a screwdriver (Sec 17)

66 With the balls removed, rotate the outer part of the cage, so that the ball guide holes line up with the projections in the inner part, the ball retainer. Then separate the two parts by sliding the cage along the axle shaft.
67 Remove the snap-ring that secures the retainer to the axle shaft.
68 Remove the retainer, the cage and the bellows from the shaft.
69 The Birfield Joint (BJ) assembly is permanently fixed to the axle shaft and cannot be removed or disassembled. To inspect the joint, remove the two bellows bands from the rubber bellows. Then slide the bellows off the axle shaft.
70 Check all of the disassembled parts for signs of wear or damage and replace them as necessary.
71 Inspect the BJ case and the center shaft for any signs of contact. Also check the bottom face of the DOJ case for signs of contact. If any are found, the steering angles should be checked after reinstallation.
72 Also check the inside face of the DOJ cage for any signs of contact with the retainer. Also check for signs of contact between the spring clip and the balls. Contact of these parts indicate transverse misalignment of the front axle assembly.
73 To begin reassembly, carefully slip the BJ assembly rubber bellows onto the axle shaft, and slide it into its groove in the shaft.
74 Apply sufficient grease to the BJ assembly so that the interior of the joint is adequately lubricated. Also fill the bellows about half-way with grease.
75 Position the bellows over the BJ case and check that it is not collapsed or distorted. Then fit the large bellows band into position on the bellows. Tighten it by first depressing the lever by hand and then bending down the hook with a hammer.
76 Next install the small bellows band in the same manner as the large one. Check again that the bellows is not distorted.
77 To assemble the DOJ assembly, first slide the large bellows band onto the axle shaft.
78 Apply a thin coat of grease to the shaft and then carefully slide the bellows onto the shaft.
79 Slide the cage onto the shaft, with the smaller diameter side toward the BJ assembly.
80 Fit the ball retainer onto the shaft and secure it by installing the snap-ring.
81 Align the ball guide holes in the cage with the protrusions in the retainer and slide them together. Then turn the cage one-half pitch, so the track on the retainer is aligned with the cage holes.
82 Install the six balls into position and pack the joint with grease.
83 Fit this assembly into the DOJ case and secure it by installing the spring clip. The open ends of the spring clip should be positioned at the inner circumference away from the ball groove.
84 Apply a sufficient amount of grease to the joint, then slide the bellows into position over the case.
85 If the bellows is at all collapsed, insert a screwdriver under the large end to allow the inside and outside air to equalize.
86 With the bellows free of distortion, install both bellows bands. As with the BJ assembly bands, these are tightened by depressing the lever by hand and then bending the hook down with a hammer.

Front differential — removal and installation

87 Remove the front axle assembly from the truck as previously described.
88 Remove the front axle shafts as previously described.
89 Remove the bolts that attach the front differential to the axle case.
90 Because of the complexity and critical nature of the differential, if there are problems with it we suggest that you either exchange it for a rebuilt one or have a Chevrolet dealer or other qualified mechanic overhaul it.
91 Prior to installation, be sure the mating faces of the differential and axle case are clean. Then apply a thin coat of liquid gasket solution and place the differential gasket into position.
92 Mount the differential into the axle case and tighten the bolts to the proper torque.
93 Reinstall the axle shaft and then fill the differential to just below the filler hole with the proper oil.
94 The remainder of the installation is the reverse of the removal procedure.

18 Front suspension

Refer to Figures 13.63 thru 13.66

Front wheel bearing adjustment

1 Apply the handbrake and raise the front of the vehicle so that the front wheel is clear of the ground.
2 Remove the hub cap and carefully tap or pry out the hub grease cap.
3 Wipe any excess grease from the end of the spindle and then withdraw the cotter pin and castellated nut retainer.
4 Spin the wheel and tighten the hub nut to a torque of 22 ft-lbs.
5 Rotate the hub two or three complete turns and loosen the nut sufficiently to allow the hub to be turned by hand and check that there is no free play in the hub. Ensure when rotating the hub that the brake linings are not in contact with the drum.
6 Pull on one of the wheel hub studs with an accurate pull scale and measure the torque needed to start it rotating. The correct torque should be 1.1 to 2.6 ft-lbs. If necessary, tighten the spindle nut until the correct torque is indicated on the scale. **Note**: *Be sure that the brake pads are not rubbing against the rotor during this operation.*
7 Refit the castellated nut retainer and insert a new cotter pin, then bend over the ends to retain it in position.
8 Check that the wheel rotates freely and then install the grease cap and hub cap. **Note**: *If the wheel does not rotate freely, then the bearings are probably due for replacement.*

Front wheel bearing (4 x 4) — adjustment

9 Raise the front of the truck and support it on jackstands. The 4WD shift lever should be in 2H
10 With the free-wheeling hub in the Free position, remove the screws that retain the hub cover assembly.
11 Remove the snap-ring and shims from the end of the spindle.
12 Remove the hub body.
13 Remove the lock washer.
14 Fit a front hub nut wrench onto the hub nut. Then, while spinning the wheel, tighten the hub nut just until the wheel can't be turned by hand.

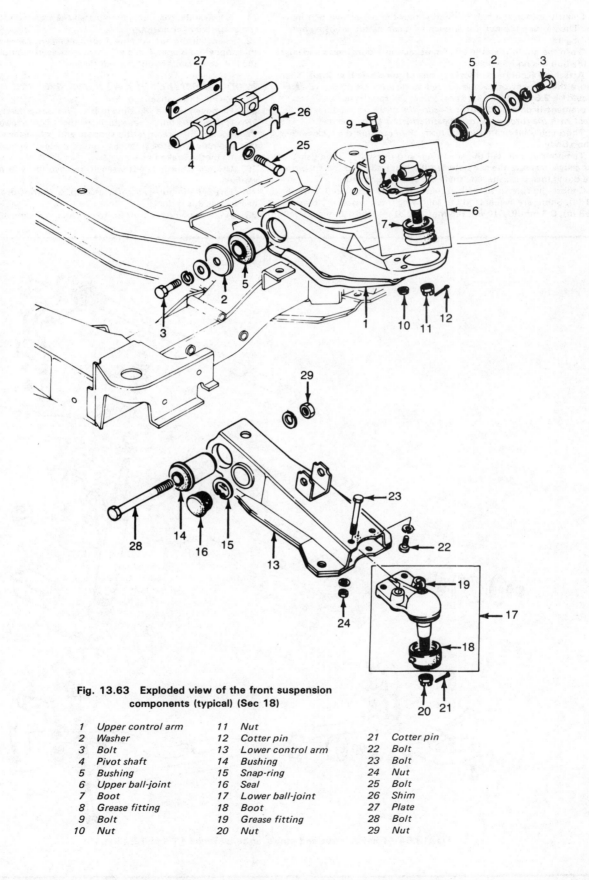

Fig. 13.63 Exploded view of the front suspension components (typical) (Sec 18)

1 Upper control arm	11 Nut	
2 Washer	12 Cotter pin	21 Cotter pin
3 Bolt	13 Lower control arm	22 Bolt
4 Pivot shaft	14 Bushing	23 Bolt
5 Bushing	15 Snap-ring	24 Nut
6 Upper ball-joint	16 Seal	25 Bolt
7 Boot	17 Lower ball-joint	26 Shim
8 Grease fitting	18 Boot	27 Plate
9 Bolt	19 Grease fitting	28 Bolt
10 Nut	20 Nut	29 Nut

15 Slightly loosen the hub nut and rotate the wheel two or three turns. Then loosen the nut just enough so it can be turned using only your fingers.
16 Turn the hub nut all the way in, using only your fingers and check that the hub has no free play.
17 Attach an accurate pull scale to one of the wheel hub studs. Then measure the amount of torque needed to begin rotating the wheel. It should be 2.6 to 4.0 ft-lbs. If not, adjust the spindle nut until this measurement is obtained. **Note**: *Be sure the brake pads are not in contact with the rotor while this measurement is being taken.*
18 Thoroughly lubricate the hub body then reinstall the lock washer and hub body.
19 Temporarily reinstall the snap-ring onto the shaft. Then using a feeler gauge, measure the gap between the hub body and the snap-ring. While doing this, push the axle shaft out with your hand.
20 Choose the correct shim(s) so that this gap is less than 0.3 mm (0.01 in). Shims are available in the following thicknesses: 0.2 mm (0.008 in), 0.3 mm (0.010 in), 0.5 mm (0.020 in), and 1.0 mm (0.040 in).
21 Remove the snap-ring, then install the selected shims and reinstall the snap-ring once again.
22 Reinstall the hub cover assembly along with a new gasket. Be sure the stopper nails are aligned with the grooves in the hub body. Tighten the hub cover bolts to the correct torque.

Front wheel hub (4 x 4) — removal and installation

23 Remove the wheel and tire.
24 Remove the free wheeling hub as previously described.
25 Remove the bolts that attach the disc brake support assembly to the adapter, then remove the support and caliper assembly. Suspend this assembly from the frame by using a piece of wire. Do not allow it to hang by the brake line.
26 Using a front hub nut wrench, remove the hub nut and lock washer.
27 Lift off the hub and rotor assembly. Do not drop the wheel bearings as this is done.
28 If either component needs to be replaced, remove the bolts that

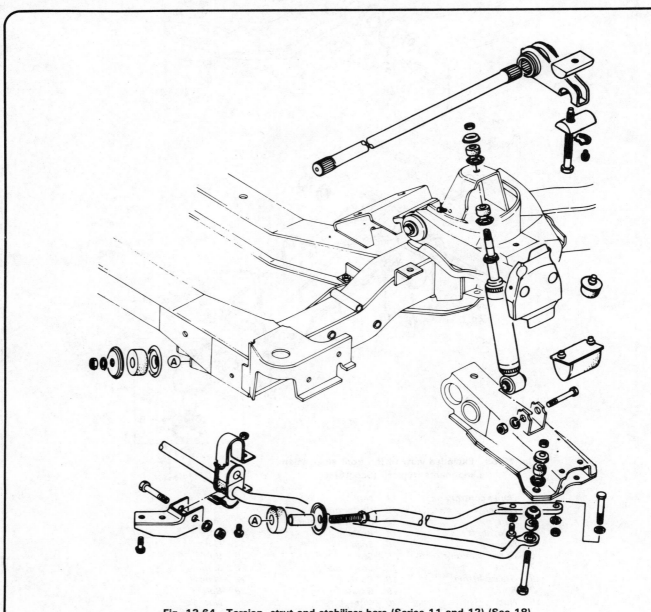

Fig. 13.64 Torsion, strut and stabilizer bars (Series 11 and 12) (Sec 18)

attach them and separate the two parts.

29 Installation is the reverse of the removal procedure, but be sure to follow the procedure under front wheel bearing adjustment.

Front wheel bearings — removal and installation

30 Remove the front hub assembly as described in either Chapter 11 (4 x 2) or this Chapter (4 x 4).

31 After removing the hub, the outer roller bearing can be lifted out with the fingers.

32 The inner bearing will remain in the hub and will have to be driven out, along with its oil seal and retaining ring with a suitable driving tool.

33 Wash all parts in cleaning solvent and blow them dry. Discard the oil seal.

34 If the bearing races need to be replaced, they can be driven out by using a brass drift placed behind the race. New races can be installed by using an appropriate sized socket or other installing tool. Be sure to start it squarely into the hub to avoid damaging the race.

35 Pack the bearings with a high-temperature wheel bearing grease.

36 Apply a light coat of grease to the spindle and to the inside surface of the hub.

37 Place the inner bearing in position against the race. Then install a new oil seal and retaining ring by using an appropriate sized socket.

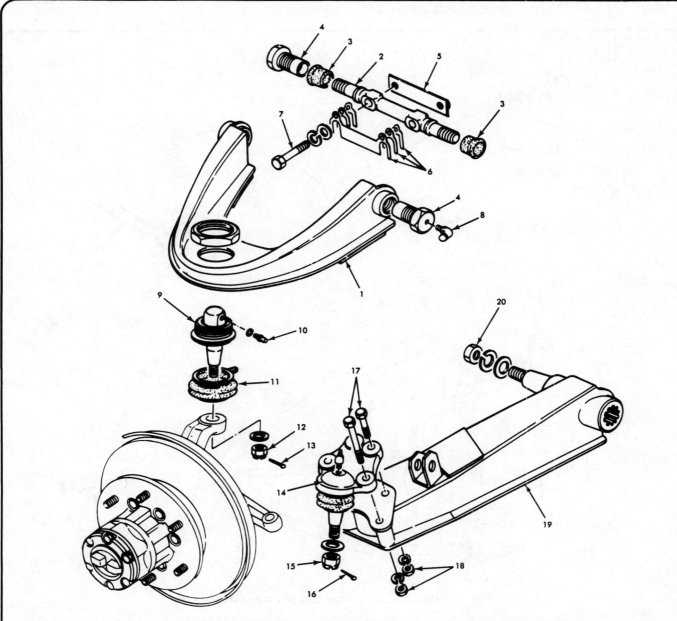

Fig. 13.65 Exploded view of the front suspension components (4 x 4 models) (Sec 18)

1	Upper control arm	6	Shim	11	Boot	16	Cotter pin
2	Pivot shaft	7	Bolt	12	Nut	17	Bolt
3	Seal	8	Grease fitting	13	Cotter pin	18	Nut
4	Bushing	9	Upper ball-joint	14	Lower ball-joint	19	Lower control arm
5	Plate	10	Grease fitting	15	Nut	20	Nut

38 Carefully install the hub assembly onto the spindle and then install the outer wheel bearing.
39 Installation is the reverse of the removal procedure, but be sure to follow the procedure for front wheel bearing adjustment as previously described.

Lower balljoint (4 x 4) — removal and installation

40 Raise the front of the truck and support it firmly on jackstands.
41 Mark the position of the torsion bar adjusting bolts and then loosen them.
42 Remove the strut rods at the lower control arm.
43 Remove the stabilizer link bolts at the lower control arm.
44 Remove the tie-rod ends.
45 Remove the cotter pin and nut that attaches the balljoint to the steering knuckle.
46 Using an appropriate tool, separate the balljoint from the knuckle.
47 Remove the bolts that secure the balljoint to the lower control arm.
48 Remove the lower shock absorber bolt.
49 Remove the control mechanism and the hub body.
50 Slide the knuckle assembly off the axle shaft.
51 Remove the balljoint.
52 Installation is the reverse of the removal procedure.

Steering knuckle (4 x 4) — removal and installation

53 Remove the hub and rotor assembly as previously described.
54 Remove the four bolts that connect the knuckle to the dust shield and adapter.

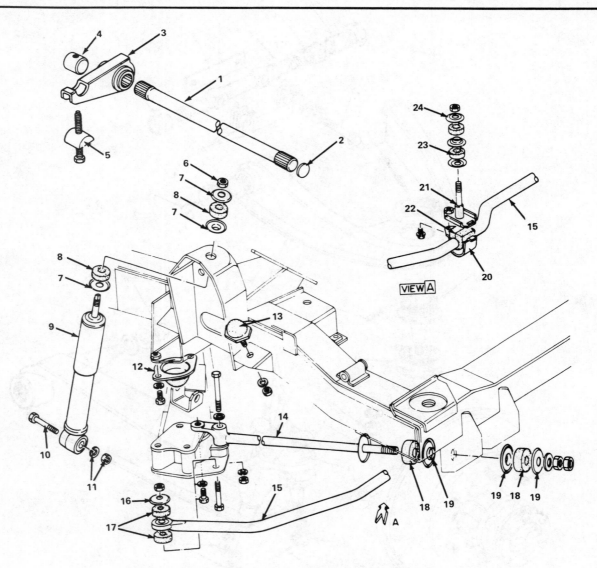

Fig. 13.66 Torsion, strut and stabilizer bars (4 x 4 models) (Sec 18)

1 Torsion bar
2 Rubber seat
3 Height control arm
4 Pivot nut
5 Height control seat & bolt
6 Nut
7 Retainer
8 Bushing
9 Shock absorber
10 Bolt
11 Nut & washer
12 Lower control arm bumper
13 Upper control arm bumper
14 Strut bar
15 Stabilizer bar
16 Washer
17 Bushing
18 Strut bar bushing
19 Strut bar washer
20 Stabilizer bar bracket
21 Stabilizer bar support
22 Stabilizer bar bushing
23 Bushing
24 Washer

55 Remove the cotter pins and nuts that secure the upper and lower balljoints to the knuckle.
56 Using the appropriate tool, separate the balljoints from the steering knuckle and lift off the knuckle.
57 If the knuckle oil seal needs replacing, pry it out with a screwdriver, then install a new one using an appropriately-sized socket.
58 The axle shaft bearing, inside the knuckle, can be removed and a new one installed, if necessary, by using special removal and installing tools.
59 Installation is the reverse of the removal procedure. Be sure to follow the procedure for adjusting the front wheel bearings, as previously described.

Upper control arm and balljoint assembly (Series 11 and 12) — removal and installation

60 Apply the parking brake and jack up the front of the vehicle. Place chassis stands or blocks under the lower control arm. Remove the wheel.
61 Withdraw the cotter pin from the nut retaining the upper control arm and upper balljoint unit. Detach the upper control arm from the steering knuckle, but support the knuckle unit by tying it with wire or cord and hang it out of the way. Do not allow it to hang from the brake hose!
62 Remove the two upper pivot shaft bolts.
63 Remove the retaining nut, retainer and rubber grommet from the upper stem of the shock absorber. Then depress the shock absorber.
64 Remove the upper control arm from the bracket, noting the number and position of the camber adjusting shims.
65 If the pivot shaft and bushings need to be removed from the control arm, first remove the bolts from each side of the pivot shaft. Then remove the lock washer, washer and plate. Finally, remove the bushings and pivot shaft.
66 If the pivot shaft and bushings have been removed, begin the installation by installing the shaft and bushings into the control arm. Then, into each side, install the bolts, complete with plates, washers and lock washers.
67 The remainder of the installation is the reverse of the removal procedure. Be sure that the camber adjusting shims are replaced in the same locations they were removed from.

Stabilizer bar (Series 11 and 12) — removal and installation

68 Raise the front of the vehicle and support it on jackstands.
69 Remove the bolts, nuts, washers and rubber grommets that connect the stabilizer bar to the lower control arms.
70 Remove the stabilizer bar bushing clamp bolts, then remove the lower clamps and lift off the stabilizer bar.
71 Remove the bolts that attach the bushings to their brackets and lift out the bushings.
72 Check the grommets and bushings for wear or damage and replace as necessary.
73 Begin the installation by installing the bushings and clamps onto the stabilizer bar, but do not tighten the clamp bolts yet.
74 Install the bushings onto their brackets and insert the attaching bolts. Torque these bolts to specs. Install the ends of the stabilizer bar to the lower control arms.
75 Tighten the bushing clamp bolts.

Ride height checking

76 The procedure is identical to that given in Chapter 11 and is only made with the vehicle at curb weight. The front checking dimension is identical, but the rear dimensions are as given in Specifications (Section 2).

19 Steering system (manual)

Refer to Figures 13.67 thru 13.70

Caster angle (Series 11 and 12) — adjustment

1 Beginning with Series 11 and 12 models, the caster angle is adjusted by varying the length of the strut bar. This is done by turning the locknuts to obtain the correct angle. Refer to Figure 13.67.

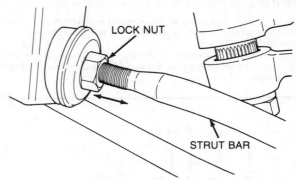

Fig. 13.67 On Series 11 and 12 models, the caster angle is adjusted by varying the length of the strut bar (Sec 19)

Steering wheel (Series 11 and 12) — removal and installation

2 Disconnect the negative battery cable.
3 Remove the two screws that retain the horn shroud, lift off the shroud and disconnect the horn electrical lead.
4 Unscrew and remove the steering wheel retaining nut, washer and lock washer.
5 Mark the relative position of the steering wheel and steering shaft.
6 Remove the five steering column cowling retaining screws and detach the cowling.
7 Using a suitable steering wheel puller, remove the steering wheel.
8 Installation of the steering wheel is the reversal of the above procedure but ensure that the wheel-to-shaft alignment marks are correct and check the horn operation when the battery is reconnected. Torque the steering wheel nut to that given in the Specifications.

Steering shaft flexible coupling (Series 11 and 12) — removal and installation

9 Jack up the front of the vehicle just until the wheels are off the ground. Place it firmly on jackstands.
10 Remove the coupling through bolts and locknut.
11 Remove the pinch bolts on the upper and lower coupling flanges.
12 Remove the two bolts that attach the steering column to the dash.
13 Pull the steering column to the rear approximately two inches (50 mm).
14 Remove the upper coupling flange and coupling. Then remove the lower coupling flange.
15 Installation is the reverse of the removal procedure. **Note:** *Do not fully tighten the steering column-to-dash bolts until the vehicle has been lowered to the ground.*

Steering column and shaft assembly (Series 11 and 12) — removal and installation

16 Disconnect the negative battery cable.
17 Remove the steering wheel as described previously.
18 Remove the combination switch as described in Section 16.
19 Remove the bolt that retains the ignition switch bracket.
20 Lift the hood and remove the upper coupling clamp pinch bolt from the steering shaft flexible coupling. The relation of the steering shaft to the coupling clamp should be marked prior to separating.
21 Remove the two bolts that attach the steering column to the dash.
22 Remove the column to the rear, being careful not to jar or damage the shaft.
23 Installation is the reverse of the removal procedure with the following notes.

  a) Be sure the marks line up when connecting the steering shaft to the coupling clamp.
  b) Leave the column-to-dash bolts loose until the vehicle is lowered onto its tires. Then tighten them to the specified torque.
  c) Following installation, check the alignment between the front wheels and the steering wheel. Also check the steering wheel free play at the outside diameter of the steering wheel. It should be about 10 mm (0.4 in).

Steering gear (Series 8 thru 10) — reassembly

24 On the initial reassembly of the steering gear worm shaft, the initial starting torque should be between 2.9 and 6.0 ft-lbs when measured with a pull scale.

25 After complete reassembly of the steering gear, the worm shaft starting torque should be between 4.9 and 11.7 ft-lbs when measured with a pull scale.

Steering gear (Series 11 and 12) — disassembly and reassembly

26 Remove the steering gear from the vehicle (Chapter 11).

27 Remove the coupling pinch bolt and separate the coupling from the worm shaft. Prior to separating the two parts a mark should be made to show the relationship between them.

28 Drain the oil from the steering gear through the filler hole.

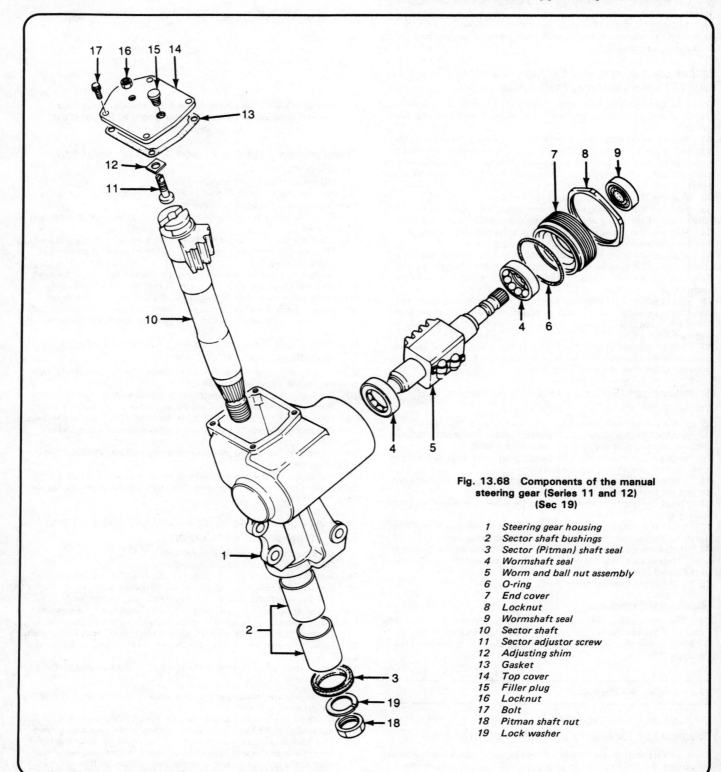

Fig. 13.68 Components of the manual steering gear (Series 11 and 12) (Sec 19)

1 Steering gear housing
2 Sector shaft bushings
3 Sector (Pitman) shaft seal
4 Wormshaft seal
5 Worm and ball nut assembly
6 O-ring
7 End cover
8 Locknut
9 Wormshaft seal
10 Sector shaft
11 Sector adjustor screw
12 Adjusting shim
13 Gasket
14 Top cover
15 Filler plug
16 Locknut
17 Bolt
18 Pitman shaft nut
19 Lock washer

29 Remove the adjusting screw locknut and then turn the adjusting screw counterclockwise to relieve the preload between the sector gear and the ball nut rack.
30 Remove the top cover bolts.
31 While holding the top cover stationary, turn the adjusting screw clockwise until the cover is free.
32 With the worm shaft positioned straight-ahead, remove the sector shaft from the gear box. Do not hammer on or use other impact tools to remove the sector shaft. Also be careful not to damage the oil seal when removing the shaft.
33 Using a chisel and hammer in the locknut slots, remove the locknut from the end cover.
34 Using a special tool as shown in Figure 13.69, turn the end cover counterclockwise to remove it. Be careful not to damage the oil seal when removing the cover. This can be prevented by taping the splines of the shaft.

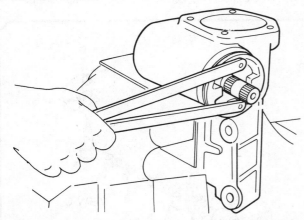

Fig. 13.69 A special tool such as shown above, must be used to remove or tighten the end cover of the Series 11 and 12 steering gear (Sec 19)

35 Extract the worm and ball nut from the gearbox and remove the lower bearing. Retain the worm and ball nut in the horizontal position and do not hold it vertically or the ball nut will fall onto the worm gear end and damage the ball tubes.
36 Clean the respective components for inspection, but do not disturb the clamp plate on the ball tube which is sealed with white paint.
37 Check the following items for wear, distortion, pitting or damage.
   a) Steering shaft
   b) Ball nut teeth
   c) Bearing
   d) Gear casing

38 If, on inspection, the worm shaft components are found to be faulty, then the worm and ball nut unit must be replaced, as they are a matched pair. Oil seals must always be replaced on assembly and bearings, seals, gearbox and covers can be individually replaced as required.
39 To check the ball nut, position the worm and nut vertically and check that the ball nut lowers in a smooth twisting motion. If the motion is not smooth or the ball nut is noisy when lowering, the complete steering shaft unit must be replaced.
40 Check that the sector shaft teeth and serrated portion are not worn or damaged. Measure the reduction in the outside diameter of the shaft. The standard diameter is 1.181 in and the wear limit is 0.001 in. Replace the sector shaft if necessary.
41 Replace any oil seals that were removed with new ones.
42 Begin reassembly by installing the lower bearing into its position in the gear box.
43 While keeping the assembly horizontal, carefully insert the ball screw and nut assembly into the gear box.
44 Be sure the mating surfaces of the gear box and the end cover are clean, then fit the O-ring into the cover.
45 Apply a little gear oil to the O-ring and the inside lip of the oil seals.
46 Wrap the splines of the worm gear shaft with tape. Then fit the cover over the worm shaft, being careful not to damage the seal.
47 Again using an appropriate tool, screw in the end cover, until the torque required to rotate the worm shaft is 2.6 to 5.2 in-lbs (0.3 to 0.5 Nm). This can be measured by using a 15 mm 12-point socket and an in-lb torque wrench on the taped worm gear spline. This measurement is the worm gear bearing preload.
48 With the worm bearing preload properly set, install the locknut and tighten it using a hammer and chisel. Again, check the worm bearing preload to be sure it hasn't changed.
49 Remove the tape from the worm shaft.
50 Fit the adjustment screw into the sector shaft T-slot so that there is no clearance between the adjuster screw head and sector shaft. The adjuster screw must slide freely in the slot. If the adjustment screw-to-sector shaft clearance exceeds 0.001 inch, adjustment can be made by shims available in thickness of 0.059 in.
51 Apply a dab of grease to the bushing and the oil seal in the gear box. Then install the sector shaft so it is aligned with the rack as shown in Figure 13.68. **Note:** *Wrap the shaft splines with tape as necessary to avoid damaging the oil seal.*
52 Apply liquid gasket to the top cover and attach a gasket to the cover.
53 Thread the adjusting screw into the top cover. Make sure the adjusting screw and shim are centered in the T-slot of the sector shaft.
54 Turn the adjusting screw in a counterclockwise direction until the top cover contacts the gear box. Then turn it two more turns and install the bolts and washers. Torque the bolts to specs.
55 Check again to see if the sector gear and ball nut are properly installed. This is done by turning the worm from lock to lock. If the worm must be turned more than five times, then it is properly installed If not, then it must be disassembled and reinstalled correctly.
56 Next the back lash between the ball nut rack and the sector gear must be adjusted. This is the torque required to begin rotating the worm shaft. With the steering gear set in a straight-ahead position, fit an in-lb torque wrench to the worm shaft. If the initial torque is not within 4.3 to 8.7 in-lb (0.5 to 1.0 Nm) turn the adjusting screw until the proper torque is obtained.
57 Install and tighten the adjusting screw locknut.
58 Connect the sector shaft to the Pitman arm and tighten the nut to the specified torque.
59 Install the coupling on the worm shaft so that the marks made during disassembly line up. If necessary, adjust the setting of the end cover so that the clearance between the coupling and the end of the end cover is 0.146 to 0.224 in (3.7 to 5.7 mm).
60 Install and tighten the pinch bolt to the correct torque.
61 Fill the steering gear with the proper grade and amount of oil.

Intermediate rod and tie-rods (Series 11 and 12) — removal and installation

62 Raise the front of the truck and support it on jackstands.
63 Remove the cotter pin, nut, washer and bolt that attaches the steering damper to the intermediate rod.
64 Remove the cotter pins and nuts from the ball studs that connect the intermediate rod and tie-rods to the idler arm, Pitman arm and adapters.
65 Separate the parts, using an appropriate puller.
66 Remove the intermediate rod, together with the tie-rods.
67 The tie-rods need to be removed from the intermediate rod only if they are to be replaced separate from the intermediate rod.
68 Installation is the reverse of the removal procedure, with the following notes:
   a) If the tie-rods have been removed from the intermediate rod, liquid gasket should be applied to the junction prior to attaching the rods. Following attachment, stake the upper and lower points with a hammer and punch.
   b) Ensure that the ball stud threads are clean before installation.
   c) Lubricate the ball studs.
   d) Tie-rods are installed with the lubrication fittings to the front.
   e) Tighten the ball stud nuts to the specified torque.

Steering damper (Series 11 and 12) — removal and installation

69 Raise the front of the vehicle and support it on jackstands.
70 Remove the cotter pin, nut and washer from the bolt that attaches the damper to the intermediate rod.

Chapter 13/Supplement: Revisions and information on later models

71 Remove the cotter pin and nut and bolt that attaches the damper to the second crossmember (4 x 2 models) or the idler arm bracket (4 x 4 models).
72 Remove the steering damper.
73 Installation is the reverse of the removal procedure.

Idler arm assembly (Series 11 and 12) — removal and installation

74 Raise the front of the vehicle and support it on jackstands.
75 Remove the nut and lock washer from the ball stud that attaches the idler arm to the intermediate rod.
76 Separate the idler arm from the intermediate rod using an appropriate puller.
77 Remove the four bolts, nuts and washers that attach the idler arm bracket to the frame and lift off the idler arm assembly.
78 If the idler arm needs to be removed from the idler arm pivot shaft, remove the nut and lock washer, then separate the two parts using an appropriate puller.
79 Installation is the reverse of the removal procedure, with the following notes.
   a) All nuts should be torqued to their proper specifications.
   b) Use new cotter pins where appropriate.
   c) The idler arm pivot shaft should be lubricated following installation.

20 Power steering system (Series 12)

Refer to Figures 13.71 thru 13.78

Power steering system — general description

1 The hydraulic components used in the power steering system

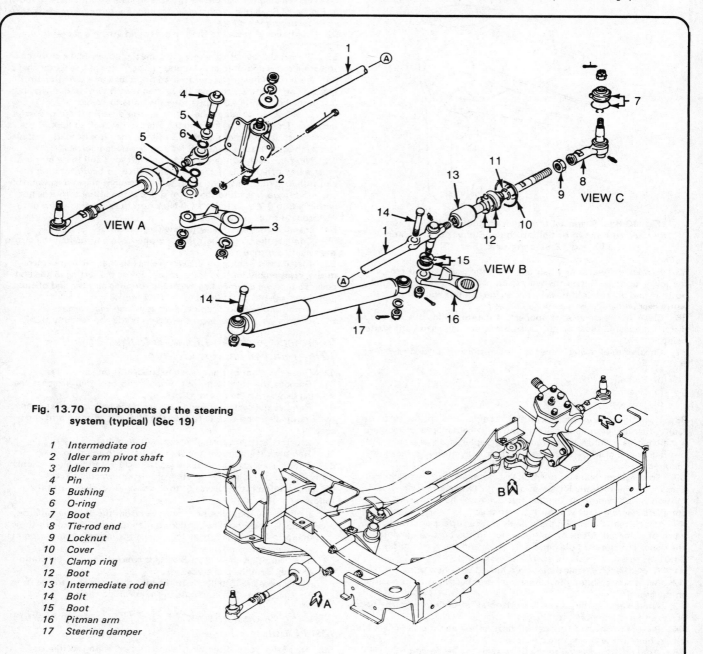

Fig. 13.70 Components of the steering system (typical) (Sec 19)

1 Intermediate rod
2 Idler arm pivot shaft
3 Idler arm
4 Pin
5 Bushing
6 O-ring
7 Boot
8 Tie-rod end
9 Locknut
10 Cover
11 Clamp ring
12 Boot
13 Intermediate rod end
14 Bolt
15 Boot
16 Pitman arm
17 Steering damper

Chapter 13/Supplement: Revisions and information on later models

include a belt-driven oil pump, separate reservoir tank, power steering gear assembly and connecting hoses and lines.

2 Normal maintenance of the power steering system consists mainly of periodically checking the fluid level in the reservoir, keeping the pump drivebelt tension correct and visually checking the hoses for any evidence of fluid leakage. It will also be necessary, after a system component has been removed, to bleed the system as described in this Section.

3 If the operational characteristics of the system appear to be suspect, and the maintenance and adjustment mentioned in this Section is in order, the vehicle should be taken to a dealer, who will have the necessary equipment to check the pressure in the system and the operational torque needed to turn the steering wheel. These two operations are considered beyond the scope of the home mechanic, in view of the high working pressure of the system and the special tools required.

4 If the checks mentioned in paragraph 3 prove that the oil pump assembly is at fault, a new component will have to be purchased, as it is not possible to overhaul it.

5 Excluding the oil pump, reservoir, power steering gear and its

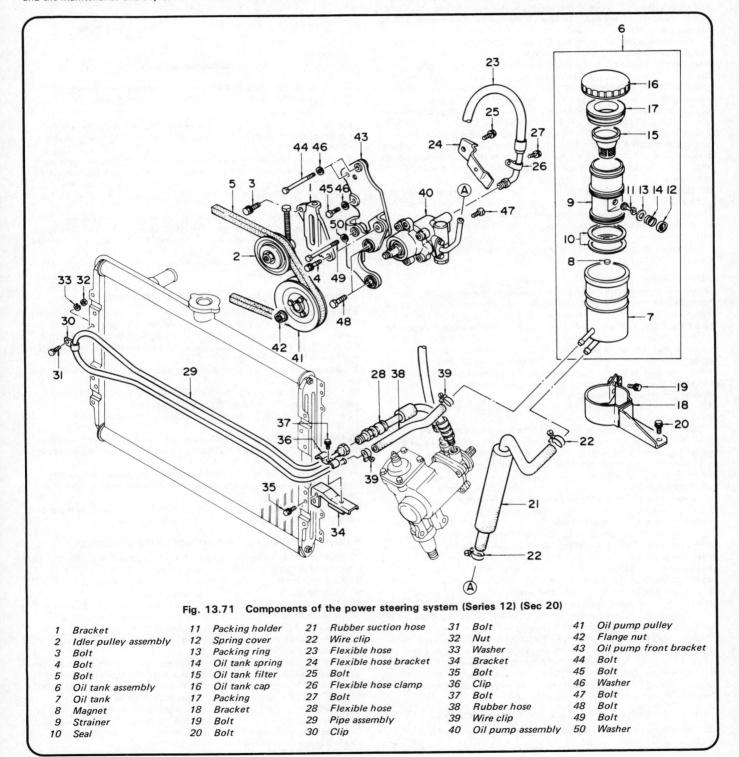

Fig. 13.71 Components of the power steering system (Series 12) (Sec 20)

1	Bracket	11	Packing holder	21	Rubber suction hose	31	Bolt	41	Oil pump pulley
2	Idler pulley assembly	12	Spring cover	22	Wire clip	32	Nut	42	Flange nut
3	Bolt	13	Packing ring	23	Flexible hose	33	Washer	43	Oil pump front bracket
4	Bolt	14	Oil tank spring	24	Flexible hose bracket	34	Bracket	44	Bolt
5	Bolt	15	Oil tank filter	25	Bolt	35	Bolt	45	Bolt
6	Oil tank assembly	16	Oil tank cap	26	Flexible hose clamp	36	Clip	46	Washer
7	Oil tank	17	Packing	27	Bolt	37	Bolt	47	Bolt
8	Magnet	18	Bracket	28	Flexible hose	38	Rubber hose	48	Bolt
9	Strainer	19	Bolt	29	Pipe assembly	39	Wire clip	49	Bolt
10	Seal	20	Bolt	30	Clip	40	Oil pump assembly	50	Washer

associated pressure hoses, the rest of the steering system components are identical to those used on the manual steering models. Servicing these components can be carried out by following the operations described in the relevant Sections of either this Chapter or Chapter 11.

Power steering pump drivebelt — checking and adjustment

6 The power steering pump is driven by a belt which is, in turn, driven off of the crankshaft. Maintaining the proper tension on this belt is very important because if the belt is either too tight or too loose it can put excessive sidewards strain on the various shafts and bearings. It should also be periodically checked for wear, to minimize the possibility of it breaking and thus immobilizing the power steering pump.
7 With the engine off, open the hood and locate the power steering pump to the left of the crankshaft pulley. Using your fingers (and a flashlight if necessary), move along the belt checking for cracks or separation. Also check for fraying and for glazing which gives the belt a shiny appearance. Both sides of the belts should be inspected, which means you will have to twist the belt to check the underside.
8 To check the tension of this belt, find the lower stretch of belt that goes between the crankshaft and the water pump pulley and, gripping it in the center of the stretch, move the belt up and down to gauge its slack. The distance that the belt can be moved up and down should be about 1/2 in (13 mm). If the slack allows the belt to move more or less than this distance, the belt should be adjusted using the following procedure.
9 Locate the drivebelt idler pulley, to the right of the pump. The idler pulley locknut, at the center of the pulley, must first be loosened. Turn the idler pulley adjusting bolt either clockwise or counterclockwise until the proper belt tension is obtained. Retighten the idler pulley nut.
10 After retightening the locknut, recheck the belt tension to make sure it is still correct.

Power steering system — bleeding

Note: *Whenever a hose in the power steering hydraulic system has been disconnected, it is quite probable, no matter how much care was taken to prevent air entering the system, that the system will need bleeding. To do this, proceed as described in the following paragraphs.*
11 First, ensure that the reservoir level is correct; if necessary, add fluid to bring the level to the proper mark. If the vehicle has not just been driven, the power steering oil should be brought up to operating temperature. This can be done either by idling the engine and turning the steering wheel from left to right for about two minutes, or by driving the vehicle for several miles.
12 Raise the front end of the vehicle until the front wheels are just clear of the ground.
13 Quickly turn the steering wheel all the way to the right lock and then the left lock. Do not allow the lock stoppers to be struck with a bang. Try to gauge the end of the lock and only lightly touch the lock stoppers. This operation should be repeated about ten times.
14 Now check the reservoir fluid level again.
15 Start the engine and allow it to idle. Again turn the steering wheel from lock-to-lock several times. Note: *Do not hold the wheel in the full lock position for longer than ten seconds at a time.*
16 Check the reservoir fluid level again and add more fluid if necessary.
17 Lower the front wheels to the ground. Then, again with the engine idling, turn the steering wheel lock-to-lock several times and then bring the front wheels to the straight-ahead position.
18 Shut off the engine and immediately check that the reservoir fluid does not rise in the reservoir. If the fluid level rises sharply, there is still air trapped in the system and the procedure described in the preceding paragraph should be repeated.
19 If it becomes obvious, after several attempts, that the system cannot be satisfactorily bled, there is quite probably a leak in the system. Have an assistant turn the wheel lock-to-lock a few times and then hold it for about five seconds at each stop. Note: *Do not hold the wheel at full lock for more than ten seconds at a time.* Visually check the hoses and their connections for leaks. If no leaks are evident, the problem could be in the steering box itself and the only solution is to have the entire system checked by a Chevrolet dealer.

Power steering system — draining and refilling

20 Raise the front of the truck so that the front wheels are just off the ground.

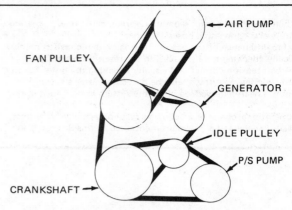

Fig.13.72 Engine drivebelt arrangement for Series 12 models equipped with power steering (Sec 20)

21 Place a drain pan under the power steering fluid reservoir. Then disconnect both hoses from the reservoir and allow the fluid to drain into the pan.
22 After the inital fluid has drained, remove the remaining fluid from the system by turning the steering wheel to its stops in both directions several times. Do not bang the stops but just lightly touch them.
23 After all fluid has drained from the hoses, reconnect them to the reservoir.
24 Remove the cap from the reservoir and remove the filter and strainer. Clean both with soapy water and allow them to air dry.
25 Reinstall the strainer and filter and refill the reservoir with Dexron II ATF to the proper level.
26 After the system has been filled, bleed the system as described in the preceding sub-section.

Power steering pump — removal and installation

27 Remove the engine rock shield.
28 Place a suitable drain pan under the power steering pump. Disconnect the hydraulic hoses from the pump. As each is disconnected, cap or tape over the hose opening and then secure the end in a raised position to prevent leakage.
29 Cover or plug the pump openings to prevent dirt from entering the unit.
30 Loosen the idler pulley locknut and turn the adjusting bolt until the drivebelt can be removed from the pump pulley.
31 Remove the bolts that attach the pump to its bracket.
32 Remove the pump.
33 Overhaul of the power steering pump is not feasible for the home mechanic and, if defective, it should be replaced with a new or rebuilt unit.
34 Installation is the reverse of the removal procedure. Note: *After installing the drivebelt, adjust its tension as described elsewhere in this Section.*
35 Following installation, bleed the power steering system as described elsewhere in this Section.

Power steering gear — removal and installation

36 Prior to removing the steering gear, wash the outside of the assembly thoroughly to remove any built-up dirt or oil.
37 Raise the front of the truck and support it on jackstands.
38 Remove the air intake silencer and duct.
39 Disconnect the hydraulic hoses from the steering gear. As each is disconnected, cap or tape over the hose opening and then secure the end in a raised position to prevent leakage of fluid.
40 Cover or plug the steering gear openings to prevent dirt from entering the assembly.
41 Mark the relationship of the steering gear's stub shaft to the steering column coupling. Then remove the pinch bolt on the universal joint.
42 Remove the two bolts that attach the steering column to the dash.
43 Separate the universal joint from the stub shaft by pulling the steering column into the cab approximately two inches.
44 Working under the truck, remove the Pitman arm nut and washer. Then use an appropriate puller to separate the Pitman arm from the steering gear's Pitman shaft.

Chapter 13/Supplement: Revisions and information on later models

45 Remove the engine rock shield.
46 Remove the bolts that attach the steering gear to the frame and lift the steering gear out.
47 Installation is the reverse of the removal procedure, with the following notes:
   a) When attaching the stub shaft to the universal joint, be sure the alignment marks made during removal are lined up.
   b) Be sure to tighten all bolts and nuts to the specified torque.
48 Following installation, bleed the power steering system as described in this Section.

Power steering gear — disassembly, inspection and reassembly

49 Remove the power steering gear as described elsewhere in this Section.

Disassembly

Note: *When securing the power steering gear in a vise, do not clamp the power cylinder portion of the steering gear.*

50 Remove the dust cover from the stub shaft. Clean the faces of the stub shaft thoroughly. Remove the retaining ring and back-up ring from

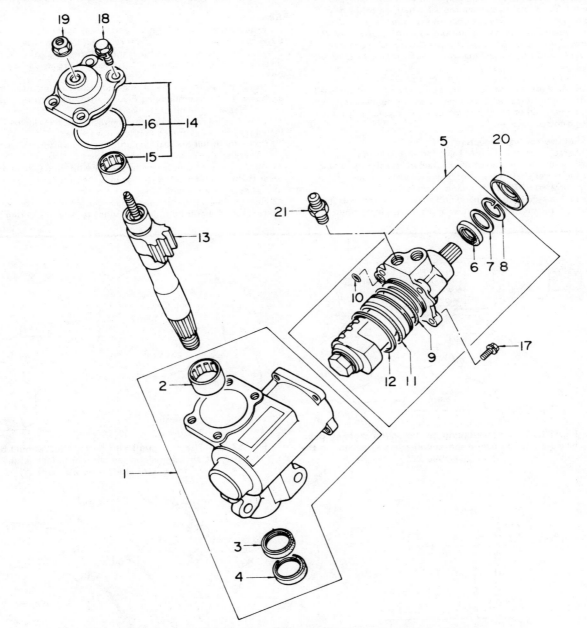

Fig.13.73 Components of the power steering gear assembly (Sec 20)

1 Gear box assembly
2 Sector shaft bearing
3 Sector shaft seal ring
4 Sector shaft dust seal ring
5 Ball screw and valve housing assembly
6 Stub shaft oil seal
7 Back up ring
8 Retaining ring
9 Valve housing O-ring
10 Fluid passage O-ring
11 Piston O-ring
12 Piston seal ring
13 Sector shaft
14 Top cover assembly
15 Sector shaft bearing
16 Top cover O-ring
17 Valve housing retaining bolt
18 Top cover retaining bolt
19 Adjusting screw locknut
20 Dust cover
21 Hose fitting

around the stub shaft.
51 Securely plug the inlet hydraulic hose fitting. Then apply compressed air into the outlet hose fitting and blow out the stub shaft oil seal.
52 Remove the adjusting screw locknut. Then turn the adjusting screw counterclockwise. This will remove the preload between the sector gear and the rack piston.
53 Remove the top cover bolts. While holding the cover stationary, turn the adjusting screw clockwise enough to raise and free the cover.
54 Clean the faces of the sector shaft. Bring the stub shaft into a straight-ahead position as shown in the accompanying Figure, then withdraw the sector shaft from the gear box. With both shafts held in the straight-ahead position, it should come out without trouble.
55 Remove the four bolts that attach the valve housing to the gear box. Separate the valve housing with ball screw from the gear box.
Note: *Keep the ball screw and valve housing assembly horizontal. If it is held vertical, the rack piston will fall off the end of the worm, causing the rack piston to slip off the worm shaft and the balls to fall out.*
56 The valve housing O-rings, piston seal ring and O-ring, top cover O-ring and sector shaft seals can be removed using a piece of wire. All of these parts should be replaced with new ones during reassembly.
57 Thoroughly wash all of the disassembled parts in clean solvent and dry them using compressed air.

Ball screw and valve housing inspection
58 Inspect the rack piston and worm shaft for excessive end play.
59 Inspect the outer circumference of the rack piston and teeth where they engage the sector shaft. Check for excessive wear or damage.
60 Check the serrated surfaces for wear or damage.
61 Hold the rack piston with one hand and slowly turn the stub shaft, checking the rack piston for smoothness of rotation. There should be no roughness or binding.
62 If any parts of the ball screw or valve housing are defective, the entire assembly must be replaced as a unit.

Gear box and sector shaft inspection
63 Carefully inspect the gear box for any signs of wear or damage from the sliding of the piston. Replace the gear box if the piston sliding surfaces are excessively worn or damaged.
64 Use your finger to slowly turn the sector shaft bearing, noting any noise, roughness, looseness or binding. Replace the bearing if necessary.
65 Inspect the teeth and the serrated surface of the sector shaft for wear or damage and replace the sector shaft if necessary.

Reassembly
66 Prior to reassembly, clean all parts thoroughly.
67 If the sector shaft bearing was removed from the gear box, install it so the face stamped with the name is turned toward you. Also when installed, the bearing should be flush with the recessed face of the gear box.
68 Install the sector shaft seal ring and dust seal ring into the gear box as shown in the accompanying Figure. After installation, apply a thin coat of grease to the lip of the seals.
69 Apply a thin coat of grease to the rack piston O-ring and carefully install it onto the piston so that it is not twisted.
70 Carefully install the seal ring onto the rack piston and then apply a coat of grease to the circumference of the seal ring.
71 Apply a thin coat of grease to the valve housing O-rings and top cover O-ring and install the O-rings into their respective grooves. Be careful not to twist them during installation.
72 Keeping the ball screw/valve housing assembly horizontal, carefully insert it into the gear box. Be careful not to drop the O-ring fitted to

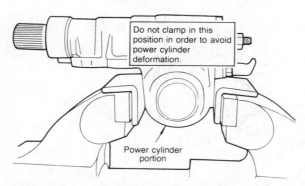

Fig. 13.74 When working on the power steering gear, be sure NOT to clamp the power cylinder portion in the vise (Sec 20)

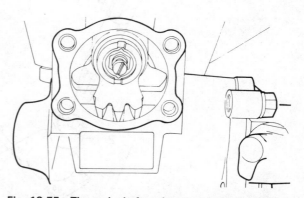

Fig. 13.75 The stub shaft and sector shaft should be in the straight ahead-position (as shown) before removing the sector shaft (Sec 20)

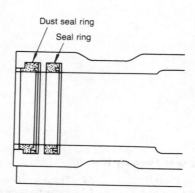

Fig. 13.76 Proper installed positions of the sector shaft seal ring and dust seal ring (Sec 20)

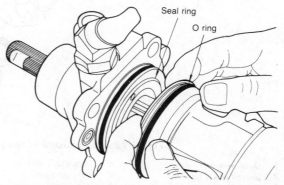

Fig. 13.77 Installed locations of the rack piston O-ring and seal ring (Sec 20)

the oil passage in the valve housing.
73 Install the valve housing attaching bolts and tighten them to the specified torque.
74 Using an appropriate sized socket, install a new stub shaft oil seal into the valve housing. Then install the back-up ring and retaining ring, each positioned so that the face with the rounded edge is turned toward the oil seal. Finally install the dust cover.
75 If the hose fitting was removed from the valve housing, install it with a new O-ring so it is set in the direction shown in the accompanying Figure. Tighten the hose fitting locknut to the specified torque.
76 Apply electrical tape to the splines of the sector shaft, then with it held in the straight-ahead position, install the shaft. The sector shaft and rack should be aligned as shown in the accompanying Figure.
77 Thread the adjusting screw into the top cover and turn the screw counterclockwise until the cover just contacts the gear box. Continue turning for two more turns, then install the top cover bolts, tightening them to the specified torque.
78 At this point, slowly turn the stub shaft from lock-to-lock counting the number of turns necessary. If it takes more than four turns, the sector shaft and rack piston are installed properly. If not, the sector shaft should be removed again and reinstalled correctly.
79 When installed correctly, remove the tape from the sector shaft.
80 Adjust the backlash between the rack piston and the sector shaft using the following procedure:
   a) If not already done, wrap the splines of the stub shaft with several layers of electrical tape, so that a 17 mm - 12 point socket can be fitted over the shaft without damaging the splines.
   b) Hold the piston and sector shaft in the straight-ahead position and use an inch-pound torque wrench with the 17 mm socket to slowly turn the stub shaft. Note the amount of torque necessary to start the stub shaft rotating.
   c) The normal amount of starting torque required is 5.2 to 7.8 in-lb. If the stub shaft required more or less torque, turn the adjusting screw until the correct torque is achieved. After the adjustment is correct, remove the tape from the shaft.
81 Install a new adjusting screw locknut and tighten it to the specified torque. Be sure the adjustment is not altered when the locknut is installed.

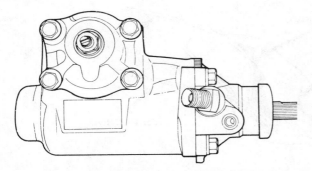

Fig. 13.78 The valve housing hose fitting should be installed in the direction shown above (Sec 20)

21 Body

Refer to Figure 13.79 and 13.80

Grille (Series 11 and 12) — removal and installation

1 The grille is retained by nine clips. The grille can be released from these clips simply by raising the interlocking tab of each clip.
2 When installing the grille, set the clips into position on the grille first, then push the grille onto the body until all the clips click into place.

Hood latch (Series 11 and 12) — removal and installation

3 Remove the clip on the outer end of the cable.
4 Remove the two screws that retain the hood latch and lift off.
5 Installation is the reverse of the removal procedure.

Hood control cable (Series 11 and 12) — removal and installation

6 Remove the grille as previously described.
7 Open the hood, then disconnect the inner cable from the slot in the hood latch.
8 Remove the cable from the retaining clip on the fender skirt.
9 Depress the grommet on the dash panel toward the cab, then pull the hood control cable into the cab.
10 Installation is the reverse of the removal procedure.

Front bumper (Series 11 and 12) — removal and installation

11 Remove the screws that retain the front combination lights and the front side marker lights to the front bumper, then lift out the lights from the bumper.
12 Remove the three bolts on each side that retain the bumper to the backbars and lift off the bumper.
13 The backbars can be removed by removing the bolts that retain them to the frame.
14 If necessary, the side sections of the bumper can be separated from the center section by removing the bolts that attach them.
15 If more convenient, the bumper and backbars can also be removed as one assembly.
16 Installation is the reverse of the removal procedure.
17 If necessary during installation, the bumper-to-fender alignment can be adjusted by using shims placed between the back bar and front side bumper. When properly set, there is a 0.65 in (16.5 mm) clearance between the lower side of the fender and the upper side of the side bumper section.

Front fender (Series 11 and 12) — removal and installation

18 Remove the screws that retain the front side marker light to the front bumper and lift out the light.
19 Remove the bolts attaching the side section of the front bumper to the center section and lift off the side section.
20 There are five screw-type rivets retaining the inner fender liner. Turn them counterclockwise to remove them.
21 Disengage the five clips located at the edge of the fender, while pulling the inner fender inward, and remove the liner.
22 Remove the bolts that attach the front fender to the deflector and rocker panel.
23 Open the door and remove the bolts on the upper part of the hinge pillar.
24 Remove the cover or the speaker on the dash side and then remove the fender mounting bolt accessible through the opening.
25 Remove the five bolts that retain the front fender to the fender skirt and lift off the fender.
26 Installation is the reverse of the removal procedure.

Door trim panel (Series 11 and 12) — removal and installation

27 Press the door trim panel in around the window regulator handle and withdraw the handle retaining clip.
28 Lift off the window handle and escutcheon.
29 With the inner door handle lever pulled inward, remove the screw that retains the bezel, then lift off the bezel.
30 Remove the screws that retain the armrest.
31 Remove the inner waist seal.
32 Remove the door lock knob escutcheon.
33 The trim panel is retained by ten plastic bullet-type fasteners. The panel can be pryed off the door by carefully using a screwdriver at each fastener. **Note:** *The upper edge of the trim panel is secured by adhesive tape.* If the trim panel does not have to be removed completely, as during replacement, do not disengage this tape. To keep the panel out of the way, swing the lower edge up and tape it to the upper edge of the door.
34 Installation is the reverse of the removal procedure, with the following notes.
   a) If the trim panel has been completely removed from the door, be sure to secure the upper edge with double-adhesion tape.
   b) Prior to installing the window regulator handle, insert the

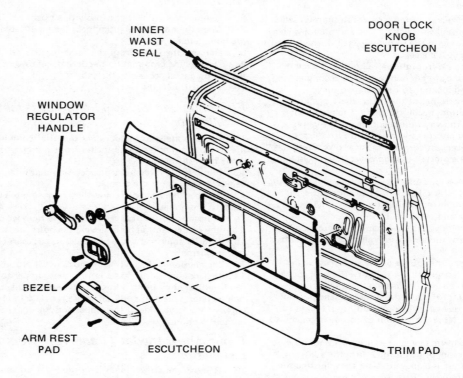

Fig. 13.79 Door trim panel and related exterior parts (Series 11 and 12) (Sec 21)

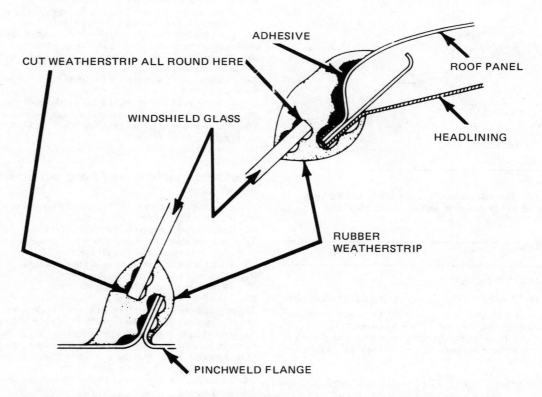

Fig. 13.80 To remove the windshield, the weatherstrip must be cut at the places shown (Sec 21)

retaining clip into its slot, then push the handle onto its shaft until the clip clicks into place.

Door lock cylinder (Series 11 and 12) — removal and installation

35 Remove the door trim panel as previously described.
36 Partially detach the inner water deflector to allow access to the door lock cylinder.
37 Raise the window all the way.
38 Disengage the door lock link from the nylon fastener.
39 Release the lock cylinder retaining clip with pliers and then push the cylinder out on the exterior side of the door.
40 Installation is the reverse of the removal procedure.

Window regulator — removal and installation

41 The procedure is similar to that described in Chapter 12 but, in addition, it will be necessary to remove the two screws retaining the bottom channel and regulator.

Door glass — removal and installation

42 Remove the door trim panel and peel back the water deflectors.
43 Remove the window regulator with reference to paragraph 41 and Chapter 12.
44 Remove the inner weatherstrip and waist seal assembly, then extract the glass by lowering it and tilting it so that it can be withdrawn.
45 Remove the bottom channel by tapping carefully with a soft-faced hammer.
46 Installation is the reverse of removal, but make sure that the glazing rubber is located in the bottom channel and that the rear of the channel is about 2-1/8 in from the rear of the glass.

Windshield glass — removal

47 If the windshield glass is to be removed complete and intact, it is recommended that the weatherstrip first cut as shown in Figure 13.80. This will reduce the possibility of breaking the glass. It will of course be necessary to obtain a new weatherstrip before installing the glass.

268

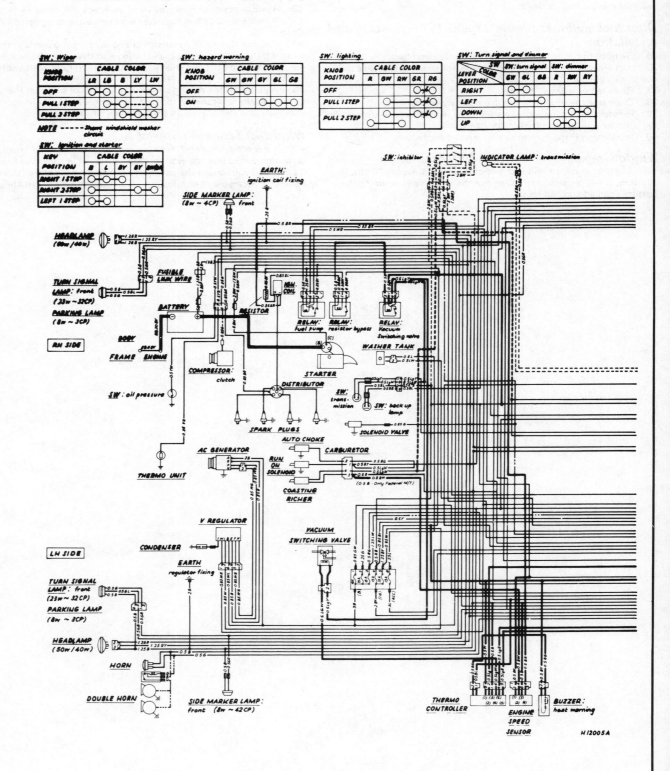

Fig. 13.81 Wiring diagram (Series 8) (1 of 2)

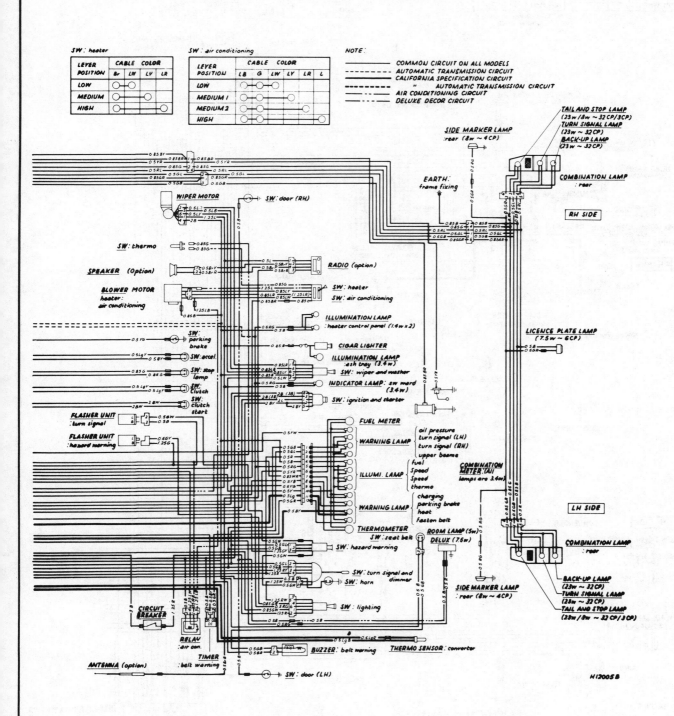

Fig. 13.82 Wiring diagram (Series 8) (2 of 2)

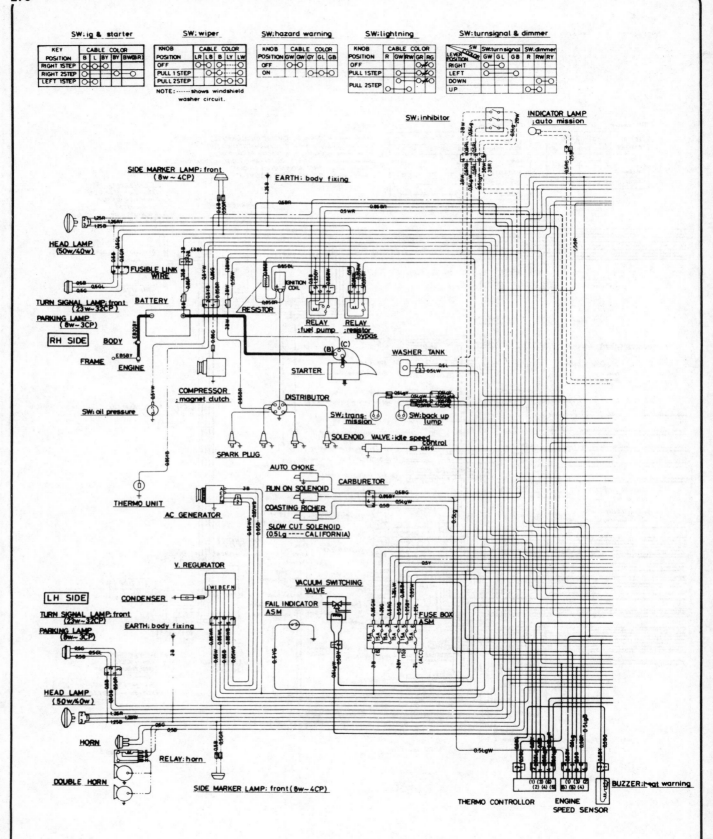

Fig. 13.83 Wiring diagram (Series 9 and 10) (1 of 2)

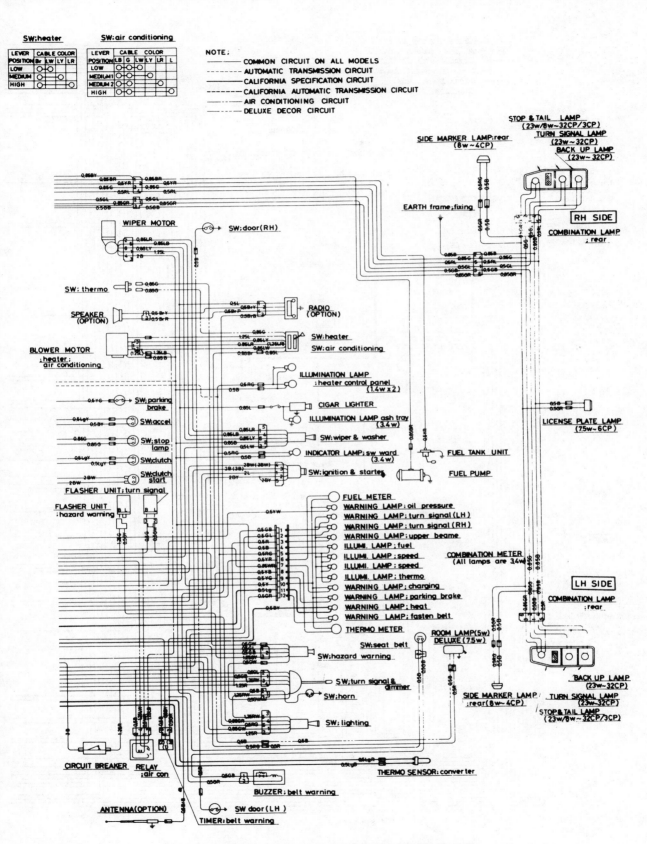

Fig. 13.84 Wiring diagram (Series 9 and 10) (2 of 2)

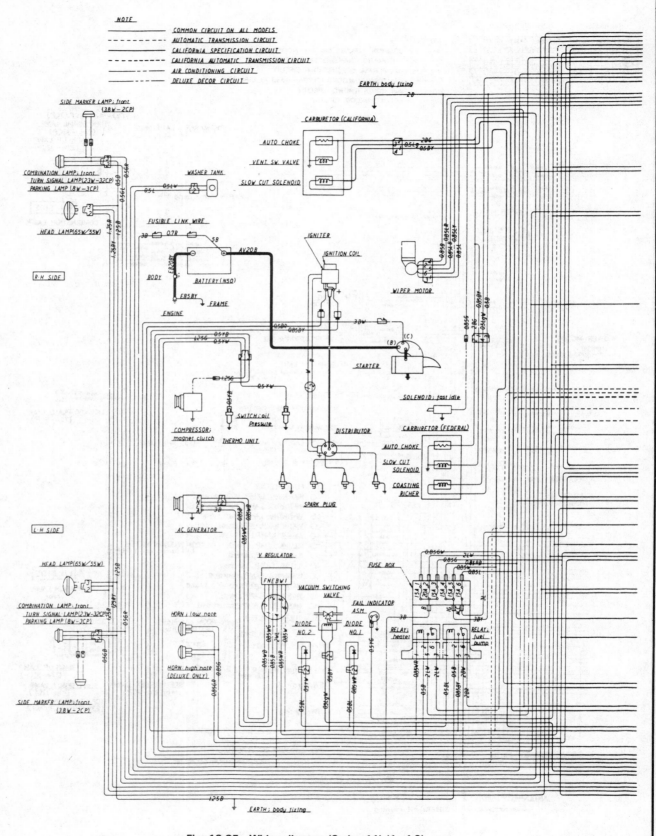

Fig. 13.85 Wiring diagram (Series 11) (1 of 2)

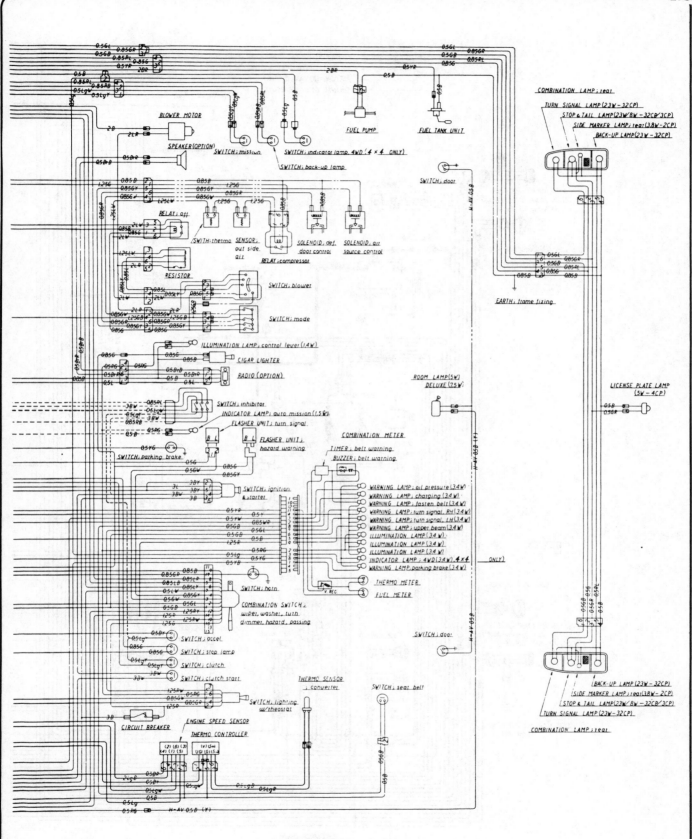

Fig. 13.86 Wiring diagram (Series 11) (2 of 2)

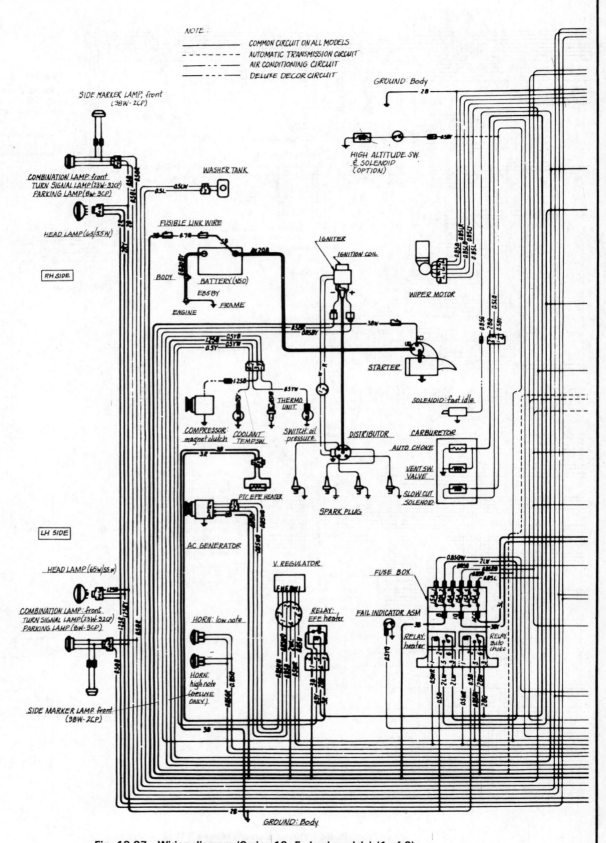

Fig. 13.87 Wiring diagram (Series 12, Federal models) (1 of 2)

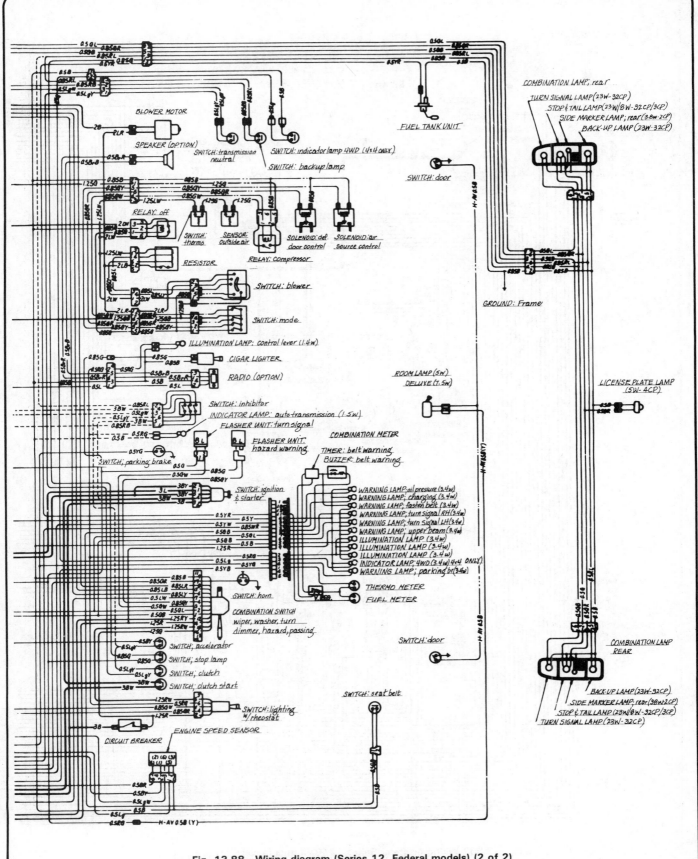

Fig. 13.88 Wiring diagram (Series 12, Federal models) (2 of 2)

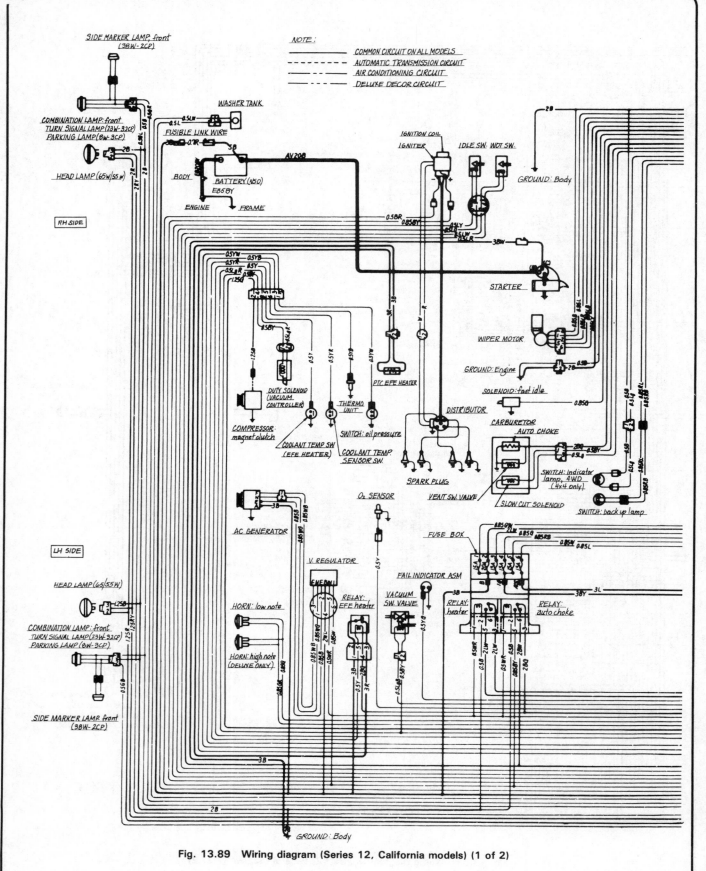

Fig. 13.89 Wiring diagram (Series 12, California models) (1 of 2)

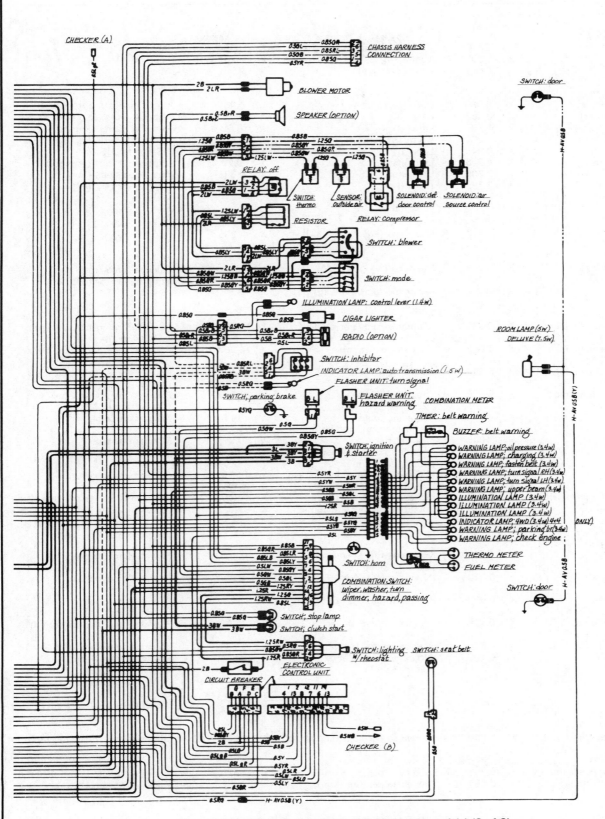

Fig. 13.90 Wiring diagram (Series 12, California models) (2 of 2)

Conversion factors

Length (distance)
Inches (in)	X	25.4	= Millimetres (mm)	X 0.0394	= Inches (in)
Feet (ft)	X	0.305	= Metres (m)	X 3.281	= Feet (ft)
Miles	X	1.609	= Kilometres (km)	X 0.621	= Miles

Volume (capacity)
Cubic inches (cu in; in³)	X	16.387	= Cubic centimetres (cc; cm³)	X 0.061	= Cubic inches (cu in; in³)
Imperial pints (Imp pt)	X	0.568	= Litres (l)	X 1.76	= Imperial pints (Imp pt)
Imperial quarts (Imp qt)	X	1.137	= Litres (l)	X 0.88	= Imperial quarts (Imp qt)
Imperial quarts (Imp qt)	X	1.201	= US quarts (US qt)	X 0.833	= Imperial quarts (Imp qt)
US quarts (US qt)	X	0.946	= Litres (l)	X 1.057	= US quarts (US qt)
Imperial gallons (Imp gal)	X	4.546	= Litres (l)	X 0.22	= Imperial gallons (Imp gal)
Imperial gallons (Imp gal)	X	1.201	= US gallons (US gal)	X 0.833	= Imperial gallons (Imp gal)
US gallons (US gal)	X	3.785	= Litres (l)	X 0.264	= US gallons (US gal)

Mass (weight)
Ounces (oz)	X	28.35	= Grams (g)	X 0.035	= Ounces (oz)
Pounds (lb)	X	0.454	= Kilograms (kg)	X 2.205	= Pounds (lb)

Force
Ounces-force (ozf; oz)	X	0.278	= Newtons (N)	X 3.6	= Ounces-force (ozf; oz)
Pounds-force (lbf; lb)	X	4.448	= Newtons (N)	X 0.225	= Pounds-force (lbf; lb)
Newtons (N)	X	0.1	= Kilograms-force (kgf; kg)	X 9.81	= Newtons (N)

Pressure
Pounds-force per square inch (psi; lbf/in²; lb/in²)	X	0.070	= Kilograms-force per square centimetre (kgf/cm²; kg/cm²)	X 14.223	= Pounds-force per square inch (psi; lbf/in²; lb/in²)
Pounds-force per square inch (psi; lbf/in²; lb/in²)	X	0.068	= Atmospheres (atm)	X 14.696	= Pounds-force per square inch (psi; lbf/in²; lb/in²)
Pounds-force per square inch (psi; lbf/in²; lb/in²)	X	0.069	= Bars	X 14.5	= Pounds-force per square inch (psi; lbf/in²; lb/in²)
Pounds-force per square inch (psi; lbf/in²; lb/in²)	X	6.895	= Kilopascals (kPa)	X 0.145	= Pounds-force per square inch (psi; lbf/in²; lb/in²)
Kilopascals (kPa)	X	0.01	= Kilograms-force per square centimetre (kgf/cm²; kg/cm²)	X 98.1	= Kilopascals (kPa)
Millibar (mbar)	X	100	= Pascals (Pa)	X 0.01	= Millibar (mbar)
Millibar (mbar)	X	0.0145	= Pounds-force per square inch (psi; lbf/in²; lb/in²)	X 68.947	= Millibar (mbar)
Millibar (mbar)	X	0.75	= Millimetres of mercury (mmHg)	X 1.333	= Millibar (mbar)
Millibar (mbar)	X	0.401	= Inches of water (inH$_2$O)	X 2.491	= Millibar (mbar)
Millimetres of mercury (mmHg)	X	0.535	= Inches of water (inH$_2$O)	X 1.868	= Millimetres of mercury (mmHg)
Inches of water (inH$_2$O)	X	0.036	= Pounds-force per square inch (psi; lbf/in²; lb/in²)	X 27.68	= Inches of water (inH$_2$O)

Torque (moment of force)
Pounds-force inches (lbf in; lb in)	X	1.152	= Kilograms-force centimetre (kgf cm; kg cm)	X 0.868	= Pounds-force inches (lbf in; lb in)
Pounds-force inches (lbf in; lb in)	X	0.113	= Newton metres (Nm)	X 8.85	= Pounds-force inches (lbf in; lb in)
Pounds-force inches (lbf in; lb in)	X	0.083	= Pounds-force feet (lbf ft; lb ft)	X 12	= Pounds-force inches (lbf in; lb in)
Pounds-force feet (lbf ft; lb ft)	X	0.138	= Kilograms-force metres (kgf m; kg m)	X 7.233	= Pounds-force feet (lbf ft; lb ft)
Pounds-force feet (lbf ft; lb ft)	X	1.356	= Newton metres (Nm)	X 0.738	= Pounds-force feet (lbf ft; lb ft)
Newton metres (Nm)	X	0.102	= Kilograms-force metres (kgf m; kg m)	X 9.804	= Newton metres (Nm)

Power
Horsepower (hp)	X	745.7	= Watts (W)	X 0.0013	= Horsepower (hp)

Velocity (speed)
Miles per hour (miles/hr; mph)	X	1.609	= Kilometres per hour (km/hr; kph)	X 0.621	= Miles per hour (miles/hr; mph)

Fuel consumption*
Miles per gallon, Imperial (mpg)	X	0.354	= Kilometres per litre (km/l)	X 2.825	= Miles per gallon, Imperial (mpg)
Miles per gallon, US (mpg)	X	0.425	= Kilometres per litre (km/l)	X 2.352	= Miles per gallon, US (mpg)

Temperature

Degrees Fahrenheit = (°C x 1.8) + 32   Degrees Celsius (Degrees Centigrade; °C) = (°F - 32) x 0.56

*It is common practice to convert from miles per gallon (mpg) to litres/100 kilometres (l/100km), where mpg (Imperial) x l/100 km = 282 and mpg (US) x l/100 km = 235

Index

A

Accelerator — 76
Air cleaners — 59
Air conditioning — 165, 248
Air injection system — 73, 220
Alternator — 150, 226
Alternator regulator — 153
Anti-roll bar — 203
Automatic temperature — 60
Automatic transmission
    extension housing seal — 124
    fault diagnosis — 124
    fluid level — 122, 232
    general discription — 121
    inhibitor switch — 123
    kick-down switch — 123
    maintenance — 122
    rear oil seal — 124
    removal and installation — 124
    selector linkage — 122
    shift lever — 123
    shift linkage adjustment — 122
    starter inhibitor — 123
    throttle valve control cable — 123
Axleshaft — 125, 127, 238
Axleshaft (4X4) — 248, 251

B

Battery
    charging — 150
    electrolyte replenishment — 150
    maintenance and inspection — 149
    removal and installation — 149
Big-end (connecting rod bearings) — 27, 45
Bodywork and fittings
    air conditioning — 165
    bumpers — 208, 265
    door glass — 204, 267
    doors — 199, 201, 265, 267
    door locks — 201, 267
    fenders — 265
    fresh air outlet — 163
    grille — 265
    heater — 161-163
    hood — 208
    hood latch — 265
    maintenance — 198, 199
    rear mirrors — 205
    repairs
        major damage — 199
        minor damage — 198
    windshield/rear window — 204, 267

Braking system
    bleeding hydraulic system — 136
    brake adjustment — 134, 239
    brake disc/hub — 141, 241
    brake hoses — 138
    brake pedal — 134, 145, 239, 243
    brake shoes — 138, 141, 239, 241
    caliper — 140, 240
    disc pads — 140, 239
    fail indicator assembly (Series 8 to 12) — 244
    fault diagnosis — 146
    general description — 134
    master cylinder, single — 136
    master cylinder, tandem — 134, 242
    parking brake (handbrake) — 136, 145, 238, 243
    power cylinder — 143, 243
    wheel cylinder — 143, 241
Bulbs — 148
Buying spare parts — 4

C

Cables
    accelerator — 76
    caliper brake — 140
    hood — 208
    parking brake — 136, 145, 243, 244
Camshaft removal — 24, 27, 45
Carburettors
    adjustments (Series 9 thru 12) — 216
    air cleaners — 59
    choke, automatic — 69
    emission controls — 73, 216
    exhaust system — 83
    fast idle — 60, 216
    fault diagnosis — 83
    float level — 67
    fuel pump — 69, 70
    overhaul — 60, 64, 65
    Series 1 and 2 models — 60
    Series 3 models — 64
    Series 4 to 6 models — 65
    Series 8 thru 12 models — 219
Charging system (Series 11 and 12) — 226
Closed loop emission control system — 220
Clutch
    adjustment — 98, 228
    bleeding — 98
    cable removal and installation — 228
    fault diagnosis — 102
    general description — 98, 228
    inspection and renovation — 101
    installation — 102, 228
    judder — 102

    master cylinder — 99
    operating cylinder — 100
    pedal removal and installation — 98, 228
    release bearing — 102, 228
    removal — 100, 228
    shift fork — 98
    slip — 102
    spin — 102
Coasting fuel cut-off system — 222
Coasting richer system — 222
Coil — 95
Condenser — 88
Connecting rods — 27, 45
Contact breaker points — 88
Cooling system
    antifreeze — 53
    draining — 53
    fan belt adjustment — 54
    fault diagnosis — 55
    filling and flushing — 53
    radiator — 54
    thermostat — 54
    water pump — 55
Crankcase emission control — 81, 83
Crankshaft and main bearings — 25, 27, 45
Cylinder head — 23, 34, 42

D

Dash pot adjustment — 222
Decarbonisation — 29
Diagram, wiring — 168, 177, 268 - 277
Differential — 131, 252
Disc pads — 140, 239
Distributor
    adjusting points — 89
    disassembly and reassembly — 89, 226
    removal and installation — 89, 222
    timing-stroboscope lamp — 94
Doors — 199, 201, 265, 267
Driveshaft (rear axle)
    balance — 127
    center bearing — 238
    description — 125
    description (long wheelbase models) — 237
    dismantling and reassembly — 125
    removal and installation — 125
    removal and installation (long wheelbase models) — 238

E

Early Fuel Evaporation (EFE) system — 223
Electrical system
    alternator/generator — 150, 226
    battery — 150
    coil — 95
    fault diagnosis — 166
    fuses — 155
    fuses (Series 8 thru 12) — 244
    horns — 160
    instrument panel meters and gauges — 157, 245, 246
    lamps — 155, 156, 244
    speedometer — 157
    starter motor — 153, 155
    switches — 157, 245
    voltage regulator — 152
    windshield washer — 160
    windshield wiper — 158, 246
    wiring diagrams — 168, 177, 268 - 277
Emission control systems (Series 8 thru 12) — 220
Engine (Series 1 to 4)
    ancillary components — 21, 25
    big-end bearings — 27
    camshaft — 24, 27
    connecting rods — 27
    crankshaft removal — 25, 27
    cylinder head — 23, 24
    decarbonising — 29
    dismantling — general — 21
    fault diagnosis — 49
    flywheel — 29, 40
    installation — 37
    oil filter — 34
    oil pan — 23, 34
    oil pump — 23, 30, 33
    pistons and rings — 24, 31
    removal — general — 19
    rocker assembly — 29
    tappets — 37
    timing chain, gear and tensioner — 23, 30
    valve clearances and adjustments — 37
    valves and seats — 29
Engine (Series 5 and 6)
    ancillary components — 42
    big-end bearings — 45
    camshaft — 45
    connecting rods — 45
    crankshaft removal — 45
    cylinder head — 42
    decarbonising — 29
    dismantling — general — 42
    fault diagnosis — 49
    flywheel — 40, 46
    installation — 43
    oil pan — 43
    oil pump — 46
    pistons and rings — 45
    removal — general — 42
    rocker assembly — 43
    tappets — 37
    timing gear, chain and tensioner — 45
    valve clearances and adjustments — 48
    valves and seats — 49
Engine (Series 8 thru 12) — 215
Evaporative emission control — 83, 225
Exhaust gas recirculation — 77, 225
Exhaust systems — 83

F

Fan belt adjustment — 54
Fault diagnosis
    automatic transmission — 124
    braking system — 146
    clutch — 102
    cooling system — 55
    electrical system — 166
    engine — 49
    fuel system — 83, 84
    ignition — 95
    manual gearbox — 120
    rear axle — 132
    suspension and steering — 196
Filters
    air — 59
    fuel — 73
    oil — 34
Flexible coupling — 191
Flywheel — 29, 40, 46
Free-wheeling hub — 248
Front-wheel drive system — 248
Fuel and exhaust systems
    accelerator cable — 76
    air cleaner — 59
    air injection system — 73
    carburettors — 60, 64, 65, 216, 219
    choke — 69

Index

exhaust systems — 83
fault diagnosis — 83, 84
fuel evaporative emission control system — 83, 225
fuel tank — 70
fuel pump — 69, 70, 219
Fuses and fusible link — 155, 244

G

Gearbox, manual
dismantling — 105, 108, 111, 232
fault diagnosis — 120
reassembly — 114, 117, 119, 232
removal — 105, 228
steering column linkage — 191
Generator (Series 11 and 12) — 226
Glass
door — 204, 267
ventilator window — 204
windshield — 204, 205, 267
Glossary — 7
Gudgeon (piston) pin removal — 25, 31, 45

H

Halfshafts — 125, 127
Handbrake — 136, 238
Hazard warning switch — 161
Headlamp units — 155, 244
Heater
air inlet door cable adjustment (Series 8 thru 12) — 248
blower motor and heater core unit — 163, 248
control cable — 163, 246, 248
controls — 163, 246, 248
dash outlet grilles — 164, 248
removal and refitting — 161, 248
ventilation air inlet valve — 163
water valve — 163
Hood (bonnet) — 208
Horn — 160
Hub (4x4) — 248, 254
Hydraulic systems
bleeding brakes — 136
bleeding clutch — 98

I

Identification numbers — 4
Idle adjustment — 215
Idle mixture adjustment — 216
Ignition system
coil — 95
condenser — 88
contact breaker — 88
distributor — 89, 226
dwell angle, checking — 93
fault diagnosis — 95
general description — 86, 225
igniter — 95
routine maintenance — 86
spark plugs and HT leads — 95
switch — 157
timing (mechanical type) — 94
Instrument panel — 157, 245, 246

J

Jacking and towing points — 12

K

Kickdown switch — 123

L

Leads, HT - 95
Lighting system
bulb renewal — 156, 157
headlamps — 155, 156, 244
switches — 157, 245
Locks, door — 201
Lubrication chart — 8
Lubrication system — 25, 45

M

Main bearings — 27, 45
Mainshaft dismantling — 110
Maintenance, routine — 9
Major operations, engine in vehicle — 21
Manifolds — 43
Manual transmission — 104, 228
Master cylinder, brake — 136, 137, 242
Master cylinder, clutch — 99
Metric conversion tables — 278
Modifications (Series 8 thru 12 models) — 210

O

Oil filter — 34
Oil pan — 23, 34, 43, 215
Oil pump — 23, 30, 33, 46
Ordering spare parts — 4

P

Parking brakes — 136, 145, 238
Pistons and rings — 24, 31, 45
Power steering system
bleeding — 262
draining and filling — 262
drive belt — 262
general description — 260
pump — 262
steering gear — 262, 263
Propeller shaft
removal and installation — 125, 237, 238
universal joints — 125
Pump
air — 73
fuel — 69, 70
oil — 23, 30, 33, 46
power steering — 262
water — 55

R

Radiator
antifreeze — 53
draining, filling, flushing — 53
removal — 53
Radio installation — 161, 245
Rear axle
axleshaft — 129, 238
differential — 131
fault diagnosis — 132
general description — 129
housing removal and installation — 131
removal and installation — 129
Rings, piston — 24, 31, 45
Routine maintenance — 9, 215

S

Shock absorbers — 186
Spark plugs and HT leads — 95
Speakers — 245
Specifications, technical
   automatic transmission — 121
   bodywork and fittings — 197, 214
   braking system — 133, 134, 213
   carburetion, fuel and emission control systems — 56, 57, 211
   clutch — 97, 98, 212
   cooling system — 52, 211
   driveshaft — 125
   electrical system — 147, 148, 149, 213
   engine — 16, 39, 210
   front axle assembly — 213
   front differential — 213
   ignition system — 85, 86, 211
   rear axle — 128
   starter motor — 153
   suspension and steering — 178, 179, 180, 213, 214
   transmission
      automatic — 121
      manual — 103, 104, 212
   transmission/transfer case — 212
Speedometer cable — 157
Steering and suspension
   balljoints — 187, 189, 256, 257
   caster angle — 257
   control arm — 257
   fault diagnosis — 196
   front hubs — 181
   front wheel alignment — 178
   front wheel bearings — 181, 252, 255
   idler arm — 191, 260
   pitman shaft seal — 194
   power steering system — 260
   rear spring — 196
   ride height checking — 257
   shock absorbers — 186
   stabilizer bar — 186, 257
   steering angles — 181, 183
   steering column — 257
   steering damper — 259
   steering gear — 189, 192, 258
   steering knuckle (4 x 4) — 256
   steering linkage — 190
   steering shaft — 191, 257
   steering wheel — 190, 257
   strut bar — 186, 187
   tie rods — 259
   torsion bar spring — 185, 189

Sump — see oil pan
Switches — 157, 245

T

Technical specifications — see Specifications, technical
Thermostat — 165
Timing, ignition — 85
Tyre pressures — 179
Tools and working facilities — 14
Transmission
   cleaning and inspection — 113
   removal (Series 1 thru 6) — 105
   removal (Series 8 thru 12) — 228
Transmission/transfer case (4 x 4)
   transfer case
      disassembly — 234
      reassembly — 235
      removal — 232
   transmission
      disassembly — 232
      reassembly — 236
      removal — 232

U

Universal joints — 125
Upholstery — 198

V

Valve
   clearances — 37, 49, 215
   grinding — 37
   seats — 37
Voltage regulator — 152

W

Water pump — 55
Wheel bearings
   front — 181, 252, 255
   rear — 130
Windshield
   glass — 204, 205, 267
   washers — 160
   wiper blades and arms — 160
   wiper motor — 160, 246
Window regulator — 267
Wiring diagrams — 168, 177, 268 to 277

Safety first!

Regardless of how enthusiastic you may be about getting on with the job at hand, take the time to ensure that your safety is not jeopardized. A moment's lack of attention can result in an accident, as can failure to observe certain simple safety precautions. The possibility of an accident will always exist, and the following points should not be considered a comprehensive list of all dangers. Rather, they are intended to make you aware of the risks and to encourage a safety conscious approach to all work you carry out on your vehicle.

Essential DOs and DON'Ts

DON'T rely on a jack when working under the vehicle. Always use approved jackstands to support the weight of the vehicle and place them under the recommended lift or support points.
DON'T attempt to loosen extremely tight fasteners (i.e. wheel lug nuts) while the vehicle is on a jack — it may fall.
DON'T start the engine without first making sure that the transmission is in Neutral (or Park where applicable) and the parking brake is set.
DON'T remove the radiator cap from a hot cooling system — let it cool or cover it with a cloth and release the pressure gradually.
DON'T attempt to drain the engine oil until you are sure it has cooled to the point that it will not burn you.
DON'T touch any part of the engine or exhaust system until it has cooled sufficiently to avoid burns.
DON'T siphon toxic liquids such as gasoline, antifreeze and brake fluid by mouth, or allow them to remain on your skin.
DON'T inhale brake lining dust — it is potentially hazardous (see *Asbestos* below)
DON'T allow spilled oil or grease to remain on the floor — wipe it up before someone slips on it.
DON'T use loose fitting wrenches or other tools which may slip and cause injury.
DON'T push on wrenches when loosening or tightening nuts or bolts. Always try to pull the wrench toward you. If the situation calls for pushing the wrench away, push with an open hand to avoid scraped knuckles if the wrench should slip.
DON'T attempt to lift a heavy component alone — get someone to help you.
DON'T rush or take unsafe shortcuts to finish a job.
DON'T allow children or animals in or around the vehicle while you are working on it.
DO wear eye protection when using power tools such as a drill, sander, bench grinder, etc. and when working under a vehicle.
DO keep loose clothing and long hair well out of the way of moving parts.
DO make sure that any hoist used has a safe working load rating adequate for the job.
DO get someone to check on you periodically when working alone on a vehicle.
DO carry out work in a logical sequence and make sure that everything is correctly assembled and tightened.
DO keep chemicals and fluids tightly capped and out of the reach of children and pets.
DO remember that your vehicle's safety affects that of yourself and others. If in doubt on any point, get professional advice.

Asbestos

Certain friction, insulating, sealing, and other products — such as brake linings, brake bands, clutch linings, torque converters, gaskets, etc. — contain asbestos. *Extreme care must be taken to avoid inhalation of dust from such products since it is hazardous to health.* If in doubt, assume that they *do* contain asbestos.

Fire

Remember at all times that gasoline is highly flammable. Never smoke or have any kind of open flame around when working on a vehicle. But the risk does not end there. A spark caused by an electrical short circuit, by two metal surfaces contacting each other, or even by static electricity built up in your body under certain conditions, can ignite gasoline vapors, which in a confined space are highly explosive. Do not, under any circumstances, use gasoline for cleaning parts. Use an approved safety solvent.

Always disconnect the battery ground (−) cable *at the battery* before working on any part of the fuel system or electrical system. Never risk spilling fuel on a hot engine or exhaust component.

It is strongly recommended that a fire extinguisher suitable for use on fuel and electrical fires be kept handy in the garage or workshop at all times. Never try to extinguish a fuel or electrical fire with water.

Torch (flashlight in the US)

Any reference to a "torch" appearing in this manual should always be taken to mean a hand-held, battery-operated electric light or flashlight. It DOES NOT mean a welding or propane torch or blowtorch.

Fumes

Certain fumes are highly toxic and can quickly cause unconsciousness and even death if inhaled to any extent. Gasoline vapor falls into this category, as do the vapors from some cleaning solvents. Any draining or pouring of such volatile fluids should be done in a well ventilated area.

When using cleaning fluids and solvents, read the instructions on the container carefully. Never use materials from unmarked containers.

Never run the engine in an enclosed space, such as a garage. Exhaust fumes contain carbon monoxide, which is extremely poisonous. If you need to run the engine, always do so in the open air, or at least have the rear of the vehicle outside the work area.

If you are fortunate enough to have the use of an inspection pit, never drain or pour gasoline and never run the engine while the vehicle is over the pit. The fumes, being heavier than air, will concentrate in the pit with possibly lethal results.

The battery

Never create a spark or allow a bare light bulb near a battery. They normally give off a certain amount of hydrogen gas, which is highly explosive.

Always disconnect the battery ground (−) cable *at the battery* before working on the fuel or electrical systems.

If possible, loosen the filler caps or cover when charging the battery from an external source (this does not apply to sealed or maintenance-free batteries). Do not charge at an excessive rate or the battery may burst.

Take care when adding water to a non maintenance-free battery and when carrying a battery. The electrolyte, even when diluted, is very corrosive and should not be allowed to contact clothing or skin.

Always wear eye protection when cleaning the battery to prevent the caustic deposits from entering your eyes.

Mains electricity (household current in the US)

When using an electric power tool, inspection light, etc., which operates on household current, always make sure that the tool is correctly connected to its plug and that, where necessary, it is properly grounded. Do not use such items in damp conditions and, again, do not create a spark or apply excessive heat in the vicinity of fuel or fuel vapor.

Secondary ignition system voltage

A severe electric shock can result from touching certain parts of the ignition system (such as the spark plug wires) when the engine is running or being cranked, particularly if components are damp or the insulation is defective. In the case of an electronic ignition system, the secondary system voltage is much higher and could prove fatal.

HAYNES AUTOMOTIVE MANUALS

NOTE: New manuals are added to this list on a periodic basis. If you do not see a listing for your vehicle, consult your local Haynes dealer for the latest product information.

ALFA-ROMEO
- 531 **Alfa Romeo Sedan & Coupe** '73 thru '80

AMC
- **Jeep CJ** – see JEEP (412)
- 694 **Mid-size models**, Concord, Hornet, Gremlin & Spirit '70 thru '83
- 934 **(Renault) Alliance & Encore** all models '83 thru '87

AUDI
- 615 **4000** all models '80 thru '87
- 428 **5000** all models '77 thru '83
- 1117 **5000** all models '84 thru '88
- 207 **Fox** all models '73 thru '79

AUSTIN
- 049 **Healey 100/6 & 3000** Roadster '56 thru '68
- **Healey Sprite** – see MG Midget Roadster (265)

BLMC
- 260 **1100, 1300 & Austin America** '62 thru '74
- 527 **Mini** all models '59 thru '69
- *646 **Mini** all models '69 thru '88

BMW
- 276 **320i** all 4 cyl models '75 thru '83
- 632 **528i & 530i** all models '75 thru '80
- 240 **1500 thru 2002** all models except Turbo '59 thru '77
- 348 **2500, 2800, 3.0 & Bavaria** '69 thru '76

BUICK
- **Century (front wheel drive)** – see GENERAL MOTORS A-Cars (829)
- *1627 **Buick, Oldsmobile & Pontiac Full-size (Front wheel drive)** all models '85 thru '90
Buick Electra, LeSabre and Park Avenue; Oldsmobile Delta 88 Royale, Ninety Eight and Regency; Pontiac Bonneville
- *1551 **Buick Oldsmobile & Pontiac Full-size (Rear wheel drive)**
Buick Electra '70 thru '84, Estate '70 thru '90, LeSabre '70 thru '79
Oldsmobile Custom Cruiser '70 thru '90, Delta 88 '70 thru '85, Ninety-eight '70 thru '84
Pontiac Bonneville '70 thru '86, Catalina '70 thru '81, Grandville '70 thru '75, Parisienne '84 thru '86
- 627 **Mid-size** all rear-drive **Regal & Century** models with V6, V8 and Turbo '74 thru '87
- **Regal** – see GENERAL MOTORS (1671)
- **Skyhawk** – see GENERAL MOTORS J-Cars (766)
- 552 **Skylark** all X-car models '80 thru '85

CADILLAC
- *751 **Cadillac Rear Wheel Drive** all gasoline models '70 thru '90
- **Cimarron** – see GENERAL MOTORS J-Cars (766)

CAPRI
- 296 **2000 MK I Coupe** all models '71 thru '75
- 283 **2300 MK II Coupe** all models '74 thru '78
- 205 **2600 & 2800 V6 Coupe** '71 thru '75
- 375 **2800 Mk II V6 Coupe** '75 thru '78
- **Mercury Capri** – see FORD Mustang (654)

CHEVROLET
- *1477 **Astro & GMC Safari Mini-vans** all models '85 thru '90
- 554 **Camaro** V8 all models '70 thru '81
- *866 **Camaro** all models '82 thru '90
- **Cavalier** – see GENERAL MOTORS J-Cars (766)
- **Celebrity** – see GENERAL MOTORS A-Cars (829)
- 625 **Chevelle, Malibu & El Camino** all V6 & V8 models '69 thru '87
- 449 **Chevette & Pontiac T1000** all models '76 thru '87
- 550 **Citation** all models '80 thru '85
- *1628 **Corsica/Beretta** all models '87 thru '90
- 274 **Corvette** all V8 models '68 thru '82
- *1336 **Corvette** all models '84 thru '89
- 704 **Full-size Sedans** Caprice, Impala, Biscayne, Bel Air & Wagons, all V6 & V8 models '69 thru '90
- **Lumina** – see GENERAL MOTORS (1671)
- 319 **Luv Pick-up** all 2WD & 4WD models '72 thru '82
- 626 **Monte Carlo** all V6, V8 & Turbo models '70 thru '88
- 241 **Nova** all V8 models '69 thru '79
- *1642 **Nova and Geo Prizm** all front wheel drive models, '85 thru '90
- *420 **Pick-ups** '67 thru '87 – Chevrolet & GMC, all V8 & in-line 6 cyl 2WD & 4WD models '67 thru '87
- *1664 **Pick-ups** '88 thru '90 – Chevrolet & GMC, all full-size (C and K) models, '88 thru '90
- *1727 **Sprint & Geo Metro** '85 thru '91
- *831 **S-10 & GMC S-15 Pick-ups** all models '82 thru '90
- *345 **Vans – Chevrolet & GMC**, V8 & in-line 6 cyl models '68 thru '89
- 208 **Vega** all models except Cosworth '70 thru '77

CHRYSLER
- *1337 **Chrysler & Plymouth Mid-size** front wheel drive '82 thru '89
- **K-Cars** – see DODGE Aries (723)
- **Laser** – see DODGE Daytona (1140)

DATSUN
- 402 **200SX** all models '77 thru '79
- 647 **200SX** all models '80 thru '83
- 228 **B-210** all models '73 thru '78
- 525 **210** all models '78 thru '82
- 206 **240Z, 260Z & 280Z** Coupe & 2+2 '70 thru '78
- 563 **280ZX** Coupe & 2+2 '79 thru '83
- **300ZX** – see NISSAN (1137)
- 679 **310** all models '78 thru '82
- 123 **510 & PL521 Pick-up** '68 thru '73
- 430 **510** all models '78 thru '81
- 372 **610** all models '72 thru '76
- 277 **620 Series Pick-up** all models '73 thru '79
- **720 Series Pick-up** – see NISSAN Pick-up (771)
- 376 **810/Maxima** all gasoline models '77 thru '84
- 124 **1200** all models '70 thru '73
- 368 **F10** all models '76 thru '79
- **Pulsar** – see NISSAN (876)
- **Sentra** – see NISSAN (982)
- **Stanza** – see NISSAN (981)

DODGE
- *723 **Aries & Plymouth Reliant** all models '81 thru '89
- *1231 **Caravan & Plymouth Voyager Mini-Vans** all models '84 thru '90
- 699 **Challenger & Plymouth Saporro** all models '78 thru '83
- 236 **Colt** all models '71 thru '77
- 419 **Colt (rear wheel drive)** all models '77 thru '80
- 610 **Colt & Plymouth Champ (front wheel drive)** all models '78 thru '87
- *556 **D50 & Plymouth Arrow Pick-ups** '79 thru '88
- *1668 **Dakota Pick-up** all models '87 thru '90
- 234 **Dart & Plymouth Valiant** all 6 cyl models '67 thru '76
- *1140 **Daytona & Chrysler Laser** all models '84 thru '89
- *545 **Omni & Plymouth Horizon** all models '78 thru '90
- *912 **Pick-ups** all full-size models '74 thru '90
- *349 **Vans – Dodge & Plymouth** V8 & 6 cyl models '71 thru '89

FIAT
- 080 **124 Sedan & Wagon** all ohv & dohc models '66 thru '75
- 094 **124 Sport Coupe & Spider** '68 thru '78
- 310 **131 & Brava** all models '75 thru '81
- 479 **Strada** all models '79 thru '82
- 273 **X1/9** all models '74 thru '80

FORD
- *1476 **Aerostar Mini-vans** all models '86 thru '90
- 788 **Bronco and Pick-ups** '73 thru '79
- *880 **Bronco and Pick-ups** '80 thru '90
- 014 **Cortina MK II** all models except Lotus '66 thru '70
- 295 **Cortina MK III** 1600 & 2000 ohc '70 thru '76
- 268 **Courier Pick-up** all models '72 thru '82
- 789 **Escort & Mercury Lynx** all models '81 thru '90
- 560 **Fairmont & Mercury Zephyr** all in-line & V8 models '78 thru '83
- 334 **Fiesta** all models '77 thru '80
- 754 **Ford & Mercury Full-size**, Ford LTD & Mercury Marquis ('75 thru '82); Ford Custom 500, Country Squire, Crown Victoria & Mercury Colony Park ('75 thru '87); Ford LTD Crown Victoria & Mercury Gran Marquis ('83 thru '87)
- 359 **Granada & Mercury Monarch** all in-line, 6 cyl & V8 models '75 thru '80
- 773 **Ford & Mercury Mid-size**, Ford Thunderbird & Mercury Cougar ('75 thru '82); Ford LTD & Mercury Marquis ('83 thru '86); Ford Torino, Gran Torino, Elite, Ranchero pick-up, LTD II, Mercury Montego, Comet, XR-7 & Lincoln Versailles ('75 thru '86)
- *654 **Mustang & Mercury Capri** all models including Turbo '79 thru '90
- 357 **Mustang V8** all models '64-1/2 thru '73
- 231 **Mustang II** all 4 cyl, V6 & V8 models '74 thru '78
- 204 **Pinto** all models '70 thru '74
- 649 **Pinto & Mercury Bobcat** all models '75 thru '80
- *1026 **Ranger & Bronco II** all gasoline models '83 thru '89
- *1421 **Taurus & Mercury Sable** '86 thru '90
- *1418 **Tempo & Mercury Topaz** all gasoline models '84 thru '89
- 1338 **Thunderbird & Mercury Cougar/XR7** '83 thru '88
- *1725 **Thunderbird & Mercury Cougar** '89 and '90
- *344 **Vans** all V8 Econoline models '69 thru '90

GENERAL MOTORS
- *829 **A-Cars** – Chevrolet Celebrity, Buick Century, Pontiac 6000 & Oldsmobile Cutlass Ciera all models '82 thru '89
- *766 **J-Cars** – Chevrolet Cavalier, Pontiac J-2000, Oldsmobile Firenza, Buick Skyhawk & Cadillac Cimarron all models '82 thru '90
- *1420 **N-Cars** – Buick Somerset '85 thru '87; Pontiac Grand Am and Oldsmobile Calais '85 thru '90; Buick Skylark '86 thru '90
- *1671 **GM: Buick** Regal, **Chevrolet** Lumina, **Oldsmobile** Cutlass Supreme, **Pontiac** Grand Prix, all front wheel drive models '88 thru '90

GEO
- **Metro** – see CHEVROLET Sprint (1727)
- **Tracker** – see SUZUKI Samurai (1626)
- **Prizm** – see CHEVROLET Nova (1642)

GMC
- **Safari** – see CHEVROLET ASTRO (1477)
- **Vans & Pick-ups** – see CHEVROLET (420, 831, 345, 1664)

(continued on next page)

---

* Listings shown with an asterisk (*) indicate model coverage as of this printing. These titles will be periodically updated to include later model years — consult your Haynes dealer for more information.

Haynes Publications Inc., P.O. Box 978, Newbury Park, CA 91320 • (818) 889-5400 • (805) 498-6703

HAYNES AUTOMOTIVE MANUALS (continued from previous page)

NOTE: New manuals are added to this list on a periodic basis. If you do not see a listing for your vehicle, consult your local Haynes dealer for the latest product information.

HONDA
- 138   360, 600 & Z Coupe all models '67 thru '75
- 351   Accord CVCC all models '76 thru '83
- *1221   Accord all models '84 thru '89
- 160   Civic 1200 all models '73 thru '79
- 633   Civic 1300 & 1500 CVCC all models '80 thru '83
- 297   Civic 1500 CVCC all models '75 thru '79
- *1227   Civic all models '84 thru '90
- *601   Prelude CVCC all models '79 thru '89

HYUNDAI
- *1552   Excel all models '86 thru '89

ISUZU
- *1641   Trooper & Pick-up, all gasoline models '81 thru '90

JAGUAR
- 098   MK I & II, 240 & 340 Sedans '55 thru '69
- *242   XJ6 all 6 cyl models '68 thru '86
- *478   XJ12 & XJS all 12 cyl models '72 thru '85
- 140   XK-E 3.8 & 4.2 all 6 cyl models '61 thru '72

JEEP
- *1553   Cherokee, Comanche & Wagoneer Limited all models '84 thru '89
- 412   CJ all models '49 thru '86

LADA
- *413   1200, 1300. 1500 & 1600 all models including Riva '74 thru '86

LAND ROVER
- 314   Series II, IIA, & III all 4 cyl gasoline models '58 thru '86
- 529   Diesel all models '58 thru '80

MAZDA
- 648   626 Sedan & Coupe (rear wheel drive) all models '79 thru '82
- *1082   626 & MX-6 (front wheel drive) all models '83 thru '90
- *267   B1600, B1800 & B2000 Pick-ups '72 thru '90
- 370   GLC Hatchback (rear wheel drive) all models '77 thru '83
- 757   GLC (front wheel drive) all models '81 thru '86
- 109   RX2 all models '71 thru '75
- 096   RX3 all models '72 thru '76
- 460   RX-7 all models '79 thru '85
- *1419   RX-7 all models '86 thru '89

MERCEDES-BENZ
- *1643   190 Series all four-cylinder gasoline models, '84 thru '88
- 346   230, 250 & 280 Sedan, Coupe & Roadster all 6 cyl sohc models '68 thru '72
- 983   280 123 Series all gasoline models '77 thru '81
- 698   350 & 450 Sedan, Coupe & Roadster all models '71 thru '80
- 697   Diesel 123 Series 200D, 220D, 240D, 240TD, 300D, 300CD, 300TD, 4- & 5-cyl incl. Turbo '76 thru '85

MERCURY
See FORD Listing

MG
- 475   MGA all models '56 thru '62
- 111   MGB Roadster & GT Coupe all models '62 thru '80
- 265   MG Midget & Austin Healey Sprite Roadster '58 thru '80

MITSUBISHI
- *1669   Cordia, Tredia, Galant, Precis & Mirage '83 thru '90
- Pick-up – see Dodge D-50 (556)

MORRIS
- 074   (Austin) Marina 1.8 all models '71 thru '80
- 024   Minor 1000 sedan & wagon '56 thru '71

NISSAN
- 1137   300ZX all Turbo & non-Turbo models '84 thru '89
- *1341   Maxima all models '85 thru '89
- *771   Pick-ups/Pathfinder gas models '80 thru '88
- *876   Pulsar all models '83 thru '86
- *982   Sentra all models '82 thru '90
- *981   Stanza all models '82 thru '90

OLDSMOBILE
- Custom Cruiser – see BUICK Full-size (1551)
- 658   Cutlass all standard gasoline V6 & V8 models '74 thru '88
- Cutlass Ciera – see GENERAL MOTORS A-Cars (829)
- Cutlass Supreme – see GENERAL MOTORS (1671)
- Firenza – see GENERAL MOTORS J-Cars (766)
- Ninety-eight – see BUICK Full-size (1551)
- Omega – see PONTIAC Phoenix & Omega (551)

PEUGEOT
- 161   504 all gasoline models '68 thru '79
- 663   504 all diesel models '74 thru '83

PLYMOUTH
- 425   Arrow all models '76 thru '80
- For all other PLYMOUTH titles, see DODGE listing.

PONTIAC
- T1000 – see CHEVROLET Chevette (449)
- J-2000 – see GENERAL MOTORS J-Cars (766)
- 6000 – see GENERAL MOTORS A-Cars (829)
- 1232   Fiero all models '84 thru '88
- 555   Firebird all V8 models except Turbo '70 thru '81
- *867   Firebird all models '82 thru '89
- Full-size Rear Wheel Drive – see Buick, Oldsmobile, Pontiac Full-size (1551)
- Grand Prix – see GENERAL MOTORS (1671)
- 551   Phoenix & Oldsmobile Omega all X-car models '80 thru '84

PORSCHE
- *264   911 all Coupe & Targa models except Turbo & Carrera 4 '65 thru '89
- 239   914 all 4 cyl models '69 thru '76
- 397   924 all models including Turbo '76 thru '82
- *1027   944 all models including Turbo '83 thru '89

RENAULT
- 141   5 Le Car all models '76 thru '83
- 079   8 & 10 all models with 58.4 cu in engines '62 thru '72
- 097   12 Saloon & Estate all models 1289 cc engines '70 thru '80
- 768   15 & 17 all models '73 thru '79
- 081   16 all models 89.7 cu in & 95.5 cu in engines '65 thru '72
- 598   18i & Sportwagon all models '81 thru '86
- Alliance & Encore – see AMC (934)
- 984   Fuego all models '82 thru '85

ROVER
- 085   3500 & 3500S Sedan 215 cu in engines '68 thru '76
- *365   3500 SDI V8 all models '76 thru '85

SAAB
- 198   95 & 96 V4 all models '66 thru '75
- 247   99 all models including Turbo '69 thru '80
- *980   900 all models including Turbo '79 thru '88

SUBARU
- 237   1100, 1300, 1400 & 1600 all models '71 thru '79
- *681   1600 & 1800 2WD & 4WD all models '80 thru '89

SUZUKI
- *1626   Samurai/Sidekick and Geo Tracker all models '86 thru '89

TOYOTA
- *1023   Camry all models '83 thru '90
- 150   Carina Sedan all models '71 thru '74
- 229   Celica ST, GT & liftback all models '71 thru '77
- 437   Celica all models '78 thru '81
- *935   Celica all models except front-wheel drive and Supra '82 thru '85
- 680   Celica Supra all models '79 thru '81
- 1139   Celica Supra all in-line 6-cylinder models '82 thru '86
- 361   Corolla all models '75 thru '79
- 961   Corolla all models (rear wheel drive) '80 thru '87
- *1025   Corolla all models (front wheel drive) '84 thru '91
- *636   Corolla Tercel all models '80 thru '82
- 230   Corona & MK II all 4 cyl sohc models '69 thru '74
- 360   Corona all models '74 thru '82
- *532   Cressida all models '78 thru '82
- 313   Land Cruiser all models '68 thru '82
- 200   MK II all 6 cyl models '72 thru '76
- *1339   MR2 all models '85 thru '87
- 304   Pick-up all models '69 thru '78
- *656   Pick-up all models '79 thru '90

TRIUMPH
- 112   GT6 & Vitesse all models '62 thru '74
- 113   Spitfire all models '62 thru '81
- 028   TR2, 3, 3A, & 4A Roadsters '52 thru '67
- 031   TR250 & 6 Roadsters '67 thru '76
- 322   TR7 all models '75 thru '81

VW
- 091   411 & 412 all 103 cu in models '68 thru '73
- 159   Beetle & Karmann Ghia all models '54 thru '79
- 238   Dasher all gasoline models '74 thru '81
- *884   Rabbit, Jetta, Scirocco, & Pick-up all gasoline models '74 thru '89 & Convertible '80 thru '89
- 451   Rabbit, Jetta & Pick-up all diesel models '77 thru '84
- 082   Transporter 1600 all models '68 thru '79
- 226   Transporter 1700, 1800 & 2000 all models '72 thru '79
- 084   Type 3 1500 & 1600 all models '63 thru '73
- 1029   Vanagon all air-cooled models '80 thru '83

VOLVO
- 203   120, 130 Series & 1800 Sports '61 thru '73
- 129   140 Series all models '66 thru '74
- 244   164 all models '68 thru '75
- *270   240 Series all models '74 thru '90
- 400   260 Series all models '75 thru '82
- *1550   740 & 760 Series all models '82 thru '88

SPECIAL MANUALS
- 1479   Automotive Body Repair & Painting Manual
- 1654   Automotive Electrical Manual
- 1480   Automotive Heating & Air Conditioning Manual
- 1763   Ford Engine Overhaul Manual
- 482   Fuel Injection Manual
- 1666   Small Engine Repair Manual
- 299   SU Carburetors thru '88
- 393   Weber Carburetors
- 300   Zenith/Stromberg CD Carburetors thru '76

See your dealer for other available titles

Over 100 Haynes motorcycle manuals also available

4-1-91

---

* Listings shown with an asterisk (*) indicate model coverage as of this printing. These titles will be periodically updated to include later model years — consult your Haynes dealer for more information.

Haynes Publications Inc., P.O. Box 978, Newbury Park, CA 91320 ● (818) 889-5400 ● (805) 498-6703

Printed by
J H Haynes & Co Ltd
Sparkford Nr Yeovil
Somerset BA22 7JJ England